Fertilizers, Crop Yield and Nutrient Management

Fertilizers, Crop Yield and Nutrient Management

Editor: Sierra Webb

Callisto Reference
www.callistoreference.com

Callisto Reference,
118-35 Queens Blvd., Suite 400,
Forest Hills, NY 11375, USA

Visit us on the World Wide Web at:
www.callistoreference.com

ISBN: 978-1-64116-822-9 (Hardback)

Cataloging-in-Publication Data

Fertilizers, crop yield and nutrient management / edited by Sierra Webb.
p. cm.
Includes bibliographical references and index.
ISBN 978-1-64116-822-9
1. Fertilizers. 2. Crop yields. 3. Soil management. 4. Soils and nutrition. 5. Agriculture. I. Webb, Sierra.
S633 .F47 2023
631.8--dc23

Table of Contents

Preface

This book has been an outcome of determined endeavour from a group of educationists in the field. The primary objective was to involve a broad spectrum of professionals from diverse cultural background involved in the field for developing new researches. The book not only targets students but also scholars pursuing higher research for further enhancement of the theoretical and practical applications of the subject.

Fertilizers are chemical substances that are applied to the soil for supplying essential nutrients required by plants. They are used as a part of nutrient management practice for enhancing the agricultural productivity or crop yield. Fertilizers can be classified into two categories, namely, organic fertilizers and inorganic fertilizers. Organic fertilizers are natural fertilizers derived from plants and animals. They have various benefits such as promoting the reproduction of microorganisms, enriching the soil with carbonic compounds, altering the physical and chemical properties of soil, and increasing the content of organic matter in the soil. Inorganic fertilizers are chemical-based fertilizers used for providing nutrients for the development of crops. These can be further classified into nitrogen fertilizers and phosphorous fertilizers. Although fertilizers are beneficial for soil and productivity, excessive application of fertilizers may also lead to soil fertility degradation. This book includes some of the vital pieces of works being conducted across the world on various topics related to fertilizers, crop yield and nutrient management. This book will serve as a reference to a broad spectrum of readers.

It was an honour to edit such a profound book and also a challenging task to compile and examine all the relevant data for accuracy and originality. I wish to acknowledge the efforts of the contributors for submitting such brilliant and diverse chapters in the field and for endlessly working for the completion of the book. Last, but not the least; I thank my family for being a constant source of support in all my research endeavours.

Editor

1

Possibility of Limiting Mineral Fertilization in Potato Cultivation by using Bio-fertilizer and its Influence on Protein Content in Potato Tubers

Dorota Wichrowska [1,*] **and Małgorzata Szczepanek** [2]

[1] Department of Microbiology and Food Technology, University of Science and Technology, Kaliskiego 7 Street, 85-796 Bydgoszcz, Poland

[2] Department of Agronomy, UTP University of Science and Technology, Kaliskiego 7 Street, 85-796 Bydgoszcz, Poland; Malgorzata.Szczepanek@utp.edu.pl

* Correspondence: wichrowska@utp.edu.pl.

Abstract: Potato protein is a valuable source of essential plant-derived amino acids, the composition of which is similar to that of chicken egg protein considering the amino acid reference. Many factors used in potato cultivation can modify its composition. The use of bio-fertilizers in potato growing offers a possibility of a better use of minerals from soil and organic sources and reducing the need for mineral fertilizers by activating minerals present in soil. The effect can be to improve not only the potato tuber yield but also the nutritional value. The aim of this study has been to determine the hanges in the content of crude protein and the composition of amino acids in potato tubers, depending on the application of the bio-fertilizer (UGmax), organic fertilizers (pea as a catch crop, straw, and farmyard manure (FYM)) as well as mineral fertilization (100% and 50% of the reference rate). The application of bio-fertilizer significantly increased the content of essential and non-essential amino acids in potato tuber protein. With the half-decreased mineral fertilization rate, bio-fertilizer most effectively increased the content of tyrosine, methionine, asparagine in potato tuber protein in the treatments with FYM or with a catch crop as well as without organic fertilization.

Keywords: UGmax; crude protein; essential amino acids; non-essential amino acids

---

1. Introduction

A modern intensive management system, based on the use of fertilizers acquired outside the farm, especially mineral fertilizers and pesticides, to achieve maximum profits, has caused environmental pollution (water pollution, intensification of soil degradation, and simplification of biodiversity). Monoculture, zero tillage, reduced organic fertilization, and the introduction of new cultivars accompanied by increased pesticides also pose a threat, which results in some deterioration of the physical, chemical and microbiological properties of the soil and, as a result, of the quantity and the quality of the crop [1,2]. This is counteracted by the implementation of the concept of sustainable agricultural production, which, however, requires a number of adjustments in terms of organization, especially the basic elements of technology, especially crop rotation, fertilization, and plant protection. In recent years there has been observed much interest in the use of various types of biostimulants, effective microorganisms, preparations, and fertilizers enriching the soil with humus, improving the health and resistance of plants to stress conditions, facilitating the intake of nutrients, and enhancing good-quality crops [3–6]. The use of composting flora can be one of the most important sources of soil regeneration, and such possibilities are provided by the use of UGmax soil fertilizer [7]. The UGmax bio-fertilizer is a microbiological preparation which consists of yeast, lactic acid bacteria-*Lactobacillus* and *Lactococcus; Pseudomonas; Actinobacteria* and *Azotobacter* as well as small amounts of micro-and macronutrients and, in favorable climate conditions, it can accelerate the decomposition of humus, crop residue and organic fertilizers, launch nutrients from minerals or insoluble compounds, restore a better soil structure, improve water relations and affect the growth and quality of potato crops [2,6,8–10].

Potato growing conditions affect the total nitrogen content in the dry matter of tuber, including the number of amino acids [11–13]. As demonstrated by the studies of various authors conducted for many years [13–16], most protein fractions building potato protein show a well-balanced amino acid composition and a high nutritional value, as evidenced by the relevant indicators: CS (Chemical score) in the range of 57–96 [13,17,18] or BV (Biological value) in the range of 45–88 [19]. Chemical indicators, such as CS, allow a simple and quick determination of the quality of the protein tested by comparing its amino acid composition with the composition of the protein adopted as a standard and an indication of a limiting amino acid. In addition to the standard protein amino acid composition modified in 1991 by the FAO/WHO (Food and Agriculture Organization/World Health Organization Expert Committee United Nations University), newer standards are proposed, e.g., developed by WHO/FAO/UNU (United Nations University) in 2007 [20], with fewer amino acids corresponding to current research on human demand for amino acids [21].

Depending on the variety, cultivation, and storage conditions, the amino acids which limit the nutritional value of potato protein can include leucine, isoleucine [17], methionine, cysteine, and less often threonine and lysine [17,22] and valine [12]. Potato protein is particularly rich in amino acids such as glutamic and aspartic acid accounting for 30-50% of all the amino acids in tubers. Leucine, valine, alanine, lysine, and arginine are present in smaller but significant amounts (4–8%), while methionine and histidine constitute, according to Danilchenko et al. [11], a low share in potato protein. According to Eppendorfer and Eggum [19], the content of exogenous amino acids in potato tubers is negatively correlated with the level of fertilization, and the changes in the share of exogenous amino acids in the sum of all the amino acids primarily due to changes in non-protein nitrogen fractions, especially free amino acids and amides, which make up about 40–60% of the total protein in tubers. According to Eppendorfer and Eggum [19], a sulphur deficiency in soil contributes to a decrease in the content of sulphur amino acids, cysteine and methionine in potato protein and it triggers deviations from the constant dependence between the content of nitrogen and amino acids, which can be due to the effect of this component on nitrogen metabolism and protein amino acid composition. Thus, a reduction in nitrogen fertilization for organic potato cultivation can contribute to an increase in the share of protein nitrogen and exogenous amino acids in the total protein.

2. Materials and Methods

2.1. Experimental Location and Treatment

Potato tubers of the medium-early potato (*Solanum tuberosum* L.) cultivar, 'Satina N', were obtained from the field experiment as a three-factor split-split-plot design at the Experiment Station of the UTP University of Science and Technology in Mochełek (53°13′ N, 17°51′ E). Cereals constituted the catch crop. Potatoes were mechanically planted at 0.75 × 0.35 m row spacing. The single plot size was 31.5 m^2. The experiment treatments are shown in Table 1. All the mineral fertilizers were applied in spring prior to potato planting (every research year), at the rates considering the soil richness (a high phosphors content and a medium low content of potassium) and the nutrition requirements of the plants. The condition of the soil was measured every year. The chemical composition of soil before the start of the field experiment in the years analyzed was similar; it demonstrated a very low richness in available forms of magnesium and also a low amount of available nitrogen. The soil with slightly acid reaction is presented in Table 2. During cultivation the following potato protection agents were applied; herbicide: linuron (3-(3,4-dichlorophenyl)-1-methoxy-1-methylurea, Agan Chemical Manufactures Ltd., Israel-Afalon 50WP; 2 dm^3 ha^{-1}); insecticide: chlorpyrifos and cypermethrin (*O*,*O*-diethyl *O*-3,5,6-trichloro-2-pyridyl phosphorothioate and (RS)-α-cyano-3-phenoxybenzyl (1RS,3RS;1RS,3SR)-3-(2,2-dichlorovinyl)-2,2-dimethylcyclopropanecarboxylate, respectively, Agriphar

S.A., Belgium-Nurelle D 550 EC; 0.6 dm^3 ha^{-1}) as well as fungicides: cymoksanil and mancozeb (2-cyano-*N*-[(ethylamino) carbonyl]-2 (methoxyimino) acetamide and zinc complex of ethylene-bis-dithiocarbamate manganese, Helm AG, Germany-Helm-Cymi 72,5 WP; 2 kg ha^{-1}), mefenoxan and mancozeb (methyl *N*-(methoxyacetyl)-*N*-2,6-xylyl-D-alaninate and zinc complex of ethylene-bis-dithiocarbamate manganese, Syngenta Crop Protection AG, Switzerland-Ridomil Gold MZ 67,8 WG; 2 dm^3 ha^{-1}). The UGmax bio-fertilizer is a microbiological preparation which consists of yeast, *Lactobacillus* and *Lactococcus*, also *Pseudomonas; Actinobacteria* and *Azotobacter* as well as potassium 3.5 g dm^{-3}, nitrogen 1.2 g dm^{-3}, sulphur 1.0 g dm^{-3}, phosphorus 0.5 g dm^{-3}, sodium 0.2 g dm^{-3}, magnesium 0.1 g dm^{-3}, zinc 0.02 g dm^{-3} and manganese 0.003 g dm^{-3}.

Table 1. Experimental treatments.

	Type of Organic Sources (1st Factor)
Control	without organic sources
Catch crop (fodder pea)	27 t ha^{-1}, in autumn, after post-harvest treatments and before pre-winter plough
Straw	4 t ha^{-1}, in autumn, after post-harvest treatments and before pre-winter plough
Farmyard manure (FYM)	25 t ha^{-1}, in autumn, before pre-winter plough
	Rate of Mineral Fertilization (2nd Factor) (Ammonium Nitrate-34%, Triple Superphosphate-46%, Potassium Sulphate-50%)
100% NPK	100 kg N ha^{-1}, 100 kg P_2O_5 ha^{-1}, 150 kg K_2O ha^{-1}
50% NPK	50 kg N ha^{-1}, 50 kg P_2O_5 ha^{-1}, 150 kg K_2O ha^{-1} applied in spring, prior to potato planting, at the rates adjusted to the soil richness and nutritive requirements of the plant
	Bio-Fertilizer Application (3rd Factor)
Control	Without treatment
Bio-fertilizer	UGmax fertilizer applied at three rates: 0.6 l ha^{-1} in autumn, on organic sources, prior to pre-winter plough; 0.3 l ha^{-1} in spring, prior to tuber planting, during soil tillage; 0.3 l ha^{-1}1 as foliar fertilization, at the plant height of 15–20 cm

Table 2. Chemical properties of soil at the experimental site before experiment.

Parameters	Unit	Value	Categories
pH H_2O	-	5.1–6.7	Slightly acid
pH KCl	-	5.7–6.1	
Organic carbon	g kg^{-1}	7.55–7.80	-
Total nitrogen	g kg^{-1}	0.69–0.75	Low richness
Phosphorus	mg kg^{-1}	190.0–210.0	High richness
Potassium	mg kg^{-1}	95.0–150.0	Medium richness
Magnesium	mg kg^{-1}	<20.0	Very low richness

As organic matter, each research year, under winter plough, FYM was applied at the rate of 25 t/ha, straw–4 t/ha and pea (stubble intercrop) at the rate of 27 t/ha. The bio-fertilizer was applied each year at two rates applied into soil: 0.6 l/ha in autumn–prior to the winter plough, onto organic matter and 0.3 l/ha in spring, prior to tuber planting. Additionally, 0.3 l/ha was provided as foliar application at the plant height of 15–20 cm (Table 1). The chemical composition of soil prior to the field experiment was analyzed in each research year and it was similar. It showed a very low content of available forms of potassium and magnesium as well as a low amount of available phosphorus in soil (Table 2). The soil reaction was slightly acid. The chemical composition of organic sources and the total weight of macroelements applied by organic sources is presented in Table 3.

Table 3. Chemical composition of organic sources and the total weight of macroelements applied by organic sources.

Component	Organic Source [kg·t^{-1}]			Organic Source [t·ha^{-1}]		
	Catch Crop	Straw	Farmyard Manure	Catch Crop	Straw	Farmyard Manure
N	35.10	5.62	5.12	488.7	151.74	138.24
P_2O_5	4.34	1.06	4.48	90.18	28.62	120.96
K_2O	25.15	10.21	6.85	247.05	275.67	184.95
CaO	2.74	0.84	4.42	46.98	22.68	119.34
Mg	7.25	2.25	1.82	141.75	33.75	49.14

To determine the periods of drought and semi-drought, the Sielianinov hydrothermal coefficient was used (Table 4). It was determined by dividing the total precipitation by the total temperature of a given month decreased ten-fold.

Table 4. Weather conditions during the plant growing season at the Mochełek Experiment Station.

Years	Rainfall						Total
	Month						
	April	May	June	July	August	September	
2009	0.4	12.4	53.4	118.0	17.6	34.4	236.1
2010	33.8	92.6	18.1	107.4	150.7	74.7	477.3
2011	13.5	35.4	100.8	132.5	67.7	37.0	386.9
Average	15.9	46.8	57.4	119.3	78.7	48.7	366.8
	Temperature						
2009	9.8	12.3	14.5	18.6	18.2	13.7	87.1
2010	7.8	11.5	16.7	21.6	18.4	12.2	88.2
2011	10.5	13.5	17.7	17.5	17.7	14.3	91.2
Average	9.4	12.4	16.3	19.2	18.1	13.4	88.8
	Hydrothermal Coefficient K						Average
2009	0.01	0.32	1.23	2.05	0.61	0.84	0.84
2010	1.44	2.60	0.36	1.60	2.64	2.04	1.78
2011	0.43	0.85	1.90	2.44	1.23	0.86	1.29
Average	0.63	1.25	1.16	2.03	1.50	1.25	1.30

K: 0–0.5 drought; 0.6–1.0 mild-drought; 1.1–2.0 moist, optimal conditions; 2.1-humid.

K–hydrothermal coefficient value,
P–total monthly precipitation,
t–total mean daily temperature in a given month.

Value K falling in the range from 0 to 0.5 stands for drought, from 0.6 to 1.0–semi-drought, and the value above 1.0–for humid conditions.

The meteorological data were collated from the records of the Weather Station at the Experiment Station of the UTP University of Science and Technology in Mochełek (53°13′ N, 17°51′ E). The 2009 season showed drought periods during the early plant growth, much moisture afterwards and a spell of semi-dry weather at the end of the growing season. During the 2010 and 2011 growing seasons, it was quite humid, thus optimal for potato development and growth.

2.2. Plant Sampling and Chemical Analysis

Right after harvest, potato tubers were transferred to the storage chamber at temperature of 10 °C and 80% relative air humidity, and then a representative sample was prepared for analysis. The tubers were cut and frozen in liquid nitrogen and stored at the temperature of at least -18 °C for further studies. The frozen material was lyophilized (model Alpha 1–4 LDplus, Donserv, Warszawa, Poland), and then it was ground until fine powder (the particles 0.3–0.5 mm in size) was produced using the ultracentrifuge Retsch mill ZM 100 (Retsch, Haan, Germany). The ground samples were stored in the dark, in the bags which were placed in desiccators for further analysis.

The laboratory analyses (nitrogen content, and free amino acids) were carried out at the Department of Quality Plant Products in the Georg August University in Göttingen. The content of crude protein was determined by multiplying the total nitrogen by factor 6.25, with the Kjeldahl method using Büchi Labortechnik B-324 apparatus, after mineralization in concentrated sulphuric acid (VI). The total nitrogen was analyzed following the Dumas combustion method by LECO CN-2000 [23]. Free amino acids were extracted from freeze-dried potato flour using hydrochloric acid. as described by Fischer et al. [24]. 0.5 g of freeze-dried potato flour was suspended in a centrifuge tube with 4 mL 1 N hydrochloric acid. The tube was horizontally shaken for one hour and centrifuged at 15,000 relative centrifugal force (RCF) for 30 min. The supernatant was collected in a 10 mL volumetric flask. The extraction was performed three times, with the last two extractions—with 3 mL of hydrochloric acid. Finally, the hydrochloric acid was added to the flask to adjust the final 10 mL. To get a clear solution, the supernatant was centrifuged again at 15,000 RFC for 30 min.

Free amino acids were assayed using HPLC. The HPLC system consists of degasser WellChrom K-5004 (Knauer, Berlin, Germany), multisolvent delivery system 600 E (Waters, Milford, MA, USA), autosampler 2157 (Parmacia LKB, Sweden), the temperature control module (Waters, USA), 5 µm column LiChroCart 250-3 and pre-column LiChroCart 4-4 (Merc, Darmstadt, Germany), fluorescens detector 474 (Waters, USA). The gradient eluents, which consisted of methanol (71/29 and 20/80, v/v) in 50 mM sodium acetate buffer (pH 7.0), were used to separate the amino acids at the flow rate of 0.6 mL min^{-1} and the temperature of 45 °C. The amino acids were standardized using the Perbio Amino Acid Standard H (USA). The HPLC separation of fluorescent o-phtaldialdehyde (OPA) derivatives was applied to the assay of free amino acids according to the method described by Fischer et al. [24]. The procedure was based on the reaction of reducing agent β-mercaptoethanol, to give a complex which can be measured by fluorescence. The OPA solution was prepared by dissolving 125 mg of OPA in 22 mL of methanol, then mixing it with 500 µL of β-mercaptoethanol and 2.5 mL 0.5 M borate buffer at pH 9.5. The extract of amino acids was derived with the fluorescens reagent (o-phtaldialdehyde) for two minutes. The microliters of the mixture were injected into the HPLC.

The Amino Acid Score (AAS) was calculated using the following equation:

$$AAS = \frac{a_p}{a_s}$$

where a is an essential amino acid (EAA), p is the test protein, *s* is the reference protein. The reference protein used was the FAO/WHO amino acid scoring pattern from the 2007 WHO/FAO report [20].

2.3. Statistical Analysis

A three-way ANOVA analysis was performed to reveal the significance of established effects of mineral fertilization, organic fertilizers and bio-fertilizer and their interactions. The tables and Figure 1 show the mean values of crude protein (CP) and amino acids for three years of research and standard deviations of the mean of 3 replications. The results are presented as mean values for three years of study plus standard deviation. All the data were reported as the means with standard errors of the mean (s.e.m.). The significance of differences (LSD–Lowest Significance Difference) was evaluated using the Tukey multiple confidence intervals for the significance level of $\alpha = 0.05$ and 0.01. The multifactorial ANOVA was computed using Statistica® 13.1 package.

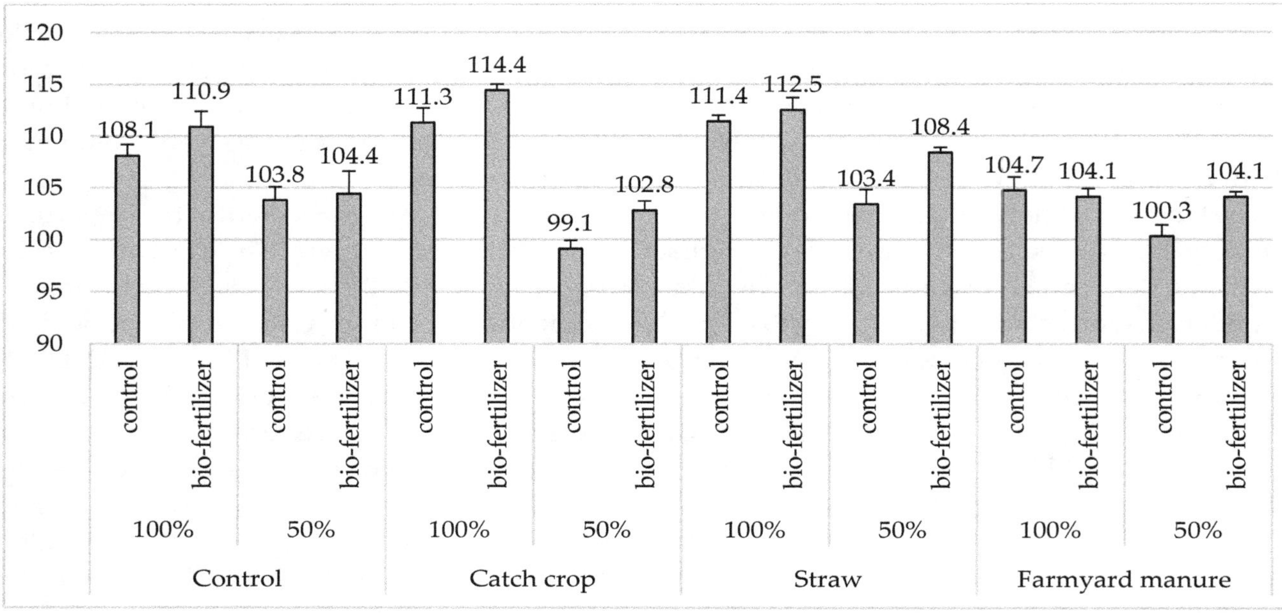

Figure 1. Content of crude protein in potato tubers as affected by organic, mineral, and bio fertilization, mean for three years of study (g·kg^{-1} of dry weight). 100%, 50%-rate of mineral fertilization with NPKLSD (Tukey-test at p $P \leq 0.05$) Organic sources = 3.595; mineral fertilizers = 1.282; Bio-fertilizers = 0.553.

3. Results

3.1. Crude Protein Content in Potato Tubers

Based on three years of research, the content of crude protein in the dry matter of edible potato tubers of the 'Satina' cultivar was determined. In our study the crude protein in tubers ranged between 99.1 and 114.4 g kg^{-1} DM (Figure 1). The effect of 100% mineral fertilization and the organic sources increased significantly the crude protein content in potato tubers. There are also positive interactions between all the factors. The highest content of crude protein in potato tubers was found after a combined application of green fertilizer (fodder pea) with a full rate of nitrogen, phosphorus and potassium (NPK) and bio-fertilizer used. A comparison of the content of crude protein in the tubers from the plots where half the rate of mineral fertilization was applied, the use of bio-fertilizer increased that component, especially after applying straw.

3.2. Essential and Non-Essential Amino Acids Content in Potato Tubers

The effect of combined organic matter, mineral fertilization, and UGmax bio-fertilizer applied on the content of essential and non-essential amino acids (Tables 5 and 6, respectively) was significant ($P \leq 0.01$), excluding mineral fertilization (significant $P \leq 0.05$) for non-essential amino acids and non-significant for essential amino acids. The content of eight essential and four non-essential amino acids identified in the potato tubers studied from the 3-year planting season are presented in Tables 7 and 8, respectively. Total free amino acids ranged between 18.00 and 64.65 g 100 g^{-1} of crude protein. The effect of mineral fertilization on methionine, valine, leucine, isoleucine, phenylalanine and glutamic acid was non-significant. The mineral fertilizer combined with organic matter with bio-fertilizer and the interaction between the three study factors affected significantly ($P \leq 0.01$) the content of tyrosine, valine, isoleucine, leucine, lysine, and phenyloalanine and significantly ($P \leq 0.05$) threonine and methionine in the crude protein content. The potato tubers derived from the experimental plot with the use of farmyard manure, 50% of the full rate of NPK and bio-fertilizer contained most tyrosine, with straw and 100% rate of NPK and bio-fertilizer; mostly threonine and lysine, in turn, 50% rate of NPK-mostly phenylalanine, with catch crop (pea) and 100% rate of NPK and bio-fertilizer–mostly valine, isoleucine, leucine. Organic fertilization increased the essential amino acids content, especially green manure (pea). A similar effect was produced after the use of bio-fertilizer. The results of the application of 100 and 50% rate of NPK were not significant for the content of methionine, valine, isoleucine, leucine, phenylalanine and a slightly significant full mineral rate increased the content of tyrosine, threonine and lysine in the protein content (Table 7). On average, for the other factors, the organic matter and bio-fertilizer resulted in a significant increase in the essential amino acid content. In turn, organic fertilization combined with bio-fertilizer had no significant influence on the aspartic acid content. Moreover, 100% rate of mineral fertilization combined with bio-fertilizer gave a non-significant result of glutamic acid content in potato protein (Table 6). The application of straw and bio-fertilizer and full rate of NPK increased aspartic acid and the alanine content (Table 8). A similar effect was found on the content of asparagine when no organic sources were applied. The highest asparagine content in crude protein of potato tubers was observed when 50% NPK rate was used, without organic fertilization and bio-fertilizer application. Green fertilizer (pea), 100% NPK mineral fertilization and bio-fertilizer increased the content of glutamic acid in potato protein (Table 8). On average for the other factors, organic fertilization and the application of bio-fertilizer resulted in the highest increase in the content of non-essential amino acids. The ASS of the potato tubers as affected by organic, mineral and bio fertilization studied. WHO/FAO adult maintenance pattern expressed as g amino acid/100 g protein shown in Figure 2. The content of leucine and valine was higher than recommended by the WHO/FAO, especially under the influence of green fertilizer and straw.

Table 5. Essential amino acids content. Significance of experimental factors and their interaction, mean for three years of study.

Essential Amino Acids	Organic Sources	Mineral Fertilizers	Bio-fertilizer	Organic Sources × Mineral Fertilizers	Organic Sources × Bio-Fertilizer	Mineral Fertilizers × Bio-Fertilizer	Organic Sources × Mineral Fertilizers × Bio-Fertilizer
Tyrosine	*	*	*	*	*	*	**
Threonine	*	*	*	**	**	*	*
Methionine	**	n.s.	**	*	*	**	*
Valine	**	n.s.	**	**	**	**	**
Isoleucine	**	n.s.	**	**	**	**	**
Leucine	*	n.s.	**	**	**	**	**
Lysine	**	**	**	**	**	**	**
Phenylalanine	**	n.s.	*	**	**	**	**
Sum	**	n.s.	**	**	**	**	**

The multifactorial ANOVA was used in order to test the single effects of organic sources, mineral fertilization, and bio-fertilizer as well as for the interaction. * significant difference $P \leq 0.05$; ** significant difference $P \leq 0.01$; n.s.-non-significant difference.

Table 6. Non-essential amino acids content. Significance of experimental factors and their interaction, the 2009-2011 mean.

Non-Essential Amino Acids	Organic Sources	Mineral Fertilizers	Bio-Fertilizer	Organic Sources × Mineral Fertilizers	Organic Sources × Bio-Fertilizer	Mineral Fertilizers × Bio-Fertilizer	Organic Sources × Mineral Fertilizers × Bio-Fertilizer
Aspartic acid	**	**	**	**	n.s.	**	**
Asparagine	**	**	**	**	**	**	**
Alanine	**	**	**	**	**	**	**
Glutamic acid	**	n.s.	**	*	*	n.s.	**
Sum	**	*	**	**	**	**	**

The multifactorial ANOVA was used to test the single effects of organic sources, mineral fertilization, and bio-fertilizer as well as the interaction. * significant difference $P \leq 0.05$; ** significant difference $P \leq 0.01$; n.s.–non-significant difference.

Table 7. Essential amino acids content (g 100 g^{-1} protein) in potato tubers. Mean for three years of study.

Organic Fertilization	Mineral NPK Fertilization	Bio-Fertilizer	Tyrosine	Threonine	Methionine	Valine	Isoleucine	Leucine	Lysine	Phenyl-Alanine	Sum
Control (without organic source)	100%	Control	0.64 ± 0.01	1.40 ± 0.08	2.19 ± 0.11	2.40 ± 0.14	1.84 ± 0.05	0.52 ± 0.04	0.17 ± 0.01	1.66 ± 0.04	10.82
		Bio-fertilizer	0.71 ± 0.02	1.32 ± 0.02	2.34 ± 0.05	2.54 ± 0.08	1.95 ± 0.11	1.72 ± 0.02	0.64 ± 0.0	1.39 ± 0.01	12.61
	50%	Control	0.83 ± 0.01	1.26 ± 0.01	2.09 ± 0.08	2.30 ± 0.01	1.57 ± 0.08	2.27 ± 0.02	0.09 ± 0.01	1.65 ± 0.01	12.06
		Bio-fertilizer	0.70 ± 0.01	1.28 ± 0.04	2.29 ± 0.03	2.31 ± 0.01	1.67 ± 0.05	2.31 ± 0.03	0.11 ± 0.03	1.26 ± 0.01	11.93
Catch crop (pea)	100%	Control	1.36 ± 0.02	2.49 ± 0.03	2.67 ± 0.07	3.89 ± 0.05	2.94 ± 0.04	4.67 ± 0.11	0.22 ± 0.02	2.39 ± 0.01	20.63
		Bio-fertilizer	1.06 ± 0.01	3.87 ± 0.11	3.05 ± 0.04	4.06 ± 0.04	3.41 ± 0.06	4.97 ± 0.08	0.89 ± 0.01	2.25 ± 0.02	23.56
	50%	Control	1.12 ± 0.05	2.21 ± 0.05	2.58 ± 0.05	3.72 ± 0.02	3.50 ± 0.07	3.92 ± 0.07	0.23 ± 0.01	2.25 ± 0.01	19.53
		Bio-fertilizer	1.35 ± 0.03	3.43 ± 0.08	2.96 ± 0.08	3.78 ± 0.01	3.57 ± 0.04	4.59 ± 0.02	0.44 ± 0.01	2.12 ± 0.01	22.24
Straw	100%	Control	1.18 ± 0.04	2.69 ± 0.02	2.34 ± 0.05	2.89 ± 0.03	3.03 ± 0.08	4.82 ± 0.03	0.31 ± 0.00	2.45 ± 0.01	19.71
		Bio-fertilizer	1.62 ± 0.01	4.52 ± 0.05	2.38 ± 0.03	2.94 ± 0.11	3.98 ± 0.11	3.45 ± 0.05	0.99 ± 0.01	2.51 ± 0.02	24.79
	50%	Control	0.91 ± 0.01	1.57 ± 0.02	2.28 ± 0.01	2.55 ± 0.08	2.15 ± 0.04	3.07 ± 0.02	0.28 ± 0.00	1.69 ± 0.01	14.50
		Bio-fertilizer	0.93 ± 0.01	3.06 ± 0.04	2.35 ± 0.05	2.98 ± 0.04	3.82 ± 0.03	3.97 ± 0.07	0.84 ± 0.00	1.78 ± 0.01	19.73
Farmyard manure (FYM)	100%	Control	1.60 ± 0.05	1.51 ± 0.02	1.98 ± 0.04	3.15 ± 0.09	2.67 ± 0.02	3.56 ± 0.01	0.14 ± 0.00	1.89 ± 0.00	16.50
		Bio-fertilizer	1.37 ± 0.04	1.16 ± 0.03	2.06 ± 0.06	3.51 ± 0.02	3.39 ± 0.05	3.30 ± 0.05	0.05 ± 0.00	2.34 ± 0.01	17.18
	50%	Control	1.04 ± 0.01	1.09 ± 0.02	1.86 ± 0.02	3.11 ± 0.03	2.39 ± 0.04	3.12 ± 0.04	0.02 ± 0.00	1.45 ± 0.00	14.08
		Bio-fertilizer	1.53 ± 0.04	2.57 ± 0.01	1.98 ± 0.03	3.47 ± 0.04	3.23 ± 0.03	3.24 ± 0.07	0.28 ± 0.00	2.11 ± 0.01	18.41
	Mean		1.12	2.21	2.34	3.10	2.82	3.34	0.36	1.94	17.24
Organic sources	Control		0.72 c	1.32 b	2.23 c	2.39 d	1.76 c	1.71 d	0.25 c	1.49 c	11.86 d
	Catch crop		1.22 b	3.00 a	2.82 a	3.86 a	3.36 a	4.54 a	0.44 b	2.25 a	21.49 a
	Straw		1.16 ab	2.96 a	2.34 b	2.84 c	3.25 ab	3.83 b	0.61 a	2.11 ab	19.08 b
	FYM		1.39 a	1.58 bc	1.97 d	3.31 b	2.92 b	3.31 c	0.12 d	1.95 bc	16.54 c
Mineral fertilization	100% NPK fertilization		1.13 a	2.26 a	2.36 a	3.09 a	2.78 a	3.06 a	0.40 a	2.05 a	17.14 a
	50% NPK fertilization		1.05 b	2.06 b	2.30 a	3.03 a	2.74 a	3.31 a	0.29 b	1.79 b	16.56 b
Bio-fertilizer (Bio-F)	Control		1.09 b	1.78 b	2.25 b	3.00 b	2.51 b	3.24 b	0.18 b	1.93 b	15.98 b
	Bio-fertilizer		1.16 a	2.65 a	2.43 a	3.20 a	3.13 a	3.44 a	0.53 a	1.97 a	18.51 a

Different letters following the mean values indicate significant differences with the Tukey test at $p \leq 0.05$ for a given feature for organic source, mineral fertilization and bio-fertilizer.

Table 8. Non-essential amino acids content (g 100 g^{-1} protein) in potato tubers. Mean for three years of study.

Organic Fertilization	Mineral NPK Fertilization	Bio-Fertilizer	Aspartic Acid	Asparagine	Alanine	Glutamic Acid	Sum
Control (without organic fertilizer)	100%	Control	6.37 ± 0.05	4.92 ± 0.04	0.22 ± 0.00	4.99 ± 0.02	16.50
		Bio-fertilizer	7.68 ± 0.07	5.83 ± 0.05	1.61 ± 0.04	6.72 ± 0.04	21.84
	50%	Control	4.47 ± 0.04	3.91 ± 0.02	0.22 ± 0.01	4.29 ± 0.05	12.89
		Bio-fertilizer	6.95 ± 0.08	7.07 ± 0.07	0.22 ± 0.00	5.95 ± 0.01	20.19
Catch crop (pea)	100%	Control	7.51 ± 0.11	4.56 ± 0.04	1.22 ± 0.01	7.56 ± 0.08	25.85
		Bio-fertilizer	8.95 ± 0.12	4.86 ± 0.05	1.24 ± 0.01	8.48 ± 0.11	28.53
	50%	Control	4.64 ± 0.04	2.24 ± 0.02	1.02 ± 0.01	7.10 ± 0.04	15.00
		Bio-fertilizer	6.62 ± 0.08	4.05 ± 0.03	1.15 ± 0.02	8.14 ± 0.03	19.96
Straw	100%	Control	6.06 ± 0.05	4.21 ± 0.05	0.27 ± 0.00	5.01 ± 0.05	15.55
		Bio-fertilizer	10.35 ± 0.14	4.97 ± 0.01	2.48 ± 0.02	8.40 ± 0.06	26.20
	50%	Control	5.86 ± 0.12	4.20 ± 0.02	0.31 ± 0.01	4.80 ± 0.05	15.17
		Bio-fertilizer	8.57 ± 0.09	1.76 ± 0.01	0.67 ± 0.01	5.81 ± 0.02	16.81
Farmyard manure (FYM)	100%	Control	4.71 ± 0.08	3.53 ± 0.02	0.22 ± 0.00	2.88 ± 0.01	11.34
		Bio-fertilizer	6.12 ± 0.11	3.37 ± 0.01	0.31 ± 0.00	5.31 ± 0.05	15.11
	50%	Control	4.12 ± 0.05	2.36 ± 0.04	0.21 ± 0.00	2.31 ± 0.04	9.00
		Bio-fertilizer	8.28 ± 0.06	1.89 ± 0.02	0.58 ± 0.01	7.50 ± 0.05	18.25
	Mean		6.70	3.98	0.75	5.95	17.39
Organic sources	Control		6.37 c	5.43 a	0.57 c	5.49 c	17.86 c
	Catch crop		6.93 b	3.93 b	1.16 a	7.82 a	19.84 a
	Straw		7.71 a	3.79 b	0.93 b	6.01 b	18.43 b
	FYM		5.81 d	2.79 c	0.33 d	4.50 d	13.43 d
Mineral fertilization	100% NPK fertilization		7.12 a	4.57 a	0.87 a	6.04 a	20.12 a
	50% NPK fertilization		6.19 b	3.43 b	0.55 b	5.74 a	17.66 b
Bio-fertilizer	Control		5.47 b	3.74 b	0.46 b	4.87 b	14.54 b
	Bio-fertilizer		7.94 a	4.23 a	1.03 a	7.08 a	20.24 a

Different letters following the mean values indicate significant differences with the Tukey test at $p \leq 0.05$ for a given feature for organic source, mineral fertilization and bio-fertilizer.

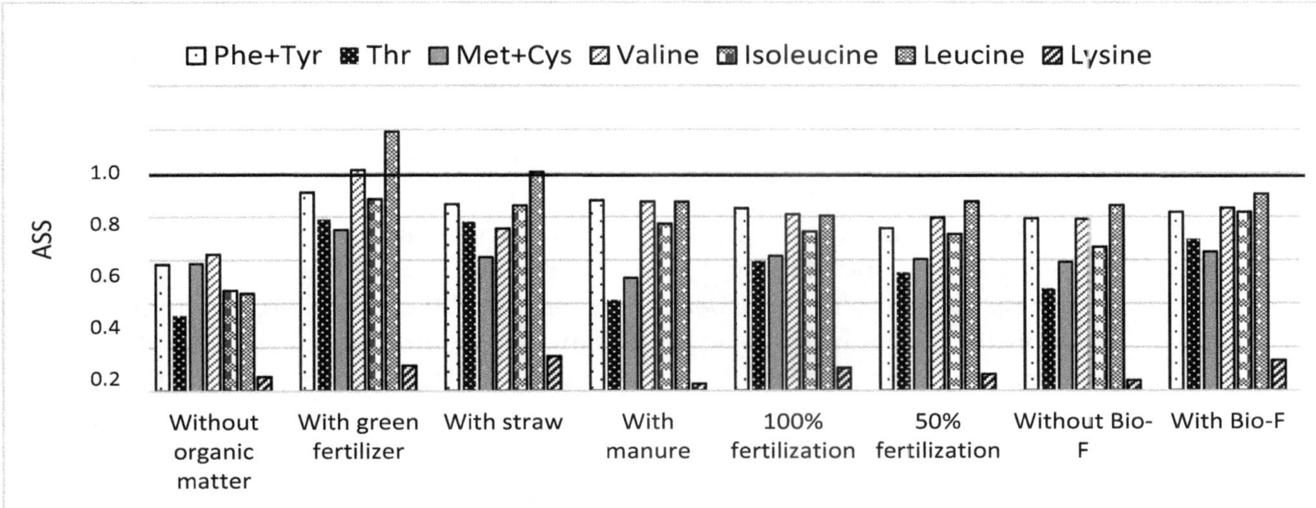

Figure 2. The ASS of the potato tubers as affected by organic, mineral and bio fertilization studied. WHO/FAO adult maintenance pattern expressed as g amino acid/100 g protein: Phe +Tyr, 3.8; Met + Cys, 2.7; Lys, 4.5; Thr, 2.3; Leu, 5.9; Val, 3.9; Ile, 3.0; His, 1.5 (WHO/FAO, 2007).

4. Discussion

The content of organic matter is the basic indicator of the soil quality and fertility and the key factor of the adequate growth, development, and efficiency of plants [25–27]. The UGmax bio-fertilizer, one of the products applied in crop cultivation, is a natural liquid concentrate, including beneficial microorganisms, including the bacteria of lactic acid, photosynthetic bacteria, Azotobacter, Pseudomonas, yeast, Actinobacteria, and some macro-and microelements [7]. However, the results of research on the effect of that product are sometimes ambiguous. A few authors did not confirm a positive effect of UGmax [28–30], while others confirmed a favorable effect of that conditioner [31,32]. Piotrowska et al. [33,34] found that the application of Ugmax increased the soil pH, the content of organic carbon (Corg), the total content of nitrogen, and the enzymatic activity of soil as compared with the control. Whereas, Dębska et al. [35] recorded an increase in the content of organic carbon in soil and in organic matter, including humins and humic acids. Many authors applied UGmax and noted a tendency of growing yields or a favorable effect on the chemical composition of the yield and an increased resistance of the plant to diseases. Sulewska et al. [31] report on the yield of maize grown for grain and silage increasing by 0.71 and 5.6 $t{\cdot}ha^{-1}$, respectively, as compared with the control. Kotwica et al. [36] noted an increase in the biomass of winter wheat and grain yield, while Górski et al. [37] point to an increased content of sugar in the sugar beet roots and a higher root efficiency. Kołodziejczyk et al. [38] report on an increase in the spring wheat grain yield. Zarzecka et al. [39,40] researching UGmax noted an increase in the amount of nitrogen and magnesium as well as in the content of starch and vitamin C in potato tubers, as compared to the control.

It is well known that protein, one of the most vital human food macronutrients, is involved in various physiological functions. The protein content in potato FM is low (2–2.5%). However, its nutritional value results from the beneficial composition of amino acids, especially the essential ones, which cannot be synthesized by the body [10,41] and, as such, they should be supplied with food. Potato protein has a high biological value; the highest one of all the plant proteins [42,43]. As one of the few plant proteins, it has a biological value corresponding to animal protein [44]. It is comparable with soybean protein and only slightly decreased in egg white [44–47]. Of the eight amino acids, people must ingest, seven are found in potato protein: leucine, lysine, isoleucine, and phenylalanine, threonine, methionine and valine [46,47]. Tyrosine does not need to be supplied with food if it contains enough phenylalanine it is made of [20]. It is included in the WHO recommendations as supplied externally (Protein standard WHO/FAO/UNU 2007). According to the American Association for Potato Research, potato has been recognized as food of excellent quality protein, gluten-free, with a number of possibilities of use. The main potato benefits are the lack of allergens, a good texture and a good nutrition value [47–50]. The daily protein requirements for an adult is 0.8 to 1.0 g per kilogram of the body weight, e.g., one with the body weight of 70 kg should consume 70 g of protein daily (including 35 g of animal and 35 g of plant protein).

In the present research, the average total protein content in the dry weight of potato tubers ranged from 99.1 to 114.4 g kg^{-1} of dry weight. In the study reported by Wierzbicka & Trawczyński [51], the protein content ranged from 75.4 to 113.7 g kg^{-1} of dry weight, whereas in the study by Mystkowska [52] it was higher, and it ranged from 129.8 g kg^{-1} to 138.2 g kg^{-1}. The differences in the protein content result from cultivar-specific properties; the capacity for accumulating that nutrient [51–57]. According to Baranowska et al. [58], biostimulants have a significant impact on the protein content, resulting in an increase in its content. A similar positive effect on crude protein content was recorded in the present research following the use of bio-fertilizer, increasing it in potato tubers. Stankiewicz et al. [12] found that changes in protein and essential amino acid contents as well as biological value of protein were mainly cultivar-specific and affected by nitrogen fertilization and weather conditions during potato growth. In the present research, both the full rate of mineral fertilization, i.e., 100 kg N ha^{-1}, 100 kg P_2O_5 ha^{-1}, 150 kg K_2O ha^{-1}, and the bio-fertilizer applied significantly increased the content of crude protein in potato tubers of 'Satina' cultivar. The research reported by Stankiewicz et al. [12] and Mitrus et al. [17] demonstrated that the protein content in potato tubers increased significantly due to nitrogen fertilization. According to Różyło and Pałys [59], potato tubers collected from combined organic and mineral fertilization contained significantly more total protein than the tubers from the plots with mineral fertilization only. In the variant with mineral-organic fertilization, the amount of nitrogen was, on average, 14.9 g kg^{-1}, with organic fertilization only–14.1 g kg^{-1} and without fertilization–13.5 g kg^{-1}. The treatments with organic fertilization at the recommended mineral fertilization rates increased the content of nitrogen, and thus protein, by 20% on average. In the studies reported by other authors [59–61], each form of fertilization resulted in a significant increase in the content of nitrogen in tubers. In addition, according to Różyło and Pałys [59], the application of effective microorganisms has contributed to an increase in the content of nitrogen in tubers by an average of 1.2 g kg^{-1}. As demonstrated in our own research, the use of bio-fertilizer enhanced not only the protein content but also the composition of amino acids, increasing the content of essential amino acids, also with half the rate of mineral fertilization combined with FYM. The use of UGmax in crops can limit the application of mineral fertilizers by facilitating the launch of elements found in soil. The UGmax microelements, enriched with a starter medium, process (compost) crop residue, straw, manure, catch crops (organic fertilizers) and, together with soil minerals, they form humus, the natural habitat of soil life and the storage of nutrients [62]. Besides, according to Piotrowska-Długosz and Wilczewski [63–65], catch crops cultivated for green manure play an important role in improving soil quality by maintaining organic matter content in soil and returning nutrients to the soil. In this study, the most positive effect in terms of the quantity and quality of protein in potato tubers was recorded for the plots where green fertilizer (fodder pea) was involved.

The main free amino acids in potato research when considering the essential amino acids were methionine, phenylalanine, isoleucine, threonine, valine, leucine, tyrosine and lysine, while when it comes to non-essential amino acids–mostly glutamic acid and, sequentially, aspartic acid, asparagine, and alanine, respectively. Many authors [66,67] report on a high metabolic activity and a great importance of microorganisms for most processes occurring in the soil environment, with the decomposition and mineralization of organic matter (crop residue, manure and other natural fertilizers, composts, catch crops) being most important. Similar results were recorded in our research when the use of organic sources and bio-fertilizer resulted in a significant increase in essential and non-essential amino acids.

AAS for essential amino acids (Figure 2) exceeded 100% in potato tubers from plots with the use of green fertilizers and straw, especially leucine and valine, the content of that amino acids was higher than recommended by the WHO/FAO. Leucine and valine, in addition to their constituent function, also play a regulatory role in the effect on the secretion of hormones and catecholamines, the formation of neurotransmitters, and stimulation of protein synthesis. They can be a source of energy, influence wellbeing, reduce physical fatigue. They are a substrate for the synthesis of other amino acids and are an important factor in the treatment of certain diseases, as obesity [68].

5. Conclusions

This article has analyzed the content of CP and the profile of amino acids in 'Satina' potato tubers from a 3-year field experiment where a full and half-decreased mineral fertilization, organic sources (pea, straw, and FYM) and UGmax bio-fertilizer were applied. The application of the bio-fertilizer increased the crude protein, essential and non-essential amino acid contents in potato tubers. The highest content of crude protein was found following the use of catch crop (pea) as organic sources with a full rate of mineral fertilization (nitrogen, phosphorus, and potassium) and the bio-fertilizer. On average for the other factors, organic sources and bio-fertilizer significantly increased the essential-and non-essential amino acid content. The reduction in the NPK rate down to 50% revealed a significant negative effect on the total crude protein content, with no negative effect on the content of individual amino acids in potato tuber protein, especially after the application of organic sources and bio-fertilizer. The content of essential amino acids depending on the rate of mineral fertilizers was comparable. Based on the AAS value, the best results were recorded after the application of stubble catch crop (pea) and straw in terms of the content of essential amino acids in potato tubers, especially valine and leucine. The results of the research presented can be useful for selecting the most valuable fertilization combination, with a possibility of using the agricultural products, e.g., the effect of straw, green fertilizers or FYM on the nutrition value of potato tubers, especially the content of essential amino acids. The application of bio-fertilizer increases the effectiveness of organic fertilizers, it allows for limiting the application of mineral fertilizers, and thus provides a possibility of eco-friendly cultivation.

Author Contributions: Conceptualization, D.W.; methodology, D.W. and M.S.; validation, D.W. and M.S.; formal analysis. D.W.; investigation, D.W.; data curation, D.W.; writing—original draft preparation, D.W and M.S.; writing—review and editing, D.W. and M.S.; visualization, D.W. and M.S., project administration. D.W. All authors have read and agreed to the published version of the manuscript.

Acknowledgments: I thank the Elke Pawelzik for the opportunity to conduct laboratory tests in Department of Quality Plant Products, Georg August University in Göttingen.

References

1. Pikuła, D. Environmental aspects of managing the organic matter in agriculture. *Econ. Reg. Stud.* **2015**, *8*, 98–112. [CrossRef]
2. Rutkowska, A.; Pikuła, D. Effect of crop rotation and nitrogen fertilization on the quality and quantity of soil organic matter. *Soil Process. Curr. Trends Qual. Assess.* **2013**, 249–267. [CrossRef]
3. Szczepanek, M.; Siwik-Ziomek, A. P and K accumulation by rapeseed as affected by biostimulant under different NPK and S fertilization doses. *Agronomy* **2019**, *9*, 477. [CrossRef]
4. Szczepanek, M.; Wszelaczyńska, E.; Pobereżny, J.; Ochmian, I. Response of onion (*Allium cepa* L.) to the method of seaweed biostimulant application. *Acta Sci. Pol. Hort. Cult.* **2017**, *16*, 113–122.
5. Dobrzański, A.; Anyszka, Z.; Elkner, K. Response of carrots to application of natural extracts from seaweed (Sargassum sp.)—Algaminoplant and from leonardite–humiplant. *J. Res. Appl. Agric. Eng.* **2008**, *53*, 53–58.
6. Wichrowska, D.; Wszelaczyńska, E.; Pobereżny, J. Effect of nutrient supply from different sources on some quality parameters of potato tubers. *J. Elementol.* **2015**, 217–230. [CrossRef]
7. Trawczyński, C.; Bogdanowicz, P. The use of Soil Fertiliser in the aspect of ecological potato cultivation. *J. Res. Appl. Agric. Eng.* **2007**, *52*, 94–97.
8. Jabłoński, K. The effect of new generation multi-component fertilization on potato yield and quality. *Zesz. Probl. Post. Nauk Rol.* **2006**, *511*, 309–315.
9. Jeske, M.; Pańka, D.; Wichrowska, D. Effect of chemical protection, organic fertilization and UGmax soil conditioner on health status of potato tubers. *Prog. Plant Prot.* **2015**, *55*, 92–97.
10. Zarzecka, K.; Gugała, M. Nutritional. dietary and medicinal value of potatoes. *Econ. Guide* **2011**, *10*, 14.

11. Danilchenko, H.; Pranaitiene, R.; Tarasieviciene, Z.; Venskutoniene, E. The effect of inhibitors on the amino acid content in the stored potato tubers. *Zesz. Probl. Post. Nauk Rol.* **2008**, *530*, 301–316.
12. Stankiewicz, C.; Bombik, A.; Rymuza, K.; Starczewski, J. The effect of selected agrotechnical treatments on the content of protein and exogenous amino acids in potato tubers of the Irga and Ekra variety during their storage. *Zesz. Probl. Post. Nauk Rol.* **2008**, *530*, 281–291.
13. Kapoor, A.C.; Desborough, S.L.; Li, P.H. Potato tuber proteins and their nutritional quality. *Potato Res.* **1975**, *18*, 469–478. [CrossRef]
14. Knorr, D. Protein quality of the potato and potato protein concentrates. *Lebensm. Wiss. Technol.* **1978**, *11*, 109–115.
15. Knoor, D. Effect of recovery methods on yield. quality and functional properties of potato protein concentrates. *J. Food Sci.* **1980**, *45*, 1183–1186. [CrossRef]
16. Knorr, D.; Kohler, G.O.; Betschart, A.A. Potato protein concentrates: The influence of various methods of recovery upon yield compositional and functional characteristics. *J. Food Process. Preserv.* **1977**, *2*, 235–247. [CrossRef]
17. Mitrus, J.; Stankiewicz, C.; Steć, E.; Kamecki, M.; Starczewski, J. The influence of selected cultivation on the content of total protein and amino acids in the potato tubers. *Plant Soil Environ.* **2003**, *4*, 131–134. [CrossRef]
18. Pęksa, A.; Rytel, E.; Kita, A.; Lisińska, G.; Tajner-Czopek, A. The properties of potato protein. *Food* **2009**, *3*, 79–87.
19. Eppendorfer, W.H.; Eggum, B.O. Dietary fibre, starch, amino acids and nutritive value of potatoes asaffected by sulfur, nitrogen, phosphorus, potassium, calcium and water stress. *Acta Agric. Scand. Sect. B Soil Plant Sci.* **1994**, *44*, 107–115.
20. World Health Organization. *Protein and Amino Acid Requirements in Human Nutrition: Report of a Joint WHO/FAO/UNU Expert Consultation*; WHO Technical Report Series 935; World Health Organization: Geneva, Switzerland, 2007; Volume 935, p. 284.
21. Gawęcki, J. *Human Nutrition. Basics of Nutrition Science*; WNP: Warszawa, Poland, 2011; pp. 210–215.
22. Galdón, B.R.; Mesa, D.R.; Rodrígez, E.M.R.; Romero, C.D. Amino acid content in traditional potato cultivars from the Canary Islands. *J. Food Compos. Anal.* **2010**, *23*, 148–153. [CrossRef]
23. Sweeney, R.A.; Rexroad, P.R. Comparison of LECO FP-228 "nitrogen determinator" with AOAC copper catalyst Kjeldahl method for crude protein. *J. Assoc. Off. Anal. Chem.* **1987**, *70*, 1028–1030. [CrossRef] [PubMed]
24. Fisher, G.H.; Arias, I.; Quesada, I.; D'Aniello, S.; Errico, F.; Di Fiore, M.M.; D'Aniello, A. A fast and sensitive method for measuring picomole levels of total free amino acids in very small amounts of biological tissues. *Amino Acids* **2001**, *20*, 163–173. [CrossRef] [PubMed]
25. Haynes, R.J. Labile organic matter fractions as central components of the quality of agricultural soils: An overview. *Adv. Agron.* **2005**, *85*, 221–268.
26. Lal, R. Sequestering carbon in soils of agro-ecosystems. *Food Policy* **2011**, *36*, S33–S39. [CrossRef]
27. Krasowicz, S.; Oleszek, W.; Horabik, J.; Debicki, R.; Jankowiak, J.; Styczyński, T.; Jadczyszyn, J. Rational management of the soil environment in Poland. *Pol. J. Agron.* **2011**, *7*, 43–58.
28. Kaczmarek, Z.; Jakubus, M.; Grzelak, M.; Mrugalska, L. Impact of the addition of various doses of effective microorganisms to arable-humus horizons of mineral soils on their physical and water properties. *J. Res. Appl. Agric. Eng.* **2008**, *53*, 118–121.
29. Dinesh, R.; Srinivasan, V.; Hamza, S.; Manjusha, A. Short-term incorporation of organic manures and biofertilizers influences biochemical and microbial characteristics of soils under an annual crop [Turmeric (*Curcuma longa* L.)]. *Bioresour. Technol.* **2010**, *101*, 4697–4702. [CrossRef]
30. Martyniuk, S. Effective and ineffective microbial preparations used in plant protection and production and methods of their evaluation. *Post. Mikrobiol.* **2011**, *50*, 321–328.
31. Sulewska, H.; Szymańska, G.; Pecio, A. Evaluation of UGmax soil additive applied in maize grown for grain and silage. *J. Res. Appl. Agric. Eng.* **2009**, *54*, 120–125.
32. Ratajczak, K.; Sulewska, H.; Szymańska, G.; Wolna-Maruwka, A.; Faligowska, A. The effect of soil type and soil additives on the selected growth parameters and yield of flowerheads of Calendula officinalis L. *Herba Pol.* **2016**, *62*, 17–30. [CrossRef]
33. Piotrowska, A.; Długosz, J.; Zamorski, R.; Bogdanowicz, P. Changes of enzymatic activity in soil supplemented with microbiological preparation UGmax. In Proceedings of the World Congress of Soil Science, Soil Solutions for a Changing World, Brisbane, Australia, 1–6 August 2010; pp. 5–8.

34. Piotrowska, A.; Długosz, J.; Zamorski, R.; Bogdanowicz, P. Changes in some biological and chemical properties of an arable soil treated with the microbial biofertilizer UGmax. *Pol. J. Environ. Stud.* **2012**, *21*, 455–463.
35. Dębska, B.; Długosz, J.; Piotrowska-Długosz, A.; Banach-Szott, M. The impact of a bio-fertilizer on the soil organic matter status and carbon sequestration–results from a field-scale study. *J. Soils Sediments* **2016**, *16*, 2335–2343. [CrossRef]
36. Kotwica, K.; Jaskulska, I.; Piekarczyk, M.; Wasilewski, P.; Gałęzewski, L.; Kulpa, D. Effect of soil conditioning and the application of biostimulants on the productivity of winter wheat in crop rotation and short-term monoculture. *Fragm. Agron.* **2013**, *30*, 55–64.
37. Górski, D.; Gaj, R.; Piszczek, J.; Ulatowska, A. Impact of soil fertilizer UGmax on leaf infection degree by leaf spot disease (Cercosporabeticola Sacc.) and yield and root quality of sugar beet. *Prog. Plant Prot.* **2015**, *55*, 195–201.
38. Kołodziejczyk, M. Effect of nitrogen fertilization and microbial preparations on potato yielding. *Plant Soil Environ.* **2014**, *60*, 379–386. [CrossRef]
39. Zarzecka, K.; Gugała, M.; Mystkowska, I.; Sikorska, A. Influence of the soil conditioner UGmax on nitrogen, phosphorus and magnesium content in potato tubers. *Acta Sci. Pol. Agric.* **2014**, *13*, 93–101.
40. Zarzecka, K.; Gugała, M.; Sikorska, A.; Mystkowska, I. The impact of the soil conditioner UGmax on selected qualitative characteristics of potato tubers. *Appl. Ecol. Environ. Res.* **2018**, *16*, 39–50. [CrossRef]
41. Pęksa, A. Potato protein characteristics and properties. *Post. Nauk Rol.* **2003**, *5*, 79–94.
42. Miedzianka, J.; Pęksa, A.; Aniołowska, M. Properties of acetylated potato protein preparations. *Food Chem.* **2012**, *133*, 1283–1291. [CrossRef]
43. Nemś, A.; Miedzianka, J.; Pęksa, A.; Kita, A. The content of health-promoting compounds in potatoes of varieties with different flesh color. *Bromatol. Toxicol. Chem.* **2015**, *48*, 473–478.
44. Emitazi, G.; Nader, A.; Etemadifar, Z. Effect of nitrogen fixing bacteria on growth of potato tubers. *Adv. Food Sci.* **2004**, *26*, 56–58.
45. Leszczyński, W. Nutritional value of potato and potato products. *Biull IHAR* **2012**, *266*, 5–20.
46. Zimnoch-Guzowska, E.; Flis, B. Genetic basis for potato quality characteristics. *Zesz. Probl. Post. Nauk Rol.* **2006**, *511*, 23–26.
47. Elfaki, A.E.; Abbsher, A.M. Nutritional situation of potato subjected to Sudanese cooking methods. *J. Appl. Sci. Res.* **2010**, *6*, 980–984.
48. Harkema, J. Potato Proteins "Free From" Texture & Nutrition. In Proceedings of the Conference Solanic Potato Proteins-Free from Food Expo, Barcelona, Spain, 4–5 June 2015; Available online: http://www.freefromfoodexpo.com/pdf/2015-conference-solanic.pdf (accessed on 1 August 2020).
49. Murray, R.K.; Granner, D.K.; Rodwell, V.W. *Harper Biochemistry Illustrated*, 6th ed.; Medical Publisher PZWL: Warsaw, Poland, 2008; pp. 295–300.
50. Pszczółkowski, P.; Sawicka, B. Protein nutritional value of selected potato varieties. In *Bioproducts—Acquisition. Properties and Application in Food Production*; Lawendowicz, G., Thanh-Blicharz, J.L., Eds.; Agriculture University Publisher of Poznań: Poznań, Poland, 2016; pp. 56–64.
51. Wierzbicka, A.; Trawczyński, C. Factors affecting the content and yield of protein in potato tubers. *Biull. IHAR* **2012**, *266*, 181–190.
52. Bártová, V.; Bárta, J.; Diviš, J.; Švajner, J.; Peterka, J. Crude protein content in tubers of starch processing potato cultivars in dependence on different agro-ecological conditions. *J. Cent. Eur. Agric.* **2009**, *10*, 57–66.
53. Mystkowska, I.T. Content of total and true protein in potato tubers in changing weather conditions under the influence of biostimulators. *Acta Agrophys.* **2018**, *25*, 475–483. [CrossRef]
54. Gugała, M.; Zarzecka, K.; Mystkowska, I.; Sikorska, A. The influence of weed control methods on total protein and true protein in table potato tubers. *Acta Sci. Pol. Agric.* **2014**, *13*, 3–11.
55. Haase, T.; Schüler, C.; Piepho, H.P.; Thöni, H.; Hess, J. The effect of preceding crop and pre-sprouting on crop growth, N use and tuber yield of maincrop potatoes for processing under conditions of N stress. *J. Agron. Crop Sci.* **2007**, *193*, 270–291. [CrossRef]
56. Wichrowska, D.; Wojdyła, T.; Rogozińska, I. Concentrations of some macroelements in potato tubers stored at 4 °C and 8 °C. *J. Elementol.* **2009**, *14*, 373–382.
57. Zarzecka, K.; Gugała, M. Content of total protein and true protein in potato tubers depending on the soil tillage and weed control methods. *Acta Sci. Pol. Agric.* **2006**, *5*, 107–115.

58. Baranowska, A.; Zarzecka, K.; Mystkowska, I.; Gugała, M. The impact of the soil fertilizer UGmax on the yield of commercial fraction, the content of protein nitrogen and nitrates in potato tubers (*Solanum tuberosum* L.). *Zesz. Probl. Post. Nauk Rol.* **2016**, *585*, 3–11.
59. Różyło, K.; Pałys, E. Effect of fertilisation and soil conditions on the chemical composition of potato tubers and their health status. *Zesz. Probl. Post. Nauk Rol.* **2006**, *511*, 279–286.
60. Gianquinto, G.; Bona, S. The significance of trends in concentrations of total nitrogen and nitrogenous compounds. In *Management of Nitrogen and Water in Potato Production*; Haverkort, A.J., MacKerron, D.K.L., Eds.; Wageningen Academic Publishers: Wageningen, The Netherlands, 2000; pp. 35–54.
61. Piikki, K.; Vorne, V.; Ojanperä, K.; Pleijel, H. Impact of elevated O_3 and CO_2 exposure on potato (*Solanum tuberosum* L. cv. Bintje) tuber macronutrients (N, P, K, Mg, Ca). *Agric. Ecosyst. Environ.* **2007**, *118*, 55–64. [CrossRef]
62. Długosz, J.; Orzechowski, M.; Piotrowska, A.; Smólczyński, S.; Bogdanowicz, P. Changes in some soil properties under the influence of the soil fertilizer UGmax. *Pol. J. Environ. Stud.* **2012**, *21*, 32–34.
63. Piotrowska-Długosz, A.; Wilczewski, E. Influence of field pea (Pisum sativum L.) as catch crop cultivated for green manure on soil phosphorus and P-cycling enzyme activity. *Arch. Agron. Soil Sci.* **2020**, 1570–1582. [CrossRef]
64. Piotrowska-Długosz, A.; Wilczewski, E. Soil phosphatase activity and phosphorus content as influenced by catch crops cultivated as green manure. *Pol. J. Environ. Stud.* **2014**, *23*, 157–165.
65. Piotrowska-Długosz, A.; Wilczewski, E. Assessment of soil nitrogen and related enzymes as influenced by the incorporation time of field pea cultivated as a catch crop in Alfisol. *Environ. Monit. Assess.* **2014**, *186*, 8425–8441. [CrossRef]
66. Dahm, H.; Wrótniak-Drzewiecka, W.; Pauter, A. Microbial biofertilizers. In *Physical, Chemical and Biological Processes in Soils*; Szajdak, L.W., Karabanow, A.K., Eds.; Prodruk: Poznań, Poland, 2010; pp. 537–547.
67. Nannipieri, P.; Falchini, L.; Landi, L.; Pietramellara, G. Management of soil microbiota. In *Biological Resource Management*; Balazs, E., Galante, E., Lynch, J.M., Schepers, J.S., Werner, D., Toutant, J.-P., Werry, P.A., Eds.; Springer: Berlin/Heidelberg, Germany, 2000; pp. 237–255.
68. Layman, D.K.; Boileau, R.A.; Erickson, D.J.; Painter, J.E.; Shiue, H.; Sather, C.; Christou, D.D. A reduced ratio of dietary carbohydrate to protein improves body composition and blood lipid profiles during weight loss in adult women. *J. Nutr.* **2003**, *133*, 411–417. [CrossRef]

2

Efficiency of Mineral Nitrogen Fertilization in Winter Wheat under Pannonian Climate Conditions

Gerhard Moitzi [1,*], Reinhard W. Neugschwandtner [2], Hans-Peter Kaul [2] and Helmut Wagentristl [1]

1 Experimental Farm Groß-Enzersdorf, Department of Crop Sciences, University of Natural Resources and Life Sciences, Vienna (BOKU), Schloßhofer Straße 31, 2301 Groß-Enzersdorf, Austria; helmut.wagentristl@boku.ac.at

2 Department of Crop Sciences, Institute of Agronomy, University of Natural Resources and Life Sciences, Vienna (BOKU), Konrad-Lorenz-Straße 24, 3430 Tulln an der Donau, Austria; reinhard.neugschwandtner@boku.ac.at (R.W.N.); hans-peter.kaul@boku.ac.at (H.-P.K.)

* Correspondence: gerhard.moitzi@boku.ac.at.

Abstract: Improvements in nitrogen (N) use efficiency in crop production are important for addressing the triple challenges of food security, environmental degradation and climate change. The three fertilizers, calcium ammonium nitrate (CAN), urea (Urea) and stabilized urea ($Urea_{stab}$), were applied at a rate of 160 kg N ha^{-1} with two or three splits to winter wheat (*Triticum aestivum* L.) in the Pannonian climate region of eastern Austria. On average, over all fertilization treatments, the grain yield (GY) increased by about a quarter and the grain N concentration (GNC) doubled compared to the control without fertilization. Consequently, the grain N yield (NY_{GRAIN}) was increased with N fertilization by 154%. The GY increased due to a higher grain density with no differences between N fertilizers but with a tendency of a higher grain yield with three compared to two splits. Three splits also slightly increased the GNC and consequently the NY_{GRAIN} of CAN and $Urea_{stab}$ in one year. The removal of N fertilizer with the NY_{GRAIN} (N surplus) was higher than the amount of applied fertilizer. Fertilization decreased the N use efficiency (NUE), the N uptake efficiency (NUpE) and the N utilization efficiency (NUtE) but increased the soil mineral nitrate (NO_3-N) at harvest and the apparent N loss (ANL). Three compared to two applications resulted in a higher NO_3-N at harvest but also a lower N surplus due to partly higher NY_{GRAIN}. Consequently, the ANL was lower with three compared to two splits. Also, the NUpE and the apparent N recovery efficiency (ANRE) were higher with three splits. The best N treatment regarding highest above-ground biomass yield with lowest N surplus, N balance and ANL was the three-split treatment (50 CAN, 50 CAN, 60 liquid urea ammonium nitrate). Three splits can, under semi-arid conditions, be beneficial when aiming high-quality wheat for bread-making and also for reducing the N loss. Whereas, two splits are recommended when aiming only at high GY, e.g., for ethanol-wheat production.

Keywords: winter wheat; mineral N fertilizer; N splitting; crop yield; N yield; apparent N loss; N use efficiency

1. Introduction

Nitrogen (N) is a major driver in crop production, but too much nitrogen is harmful for the ecosystems and potentially also to human health [1]. Nitrogen management in agriculture aims to achieve agronomic objectives (farm income, high crop and livestock productivity) and environmental objectives (minimal N losses) simultaneously. Nitrogen use efficiency (NUE) and their components are indirect measurements of the sustainability of cereals production systems [2–4]. Improving the NUE, especially the fraction of N input harvested as product, is one of the most effective means of increasing crop productivity while decreasing environmental degradation [5]. In the global analysis of cropping systems (also including mixed crop-livestock production systems, which use manure N) using the FAO (Food and Agriculture Organization of the United Nations) database, Zhang et al. [5] reported a N output-input ratio (the fraction of N input harvested as product) of 0.42 kg harvest N kg^{-1} input N, and for Europe, 0.52 kg harvest N kg^{-1} input N. The global target annual output-input ratio of N for 2050, using FAO projections for food demand and the allowable N pollution, was calculated with 0.67 kg harvest N kg^{-1} input N, and for Europe, 0.75 kg harvest N kg^{-1} input N.

Nitrogen fertilization is increasing grain yield (GY) and grain quality (protein concentration) of wheat [6]. But, fertilizer production energy is also the highest portion of the energy input in arable farming, even ahead of soil tillage [7]. The amount of fertilizer and the fertilizing strategy are influencing the energy efficiency, as shown for maize [8] and wheat [7]. An efficient N use is therefore a key indicator in sustainable plant production. Basically, there are two primary efficiency aspects: N fertilizer efficiency and crop N efficiency. N fertilizer efficiency is the fraction of applied fertilizer N that is recovered in the crop [2]. Crop N efficiency can be partitioned into the capture of N by roots (N uptake efficiency (NUpE) = crop N uptake/N available) and its conversion to grain by shoots (N utilization efficiency (NUtE) = GY/crop N uptake) [9]. NUpE is a breeding goal for wheat, whereas there was no need in breeding independently for increasing the NUtE because this automatically went up with improved yield [10]. New varieties encouraged excessive use of fertilizer with consequences for the environment which have become all too apparent according to Barraclough et al. [2], who described four key variables determining N efficiency in a wheat crop—GY, grain N concentration (GNC), N yield of the above-ground biomass (NY_{AGB}) and nitrogen harvest index (NHI), which are ultimately constrained by the law of conservation of matter. Improving the NUtE at a given N uptake and NHI can only be achieved at the expense of GNC. A high GY with a high grain quality (i.e., a high GNC) requires a high input and uptake of nitrogen. A comparison of high-yielding European hybrid vs. line cultivars of winter wheat showed on average a higher GY but a lower GNC, mainly due to higher harvest index for hybrid cultivars compared to line cultivars. The current wheat hybrids seem to be more efficient in overall N use because they are better converting comparable amounts of N taken up into grain biomass, thus they have a higher NUtE [11].

Besides genotype, the N fertilization management also determines the N efficiency in plant production [3,12,13]. The N management variables are amount of fertilized N, the source of N fertilizer (organic N, inorganic N: nitrate N, ammonium N, carbamide N), splitting (one, two or three doses) and timing of application.

N efficiency can be improved by matching N applications to crop N demand throughout the growing season by applying N in several smaller doses at periods of high N demand during rapid crop growth [6,14]. Many studies (e.g., References [14,15]) showed an increase of GY and N efficiency and a reduction of N losses through splitting of applied N fertilizer. A split application reduces lodging and N losses by leaching as N application and N uptake can be balanced both in time and amount [16]. Other research [17–19] indicates that weather conditions and total N supply are more important for yield and crude protein contents in cereals than splitting and the timing of applications. The effect of N splitting is small, if the total amount of N is sufficient [20] and the last N application cannot be taken up efficiently by the plants, particularly under dry weather conditions in May and June during the heading stage [21]. Especially under climate conditions with dry periods, N fertilization is more challenging.

Studies on efficiency of N fertilization were often carried out in wet climate regions, with the aim of mitigating environmental pollution (N leaching, gaseous emission of ammonia and nitrous oxide). There is a lack of studies on N fertilizer efficiency under climate conditions of the Pannonian region (Eastern part of Austria, Hungary, parts of Slovakia, the Czech Republic, Romania, Serbia, Croatia and the Ukraine). The aim of this study was therefore to quantify the effect of N fertilizer (calcium ammonium nitrate, urea, stabilized urea, incorporated urea, liquid urea ammonium nitrate) applied in various doses and phases of plant growth (tillering, stem elongation and ear emergence) on (i) crop yield and yield components, (ii) plant N concentrations and N yield, (iii) N balance and apparent N loss and (iv) different N efficiency parameters of winter wheat under Pannonian climate conditions.

2. Materials and Methods

2.1. Experimental Site and Climatic Conditions

A two-year experiment was performed in the growing seasons 2016/2017 (2017) to 2017/2018 (2018) at the Experimental Station of the University of Natural Resources and Life Sciences, Vienna (BOKU), in Groß-Enzersdorf (48°11′59.13″ N, 16°33′42.45″ E, 154 m elevation). The field site was located in the east of Vienna (Austria) on the edge of the Marchfeld plain, which is an important crop production region in the north-western part of the Pannonian Basin. The region is in the transition zone between the semi-humid Western-European climate and the continental East-European one. According to the phytogeographical and climatological aspect, Marchfeld is part of the Pannonicum, with high levels of sunshine, high average temperature during the growing period and low precipitation. The silt loam soil (pH_{CaCl_2}: 7.6, soil organic carbon: 16.3 g kg^{-1}) is classified as a Calcaric Chernozem of alluvial origin [22]. The Pannonian climate area is characterized by hot summers with low rainfall and cold winters with little snow. The mean annual temperature is 10.8 °C, the mean annual precipitation is 568 mm (1994–2018; [23]). Long-term precipitation pattern shows that most rainfall occurs from May to September, with monthly values above 60 mm and with the highest rainfall in July (76 mm). The mean monthly temperature and mean total precipitation during the growing season of winter wheat from October to June are shown in Table 1. Long-term average temperature and precipitation during the growing season of winter wheat (from October until June) were 8.0 °C and 361 mm (1994–2018). The deviations in the growing season of the experimental years were 0.0 °C and –48 mm (2016/2017) and +1.1 °C and +10 mm (2017/2018). The growing season of 2017 was thereby dryer and the growing season of 2018 was warmer than the long-term average. Due to a high rainfall deficiency in the year 2017 (Table 1), a supplemental irrigation with 30 mm of water was performed on the 23 May.

Table 1. Long-term average monthly temperature and precipitation (1994–2018) and deviations during the growing seasons 2017 und 2018 (weather station: Groß-Enzersdorf).

	Temperature (°C)			Precipitation (mm)		
	1994–2018	**2017**	**2018**	**1994–2018**	**2017** [1]	**2018**
October	10.4	−0.8	+1.7	39	+25	+13
November	5.8	−1.0	+0.4	33	+3	−2
December	1.2	−0.4	+1.7	30	−17	−3
January	0.3	−4.1	+3.6	27	−15	+7
February	2.0	+0.9	−2.9	22	+1	−3
March	5.7	+3.2	−2.4	35	−6	+5
April	11.2	−1.2	+4.1	41	+14	−28
May	16.0	+0.5	+2.5	63	−18	+18
June	19.5	+2.5	+1.6	71	−35	+3

[1] A supplemental irrigation with 30 mm of water was performed on 23 May 2017.

2.2. Experimental Design and Management

The experiment was arranged in a randomized complete block design with three replications. Plot size was 18 m^2 (6 × 3 m). Treatments consisted of nine N fertilization variants with a total N application of 160 kg ha^{-1} and an unfertilized control (Table 2). The three nitrogen fertilizers, calcium ammonium nitrate (CAN), urea (Urea) and stabilized urea ($Urea_{stab}$), were compared with two different fertilization regimes, giving the full N doses in either two splits (at tillering and stem elongation) or in three splits (at tillering, stem elongation and ear emergence) in an orthogonal matrix. An additional three N treatments were tested: $Urea_{stab}$, where the full dose was given at the first split, incorporated urea ($Urea_{incorp}$), with harrowing between plant rows after the first split of 100 kg N ha^{-1} and 60 kg

N ha^{-1} at the second split without harrowing, and CAN + liquid urea ammonium nitrate (CAN + UAN_{liquid}), with CAN given at the first two splits and UAN_{liquid} at the third. Urea-based fertilizers are susceptible to N losses through volatilization, denitrification and leaching, but nitrogen stabilizers (urease and nitrification inhibitors) can improve its efficiency [24]. We used in our study a stabilized urea (46% N), where a nitrification inhibitor decelerates the nitrification of ammonium-N to nitrate-N (Table 2).

Table 2. Fertilizer types and splitting regimes with growth stages (BBCH [**B**iologische **B**undesanstalt, Bundessortenamt and **CH**emical industry] -code according to Witzenberger et al. [25]) at the application dates.

Type of N Fertilizer	Percentage of N in the Fertilizer	Fertilization Schedule	Dose of Fertilizer in the Plant Development Stage (kg N ha^{-1})		
Splits Growth stage			First split [5] Tillering (BBCH 22/23)	Second split [6] Stem elongation BBCH 31/32)	Third split [7] Ear emergence (BBCH 52)
Control (N0)	-	0:0:0	0	0	0
CAN [1]	27	50:50:60	50	50	60
CAN	27	100:60:0	100	60	0
Urea	46	50:50:60	50	50	60
Urea	46	100:60:0	100	60	0
$Urea_{stab}$ [2]	46	50:50:60	50	50	60
$Urea_{stab}$	46	100:60:0	100	60	0
$Urea_{stab}$	46	160:0:0	160	0	0
$Urea_{incorp}$ [3]	46	100:60:0	100 ($Urea_{incorp}$)	60	0
CAN+UAN_{liquid} [4]	27/32	50:50:60	50 (CAN)	50 (CAN)	60 (UAN_{liquid})

[1] Calcium ammonium nitrate (27% N), [2] Stabilized urea (46% N) with nitrification-inhibitor (DCD—Dicyandiamid and 1 H-1,2,4 Triazol)—Alzon®, [3] Incorporated urea, [4] Liquid urea ammonium nitrate (8.0% nitrate N, 8.0% ammonium N, 16.0% N carbamide), [5] 9 March 2017 or 23 March 2018, [6] 25 April 2017 or 27 April 2018, [7] 23 May 2017 or 24 May 2018.

No potassium, phosphorus, magnesium and sulfur fertilization were carried out, because these nutrients were sufficiently plant-available according to former soil analysis.

The winter wheat cultivar Bernstein (quality bread wheat without awns) was sown on the 20 October 2016 and on the 17 October 2017 with a disc seed drill (row distance: 12.5 cm, seeding depth: 3–4 cm) with 340 germinable seeds m^{-2}. Pre-crop was spring barley. Plants were sprayed against weeds in one pass-over (130 g ha^{-1} of Broadway®; active ingredients: Florasulam—22.8 g kg^{-1}, Pyroxsulam—68.3 g kg^{-1} and Cloquintocet—68.3 g kg^{-1}) in mid-April of each year.

2.3. Plant and Soil Sampling and Measurements

Harvest was done at physiological maturity manually on 1 m^2 from the center of each plot on 5 July 2017 and on 27 June 2018. The yield of above-ground biomass (Y_{AGB}) and the yields of grain (GY) and straw (SY) are given at 14% moisture. The ear density (ED, ears m^{-2}) was counted and the thousand kernel weight (TKW) was determined with a seed counter (Condator, Pfeuffer GmbH, Kitzingen, Germany). Thereof, the grains ear^{-1}, the grain density (GD, grains m^{-2}) and the single ear yield (SEY, g) were calculated.

Soil samples for analyzing soil nitrate (NO_3-N) were taken with a soil auger to a depth of 0.9 m, separated into the soil horizons of 0–30, 30–60 and 60–90 cm at start of the growing season in spring (9 March 2017 and 23 March 2018) and at harvest. For each plot, a composite sample consisting of four sub-samples was taken. Soil samples were deep-frozen immediately after sampling. For analyzing NO_3-N, the soil samples were extracted with 0.01 mol L^{-1} $CaCl_2$ solution using a soil extraction ratio of 1:4 (*w*:*v*) for 30 min (shaking frequency 180 min^{-1}) using an over-head shaker. The extracts were filtrated and analyzed with a photometric flow analyzer (FIAstar™ 5000 Analyzer, FOSS, Höganäs, Sweden) [26].

The nitrogen concentrations of grain (GNC) and straw (SNC) were determined by oxidative combustions of the samples followed by reduction of nitrogen oxide to elementary nitrogen and detection with thermo-conductivity detector according Dumas [27] using an elemental analyzer (vario MAX cube CNS, Elementar Analysensysteme GmbH, Germany).

2.4. Calculations

The harvest index (HI) was calculated using Equation (1):

$$\text{HI }(\%) = \frac{\text{GY}}{\text{Y}_{\text{AGB}}} \times 100. \quad (1)$$

where GY or Y_{AGB} are dry yields of grain or above-ground biomass (kg ha^{-1}).

Grain N yield (NY_{GRAIN}) and straw N yield (NY_{STRAW}) were calculated using Equations (2) and (3) according to Liu et al. [13]:

$$\text{NY}_{\text{GRAIN}}\left(\text{kg ha}^{-1}\right) = \frac{\text{GY} \times \text{GNC}}{100} \quad (2)$$

$$\text{NY}_{\text{STRAW}}\left(\text{kg ha}^{-1}\right) = \frac{\text{SY} \times \text{SNC}}{100} \quad (3)$$

where GY and SY are dry grain and straw yields (kg ha^{-1}) and GNC and SNC are the N concentrations (%) in grain and straw. The N yield in the above-ground biomass (NY_{AGB}) and N harvest index (NHI) were calculated using Equations (4) and (5):

$$\text{NY}_{\text{AGB}}\left(\text{kg ha}^{-1}\right) = \text{NY}_{\text{GRAIN}} + \text{NY}_{\text{STRAW}} \quad (4)$$

$$\text{NHI }(\%) = \frac{\text{NY}_{\text{GRAIN}}}{\text{NY}_{\text{AGB}}} \times 100 \quad (5)$$

The N surplus (Equation (6)) or the N balance (Equation (7)) were calculated as the difference of the fertilized N (N_{FERT}, kg ha^{-1}) minus the NY_{GRAIN} or NY_{AGB} according to Hartmann et al. [28]. These simple balance calculations do not include the gaseous losses as well as the atmospheric deposition of N.

$$\text{N surplus}\left(\text{kg ha}^{-1}\right) = \text{N}_{\text{FERT}} - \text{NY}_{\text{GRAIN}} \quad (6)$$

$$\text{N balance}\left(\text{kg ha}^{-1}\right) = \text{N}_{\text{FERT}} - \text{NY}_{\text{AGB}} \quad (7)$$

The apparent net N mineralization (ANM) of the control was calculated as the difference between N recovered at harvest (sum of NY_{AGB} and NO_3-N_{END}) and N supplied (NO_3-N_{START}) using Equation (8) according to Cui et al. [29] and Hartmann et al. [28]:

$$\text{ANM}\left(\text{kg ha}^{-1}\right) = \text{NY}_{\text{AGB}} + \text{NO}_3\text{-N}_{\text{END}} - \text{NO}_3\text{-N}_{\text{START}} \quad (8)$$

where NO_3-N_{END} or NO_3-N_{START} are the soil mineral nitrate contents at harvest or at the start of the growing season in spring in a depth of 0–90 cm. The NO_3-N (0–90 cm) at the start of the growing season was, in 2017 significantly higher than in 2018, with no differences between plots (Table 3). The ANM in the control differed between years, with a lower value in 2017 (with 15.9 kg ha^{-1}), which was the drier year, than in 2018 with 44.6 kg ha^{-1}.

Table 3. Above-ground biomass yield (Y_{AGB}), grain yield (GY), straw yield (SY), harvest index (HI), ear density (ED), grains ear^{-1}, thousand kernel weight (TKW), grain density (GD) and single ear yield (SEY) depending on fertilizer treatment and year.

N Fertilizer	N Splitting	Y_{AGB}	GY	SY	HI	ED	Grains	TKW	GD	SEY
		(kg ha^{-1})	(kg ha^{-1})	(kg ha^{-1})	(%)	(m^{-2})	(ear^{-1})	(g)	(m^{-2})	(g)
Control		10,236 a [6]	4604 a	5632	44.7	380	36.3	35.9	13,013 a	1.27
CAN [1]	50:50:60	12,900 b	5855 ab	7045	45.0	431	39.0	35.8	16,711 b	1.37
CAN	100:60:0	12,267 b	5354 ab	6913	44.0	428	36.8	35.9	15,586 b	1.26
Urea	50:50:60	12,518 b	6052 b	6466	48.1	424	39.0	38.1	16,325 b	1.43
Urea	100:60:0	12,762 b	5786 ab	6976	45.0	439	38.0	35.8	16,777 b	1.32
$Urea_{stab}$ [2]	50:50:60	12,713 b	6236 b	6476	49.4	425	41.8	36.0	17,858 b	1.47
$Urea_{stab}$	100:60:0	12,516 b	5971 b	6545	48.0	452	38.5	35.2	17,355 b	1.33
$Urea_{stab}$	160:0:0	12,846 b	5375 ab	7471	45.0	432	34.0	37.0	14,972 ab	1.22
$Urea_{incorp}$ [3]	100:60:0	12,516 b	5699 ab	6817	45.6	430	38.8	34.9	16,818 b	1.32
CAN + UAN_{liquid} [4]	50:50:60	13,761 b	5897 ab	7864	42.5	449	38.3	35.6	17,051 b	1.32
Year										
2017		14,194 b	6818 b	7376 b	48.3 b	430	49.0 b	32.6 a	20,948 b	1.59 b
2018		10,813 a	4547 a	6265 a	42.4 a	428	27.1 a	39.5 b	11,545 a	1.07 a
ANOVA [5]										
N treatment (N)		*	*	n.s.	n.s.	n.s.	n.s.	n.s.	***	n.s.
Year (Y)		***	***	***	***	n.s.	***	***	***	***
N × Y		n.s. [7]	n.s.	n.s.	*	n.s.	n.s.	n.s.	n.s.	*

[1] Calcium ammonium nitrate, [2] Urea with nitrification inhibitor, [3] Incorporated Urea, [4] Liquid urea ammonium nitrate, [5] Analysis of variance, [6] Different letters indicate significant differences. Significant effects: $p < 0.05$ (*) and $p < 0.001$ (***), [7] Not significant.

The supplied N (N_{SUPP}), which is the total N supply through both fertilizer and N from the soil, was calculated using Equation (9):

$$N_{SUPP}\left(\text{kg ha}^{-1}\right) = N_{FERT} + NO_3\text{-}N_{START} + ANM \quad (9)$$

The apparent N loss (ANL) was calculated as the difference between the total N in the soil-plant system at harvest and the N_{SUPP} using Equation (10) according to Hartmann et al. [28]:

$$ANL\left(\text{kg ha}^{-1}\right) = NY_{AGB} + NO_3\text{-}N_{END} - N_{SUPP} \quad (10)$$

The N use efficiency (NUE) gives the amount of grain which is produced by the N_{SUPP}. It was calculated using Equation (11) according to Barraclough et al. [2]:

$$NUE\left(\text{kg kg}^{-1}\right) = \frac{GY}{N_{SUPP}} \quad (11)$$

The N uptake efficiency (NUpE) was calculated using Equation (12) according to Barraclough et al. [2]:

$$NUpE\ (\%) = \frac{NY_{AGB}}{N_{SUPP}} \times 100 \quad (12)$$

The agronomic N efficiency (ANE) is defined as the increase of GY per unit nitrogen applied and was calculated using Equation (13) according to Yang et al. [3] and Tian et al. [12]:

$$ANE\left(\text{kg kg}^{-1}\right) = \frac{(GY_{FERT} - GY_{CONTROL})}{N_{FERT}} \quad (13)$$

where GYFERT or GYCONTROL are the GY in N treatments or the control.

The apparent N recovery efficiency (ANRE) of the applied N was calculated using Equation (14) according to Tian et al. [12] and Liu et al. [14]:

$$ANRE\ (\%) = \left(\frac{NY_{AGB-FERT} - NY_{AGB-CONTROL}}{N_{FERT}}\right) \times 100 \quad (14)$$

where NYAGB-FERT or NYAGB-CONTROL are the NYAGB (kg ha−1) of the fertilized treatments or the control. A precise method to measure the fertilizer N recovery efficiency is by labelling the fertilizer with 15N to differentiate fertilizer N from the indigenous soil N [2]. The apparent N recovery efficiency is an easier approach but the recovery of the applied N in the fertilized treatments is just 'apparent' as the amount of fertilization is also affecting mineralization, and thereby, the difference in the N uptake of the fertilized and the control treatments might not be solely explained by N fertilization. Grain N utilization efficiency (NUtE) was calculated using Equation (15) according to Barraclough et al. [2]:

$$\text{NUtE}\left(\text{kg kg}^{-1}\right) = \frac{\text{GY}}{\text{NY}_{\text{AGB}}} \quad (15)$$

2.5. Statistical Analysis

Statistical analyses were conducted using IBM®SPSS®Statistics 21. The requirements for analysis of variance (ANOVA) were tested with the Levene test for homogeneity of variances and the Shapiro–Wilk test for normal distribution of residuals. ANOVA tests were carried out to detect growing season (year) and N treatment effects using a general linear model. Multiple comparisons to separate means were carried out with the Student-Newman-Keuls test ($p < 0.05$). For testing the split effect, the orthogonal dataset was used (i.e., two and three splits of CAN, Urea and Ureastab).

3. Results

3.1. Crop Yields and Yield Components

The means for crop yields, HI and yield components (over all fertilization treatments and both years) were: Y_{AGB}—12,504 kg ha^{-1}, GY—5683 kg ha^{-1}, SY—6821 kg ha^{-1}, HI—45.3%, ED—429 m^{-2}, grains ear^{-1}—38.1, TKW—36.0 g, GD—16,247 m^{-2} and SEY—1.33 g (Table 3).

The Y_{AGB} in all N fertilized treatments was higher than in the unfertilized control (by a mean of +24.9% over both years) with no differences between N fertilization treatments.

The GY was highest with $\text{Urea}_{\text{stab}}$ (50:50:60), $\text{Urea}_{\text{stab}}$ (100:60:0) and Urea (50:50:60), and lowest for the control with other treatments showing intermediate values. The GY of N fertilized treatments was +23.4% higher than that of the control (mean over all N fertilization treatments and both years). Splitting of CAN, $\text{Urea}_{\text{stab}}$ and Urea three times (50:50:60) resulted in an increase in GY of 344 kg ha^{-1} compared to two splits (100:60:0) (6048 vs. 5704 kg ha^{-1}, means over fertilizers and years, $p = 0.099$, not significant (n.s.)).

The SY did not differ between treatments and also not between years. The SY of N fertilized treatments was +23.4% higher than that of the control (mean over all N fertilization treatments and both years).

There was a significant N fertilization × year interaction for the HI (Table 4). In 2017, the HI was highest in the control and lowest in CAN 100:60:0, Urea 100:60:0 and $\text{Urea}_{\text{incorp}}$ 100:60:0, with other treatments showing intermediate values. Whereas in 2018, it was highest in $\text{Urea}_{\text{stab}}$ 50:50:60 and $\text{Urea}_{\text{stab}}$ 100:60:0, and lowest in the control. The HI did not differ between the splitting regimes of the individual fertilizers in 2017. Whereas in 2018, applying the fertilizers just once (160:0:0) significantly reduced the HI of $\text{Urea}_{\text{stab}}$ compared to two- and three-times applications, and two applications of Urea also tended to cause a lower HI than three applications (n.s.). The HI was higher for some fertilization treatments in 2017 than in 2018.

Table 4. Harvest index (HI), single ear yield (SEY), grain N yield (NY_{GRAIN}), N harvest index (NHI), N surplus, N use efficiency (NUE) and N utilization efficiency (NUtE) as affected by N treatment × year.

N Fertilizer	N Splitting	HI		SEY		NY_{GRAIN}		NHI		N Surplus		NUE		NUtE	
		(%)		(g)		(kg ha^{-1})		(%)		(kg ha^{-1})		(kg grain kg^{-1} N_{SUPP})		(kg grain kg^{-1} NY_{AGB})	
		2017	2018	2017	2018	2017	2018	2017	2018	2017	2018	2017	2018	2017	2018
Control		54.7 ^{c} [6]	34.4 ^{a}	1.78 ^{c}	0.77 ^{a}	63.8 ^{a}	41.5 ^{a}	91.3 ^{b}	71.5 ^{a}	63.8 ^{a}	−41.5 ^{a}	74.7 ^{b}	40.9 ^{c}	86.7 ^{d}	56.0 ^{d}
CAN [1]	50:50:60	48.3 ^{abc}	41.7 ^{abc}	1.66 ^{abc}	1.07 ^{bcd}	166.6 ^{de}	126.2 ^{cd}	87.0 ^{ab}	82.3 ^{cd}	−6.6 ^{bc}	33.8 ^{bc}	28.2 ^{a}	19.9 ^{ab}	36.7 ^{abc}	30.7 ^{abc}
CAN	100:60:0	42.3 ^{a}	44.3 ^{bbcd}	1.48 ^{ab}	1.04 ^{bcd}	131.2 ^{b}	124.4 ^{cd}	84.0 ^{a}	79.0 ^{bc}	28.8 ^{d}	35.6 ^{bc}	24.7 ^{a}	19.6 ^{ab}	39.7 ^{abc}	29.1 ^{a}
Urea	50:50:60	50.7 ^{bc}	46.0 ^{bcd}	1.69 ^{bc}	1.18 ^{cd}	163.4 ^{cde}	126.6 ^{cd}	89.3 ^{ab}	84.9 ^{cd}	−3.4 ^{c}	33.4 ^{bc}	28.7 ^{a}	21.6 ^{ab}	39.1 ^{abc}	33.4 ^{abc}
Urea	100:60:0	51.3 ^{bc}	38.3 ^{ab}	1.59 ^{abc}	1.04 ^{bcd}	147.9 ^{bcd}	109.9 ^{c}	88.0 ^{ab}	73.1 ^{ab}	12.1 ^{cd}	50.1 ^{c}	28.8 ^{a}	19.2 ^{ab}	42.8 ^{bc}	29.5 ^{ab}
$Urea_{stab}$ [2]	50:50:60	48.0 ^{abc}	50.7 ^{d}	1.67 ^{abc}	1.27 ^{d}	172.5 ^{e}	135.5 ^{d}	87.7 ^{ab}	86.5 ^{e}	−12.5 ^{bc}	24.5 ^{b}	28.5^{a}	21.3 ^{ab}	37.4 ^{abc}	32.9 ^{abc}
$Urea_{stab}$	100:60:0	46.0 ^{ab}	50.0 ^{d}	1.49 ^{ab}	1.16 ^{bcd}	140.4 ^{b}	125.7 ^{cd}	87.0 ^{ab}	85.3 ^{e}	19.6 ^{d}	34.3 ^{bc}	26.5 ^{a}	22.9 ^{b}	41.2 ^{abc}	36.5 ^{bc}
$Urea_{stab}$	160:0:0	47.0 ^{abc}	35.0 ^{a}	1.50 ^{ab}	0.94 ^{ab}	138.8 ^{b}	82.6 ^{b}	86.0 ^{ab}	71.0 ^{a}	21.2 ^{d}	77.4 ^{e}	29.2 ^{a}	15.0 ^{a}	43.3 ^{c}	32.2 ^{abc}
$Urea_{incorp}$ [3]	100:60:0	44.7 ^{ab}	47.0 ^{cd}	1.44 ^{a}	1.20 ^{cd}	146.1 ^{bc}	107.9 ^{c}	82.7 ^{a}	80.9 ^{bcd}	13.9 ^{cd}	52.1 ^{c}	23.7 ^{a}	21.0 ^{ab}	35.9 ^{ab}	38.0 ^{c}
CAN + UAN_{liquid} [4]	50:50:60	48.3 ^{abc}	36.3 ^{a}	1.64 ^{abc}	1.00 ^{abc}	183.5 ^{e}	119.4 ^{cd}	84.7 ^{a}	75.5 ^{ab}	−23.5 ^{b}	40.6 ^{bc}	28.9 ^{a}	18.7 ^{ab}	34.3 ^{a}	27.5 ^{a}
LSD [5]		7.8		0.23		20.2		6.0		20.2		6.6		7.3	

[1] Calcium ammonium nitrate, [2] Urea with nitrification inhibitor, [3] Incorporated Urea, [4] Liquid urea ammonium nitrate, [5] The least significant difference (LSD) is separating means of N treatment × year interactions. [6] Different letters indicate significant differences ($p < 0.05$) within years, N_{SUPP} = supplied N, NY_{AGB} = N yield of above-ground biomass.

The ED did not differ between treatments and between years. Anyhow, it was +14.3% higher in N fertilized treatments compared to the control (mean over all N fertilization treatments and both years).

The number of grains ear^{-1} did not differ between fertilization treatments and between years, but the control had the second lowest value after Urea$_{stab}$ (100:60:0) (n.s.). The TKW did not differ between fertilization treatments.

The GD was lower in the control than in the N fertilized treatments, except for Urea$_{stab}$ (100:60:0), which showed intermediate values. The N fertilized treatments, except for Urea$_{stab}$ (100:60:0), had a GD of about 30% higher than the control (mean of both years).

There was a significant N fertilization × year interaction for the SEY (Table 4). In 2017, the SEY was highest in the control and lowest with Urea$_{incorp}$, with other treatments showing intermediate values. Whereas in 2018, the SEY was lowest in the control and highest with Urea$_{stab}$ 50:50:60, with all fertilized treatments except Urea$_{stab}$ 160:0:0 and CAN + UAN$_{liquid}$ showing higher values than the control. The SEY was higher in all treatments in 2017 than in 2018. Splitting of individual fertilizers tended to decrease the SEY with all fertilizers in 2017 and with Urea in 2018 (n.s.), and significantly decreased the SEY with Urea$_{stab}$ from three to one split in 2018.

On average over all treatments, yield and all yield components except for ED and TKW were higher in 2017 than in 2018: Y$_{AGB}$—+3382 kg ha^{-1} (=31%), GY—+2271 kg ha^{-1} (=50%), SY—+1111 kg ha^{-1} (=18%), HI—+5.9% (=14%), grains ear^{-1}—+21.9 (=81%), GD—+9403 m^{-2} (=82%) and SEY—+0.52 g (=49%). The ED did not differ between the years. The TKW was +6.9 g (=21%) higher in 2018 than in 2017.

3.2. N Concentration, N Yield and N Harvest Index

The means for N concentration, N yield and NHI (over all treatments and both years) were: GNC—2.62%, SNC—0.43%, NY$_{GRAIN}$—127.7 kg N ha^{-1}, NY$_{STRAW}$—25.6 kg N ha^{-1}, NY$_{AGB}$—153.3 kg N ha^{-1} and NHI—82.9% (Table 5).

Table 5. N concentration, N yield and N harvest index depending on fertilizer treatment and year.

N Treatment	N Splitting	N Concentration		N Yield			NHI
		GNC	SNC	NY$_{GRAIN}$	NY$_{STRAW}$	NY$_{AGB}$	
		(%)		(kg ha^{-1})			(%)
Control		1.37 ^{a} [6]	0.23 ^{a}	52.6 ^{a}	11.7 ^{a}	64.3 ^{a}	81.4
CAN [1]	50:50:60	2.94 ^{cd}	0.44 ^{b}	146.4 ^{cd}	26.1 ^{b}	172.5 ^{cd}	84.7
CAN	100:60:0	2.85 ^{cd}	0.52 ^{b}	127.8 ^{bc}	29.7 ^{b}	157.5 ^{bcd}	81.5
Urea	50:50:60	2.82 ^{cd}	0.39 ^{b}	145.0 ^{cd}	21.4 ^{ab}	166.4 ^{bcd}	87.1
Urea	100:60:0	2.67 ^{bcd}	0.50 ^{b}	128.9 ^{bc}	30.5 ^{b}	159.4 ^{bcd}	80.6
Urea$_{stab}$ [2]	50:50:60	2.91 ^{cd}	0.42 ^{b}	154.0 ^{d}	22.4 ^{ab}	175.4 ^{cd}	87.1
Urea$_{stab}$	100:60:0	2.61 ^{bc}	0.39 ^{b}	133.1 ^{bcd}	21.5 ^{ab}	154.6 ^{bc}	86.2
Urea$_{stab}$	160:0:0	2.44 ^{b}	0.45 ^{b}	110.7 ^{b}	28.2 ^{b}	138.9 ^{b}	78.5
Urea$_{incorp}$ [3]	100:60:0	2.61 ^{bc}	0.48 ^{b}	127.0 ^{bc}	28.6 ^{b}	155.6 ^{bc}	81.8
CAN + UAN$_{liquid}$ [4]	50:50:60	3.03 ^{d}	0.53 ^{b}	151.4 ^{cd}	35.7 ^{b}	187.1 ^{d}	80.1
Year							
2017		2.47 ^{a}	0.35 ^{a}	145.4 ^{b}	23.0 ^{a}	168.5 ^{b}	86.7 ^{b}
2018		2.78 ^{b}	0.52 ^{b}	110.0 ^{a}	28.1 ^{b}	138.1 ^{a}	79.0 ^{a}
ANOVA [5]							
N treatment (N)		***	***	***	***	***	n.s.
Year (Y)		***	***	***	***	***	***
N × Y		n.s. [7]	n.s.	*	n.s.	n.s.	**

[1] Calcium ammonium nitrate, [2] Urea with nitrification inhibitor, [3] Incorporated Urea, [4] Liquid urea ammonium nitrate, [5] Analysis of variance, [6] Different letters indicate significant differences. Significant effects: $p < 0.05$ (*), $p < 0.01$ (**) and $p < 0.001$ (***), [7] Not significant. GNC = N concentration of grain, SNC = N concentration of staw, NY$_{GRAIN}$ = grain N yield, NY$_{STRAW}$ = straw N yield, NY$_{AGB}$ = N yield of above-ground biomass, NHI = N harvest index.

The GNC was +102% (mean over both years) higher in all N fertilized treatments than in the control. The highest value was observed with CAN + UAN$_{liquid}$, with +122% (mean over both years) compared to the control. The GNC was affected by splitting individual fertilizers as follows: CAN and Urea—no differences between three and two splits, and $Urea_{stab}$—higher values with three compared to one split, with two splits showing intermediate values. The lowest GNC of the N fertilized treatments was found with $Urea_{stab}$ 160:0:0.

The SNC was +104% (mean over both years) higher in all N fertilized treatments than in the control, with no differences between N treatments. The GNC and SNC did not differ between three or two splits for CAN, Urea or $Urea_{stab}$. The GNC was 13% higher and the SNC 49% higher in 2018 than in 2017.

The NY_{GRAIN} was +154% (mean over both years) higher in the N fertilized treatments than in the control. There was a significant N fertilization × year interaction for the NY_{GRAIN} (Table 4). All fertilized treatments had a significantly higher uptake than the control. Splitting of individual fertilizers significantly increased the NY_{GRAIN} from two to three splits of CAN in 2017 but not in 2018. No significant differences but higher values with three than two splits were observed for Urea. Whereas, the NY_{GRAIN} with $Urea_{stab}$ increased from one and two splits to three splits in 2017 and one to three splits in 2018. The highest NY_{GRAIN} was obtained by CAN+UAN_{liquid} in 2017 and by $Urea_{stab}$ 50:50:60 in 2018. The NY_{GRAIN} was higher in 2017 than in 2018 in all treatments.

The NY_{STRAW} was +132% (mean over both years) higher in all N fertilized treatments than in the control. It was higher with N fertilization than in the control, except for $Urea_{stab}$ (50:50:60), $Urea_{stab}$ (100:60:0) and Urea (50:50:60), which showed intermediate values. The NY_{STRAW} of individual fertilizers did not differ between the splitting regimes. The NY_{STRAW} was 22% higher in 2018 than in 2017.

The NY_{AGB} was +154% (mean over both years) higher in all N fertilized treatments than in the control. The highest value was observed for CAN + UAN_{liquid}. The NY_{AGB} was affected by splitting individual fertilizers as follows: CAN and Urea—no differences between three and two splits and $Urea_{stab}$—higher values with three compared to one split, with two splits showing intermediate values. The NY_{AGB} was 20% higher in 2017 than in 2018.

There was a significant N fertilization × year interaction for the NHI (Table 4). It was highest in 2017 in the control and lowest with $Urea_{incorp}$, the other treatments showed generally intermediate values. In 2018, the NHI was lowest in the control and highest in $Urea_{stab}$ (50:50:60). In 2017, splitting of individual fertilizers caused no differences of the NHI. Whereas in 2018, the NHI of Urea was significantly lower with two than with three splits and of $Urea_{stab}$ with one than with two and three splits. The NHI was higher in 2017 than in 2018 except for the control, CAN + UAN_{liquid}, $Urea_{stab}$ (160:0:0) and Urea (100:60:0).

3.3. N Surplus, N Balance, Soil Mineral Nitrate and Apparent N Loss

The means for N surplus, N balance, NO_3-N and ANL (over all treatments and both years) were: N surplus—16.3 kg N ha^{-1}, N balance—−9.3 kg N ha^{-1}, NO_3-N at the start of the growing season in spring—54.0 kg N ha^{-1}, NO_3-N at the end of the growing season—31.5 kg N ha^{-1} and ANL—48.3 kg ha^{-1} (Table 6).

The N surplus was between 6.0 kg N ha^{-1} for $Urea_{stab}$ (50:50:60) and 49.3 kg N ha^{-1} for $Urea_{stab}$ (160:0:0). Consequently, all values of fertilized treatments were positive. The N balance was between −27.1 kg N ha^{-1} for CAN + UAN_{liquid} (50:50:60) and 21.1 kg N ha^{-1} for $Urea_{stab}$ (160:0:0), with negative values for all treatments, where the fertilizers were applied in three splits. The mean N surplus or N balance were, in all N fertilized treatments, at 24.0 kg N ha^{-1} or at −3.1 kg N ha^{-1}, and were thereby 146% or 95% higher than in the control (means over both years). The N surplus and N balance were, in 2018, 25% or 124% higher than in 2017. The three doses application regime (50:60:60) of CAN, Urea and $Urea_{stab}$ resulted in a 18.6 kg N ha^{-1} lower N surplus than the two doses application regime (100:60:0) (11.5 vs. 30.1 kg N ha^{-1}, means over fertilizers and years, $p < 0.001$). The N balance was, in the three doses application, 14.7 kg N ha^{-1} lower than in the two doses application (−11.8 vs. 2.9 kg N ha^{-1}, means over fertilizers and years, $p = 0.024$).

Table 6. N surplus, N balance, soil mineral nitrate and apparent N loss (ANL) depending on fertilizer treatment and year.

N Treatment	N Splitting	N Surplus	N Balance	Soil NO_3-N					ANL
				Initial (0–90 cm)	At Harvest (0–30 cm)	At Harvest (30–60 cm)	At Harvest (60–90 cm)	At Harvest (0–90 cm)	
		(kg ha^{-1})	(kg ha^{-1})	(kg ha^{-1})					(kg ha^{-1})
Control		−52.6 ^{a} [6]	−64.3 ^{a}	49.7	8.6 ^{a}	3.9 ^{a}	3.0 ^{a}	15.6 ^{a}	-
CAN [1]	50:50:60	13.6 ^{bc}	−12.6 ^{bc}	52.2	33.1 ^{c}	5.6 ^{ab}	3.9 ^{ab}	42.6 ^{b}	−27.3 ^{cd}
CAN	100:60:0	32.2 ^{cd}	2.5 ^{cd}	53.1	22.4 ^{abc}	6.5 ^{ab}	4.4 ^{ab}	33.3 ^{ab}	−52.7 ^{abcd}
Urea	50:50:60	15.0 ^{bc}	−6.4 ^{bcd}	49.6	22.9 ^{abc}	5.7 ^{ab}	4.2 ^{ab}	32.9 ^{ab}	−40.6 ^{bcd}
Urea	100:60:0	31.1 ^{cd}	0.7 ^{bcd}	48.6	17.9 ^{abc}	6.0 ^{ab}	3.9 ^{ab}	27.8 ^{ab}	−51.8 ^{abcd}
$Urea_{stab}$ [2]	50:50:60	6.0 ^{b}	−16.4 ^{bc}	59.4	26.6 ^{bc}	5.2 ^{ab}	3.5 ^{ab}	35.2 ^{b}	−38.0 ^{bcd}
$Urea_{stab}$	100:60:0	26.9 ^{bcd}	5.5 ^{cd}	52.2	18.1 ^{abc}	5.5 ^{ab}	5.0 ^{b}	28.6 ^{ab}	−59.3 ^{abc}
$Urea_{stab}$	160:0:0	49.3 ^{d}	21.1 ^{d}	55.7	14.2 ^{ab}	5.6 ^{ab}	4.1 ^{ab}	23.8 ^{ab}	−83.3 ^{a}
$Urea_{incorp}$ [3]	100:60:0	33.0 ^{cd}	4.5 ^{cd}	65.0	18.1 ^{abc}	8.4 ^{b}	5.3 ^{b}	31.6 ^{ab}	−68.1 ^{ab}
CAN + UAN_{liquid} [4]	50:50:60	8.6 ^{bc}	−27.1 ^{b}	54.3	32.7 ^{c}	6.4 ^{ab}	4.3 ^{ab}	43.5 ^{b}	−14.0 ^{d}
Year									
2017		−1.4 ^{a}	−24.5 ^{a}	75.0 ^{b}	23.7	5.5	3.3 ^{a}	32.5	−49.6 ^{b}
2018		34.0 ^{b}	+5.9 ^{b}	33.0 ^{a}	19.2	6.3	5.0 ^{b}	30.5	−86.6 ^{a}
ANOVA [5]									
N treatment (N)		***	***	n.s.	**	*	*	*	***
Year (Y)		***	***	***	n.s.	n.s.	*	n.s.	***
N × Y		*	n.s. [7]	n.s.	n.s.	n.s.	n.s.	n.s.	n.s.

[1] Calcium ammonium nitrate, [2] Urea with nitrification inhibitor, [3] Incorporated Urea, [4] Liquid urea ammonium nitrate, [5] Analysis of variance, [6] Different letters indicate significant differences. Significant effects: $p < 0.05$ (*), $p < 0.01$ (**) and $p < 0.001$ (***), [7] Not significant.

The NO_3-N (0–90 cm) at the end of the growing season was highest in CAN (50:50:60) and CAN + UAN_{liquid} and lowest in the control, with other treatments showing intermediate values. There were no differences between the years. The NO_3-N (0–90 cm) of N fertilized treatments was +113% higher than that of the control (mean over all N fertilization treatments and both years). Three compared to two splits of CAN, Urea and $Urea_{stab}$ resulted in higher NO_3-N content in 0–90 cm (36.9 kg ha^{-1} vs. 29.9 kg ha^{-1}, means over fertilizers and years, $p = 0.080$, n.s.) and in 0–60 cm (33.0 kg ha^{-1} vs. 25.5 kg ha^{-1} ($p = 0.047$).

The NO_3-N at the end of the growing season was significantly influenced by the treatment in all three soil horizons. The mean values of fertilized treatments were: 0–30 cm—22.9 kg ha^{-1}, 30–60 cm—6.1 kg ha^{-1}, 60–90 cm—4.3 kg ha^{-1} (means over all fertilizers and years). Thereby, the NO_3-N was: in 0–30 cm, 2.66-fold, in 30–60 cm, 1.56-fold, and in 60–90 cm, 1.38-fold higher than in the control. In 0–30 cm, the highest NO_3-N was found in CAN (50:50:60) and CAN + UAN_{liquid}. In these two treatments, the NO_3-N was higher than in the control and in $Urea_{stab}$ (160:0:0). Three compared to two splits of CAN, Urea and $Urea_{stab}$ resulted in a higher NO_3-N content in 0–30 cm (27.6 kg ha^{-1} vs. 19.4 kg ha^{-1}, means over fertilizers and years, $p = 0.024$). In 30–60 cm, $Urea_{incorp}$ had a higher NO_3-N than the control, with other treatments showing intermediate values. Three compared to two splits of CAN, Urea and $Urea_{stab}$ resulted in a lower NO_3-N content in 30–60 cm (5.49 kg ha^{-1} vs. 6.02 kg ha^{-1}, means over fertilizers and years, n.s.)

In 60–90 cm, $Urea_{stab}$ (100:60:0) and $Urea_{incorp}$ had a higher NO_3-N content than the control, whereas the other treatments showed intermediate values. Three compared to two splits of CAN, Urea and $Urea_{stab}$ resulted in a lower NO_3-N content in 60–90 cm (3.87 kg ha^{-1} vs. 4.40 kg ha^{-1}, means over fertilizers and years, n.s.).

In 2018, the NO_3-N content at harvest was significantly higher than in 2017 in this soil. The ANL was highest in $Urea_{stab}$ (160:0:0) with −83.1 kg N ha^{-1} and lowest in CAN + UAN_{liquid} (50:50:60) with −14.0 kg N ha^{-1}. Two times application (100:60:0) of CAN, Urea and $Urea_{stab}$ resulted in a significantly higher ANL than the three times application (50:50:60) strategy (−54.6 kg ha^{-1} vs. −35.3 kg N ha^{-1}, means over fertilizers and years, $p = 0.012$). The ANL was significantly higher in 2018 than in 2017.

The correlation between the apparent N loss and the N surplus is shown in Figure 1 and that of the apparent N loss and the N balance in Figure 2. The graphs show that the apparent N loss was negatively correlated with both a higher N surplus and a higher N balance. From the graphs, it can be estimated that a zero-loss of ANL can be reached with a N surplus of −29.5 kg ha^{-1} and a N balance of −51.2 kg ha^{-1}. A theoretical zero N surplus and zero N balance showed an estimated ANL of −26.7 and −51.5 kg N ha^{-1}.

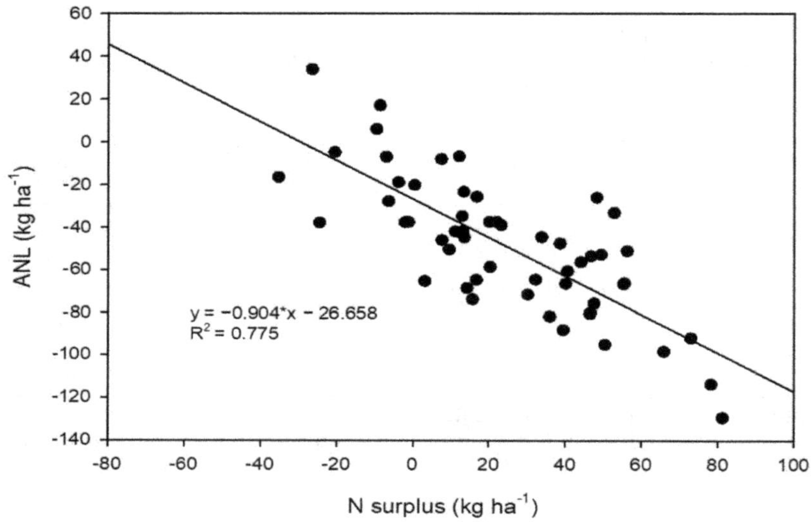

Figure 1. Correlation between apparent N loss (ANL) and N surplus.

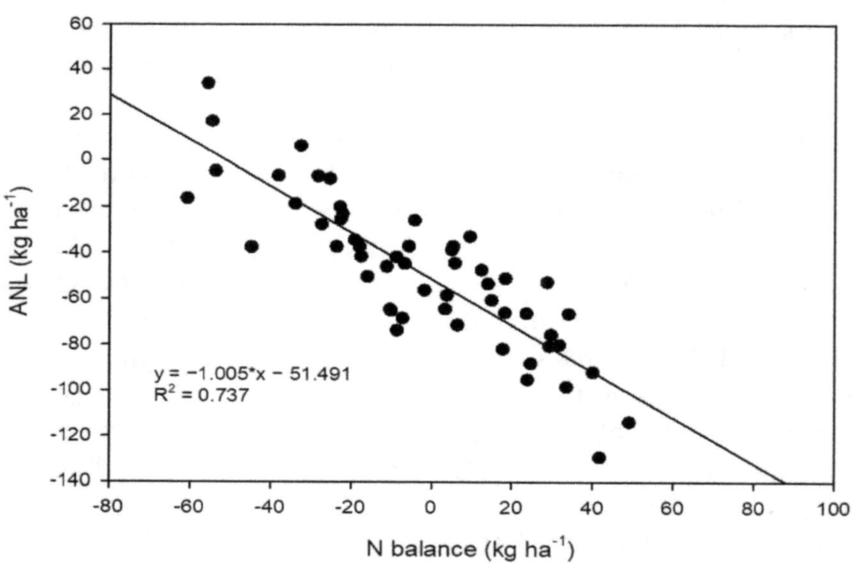

Figure 2. Correlation between apparent N loss (ANL) and N balance.

3.4. Nitrogen Efficiency

The means for the N efficiency parameters (over all fertilization treatments and both years) were: NUE—27.1 kg grain kg^{-1} N_{SUPP}, NUpE—68.1%, ANE—6.57 kg grain kg^{-1} N_{FERT}, ANRE—61.8% and NUtE—39.1 kg grain kg^{-1} NY_{AGB} (Table 7).

Table 7. N use efficiency (NUE), N uptake efficiency (NUpE), agronomic N efficiency (ANE), apparent N recovery efficiency (ANRE) and N utilization efficiency (NUtE) depending on fertilizer treatment and year.

N Fertilizer	N Splitting	NUE	NUpE	ANE	ANRE	NUtE
		(kg grain kg^{-1} N_{SUPP})	(%)	(kg grain kg^{-1} N_{FERT})	(%)	(kg grain kg^{-1} NY_{AGB})
Control		57.8 ^{b} [6]	81.1 ^{b}			71.3 ^{b}
CAN [1]	50:50:60	24.0 ^{a}	70.9 ^{ab}	6.86	67.6 ^{bc}	33.7 ^{a}
CAN	100:60:0	22.1 ^{a}	65.1 ^{ab}	4.12	58.2 ^{abc}	34.4 ^{a}
Urea	50:50:60	25.1 ^{a}	69.3 ^{ab}	7.94	63.8 ^{abc}	36.2 ^{a}
Urea	100:60:0	24.0 ^{a}	66.7 ^{ab}	6.48	59.4 ^{abc}	36.1 ^{a}
$Urea_{stab}$ [2]	50:50:60	24.9 ^{a}	70.4 ^{ab}	8.95	70.1 ^{bc}	35.2 ^{a}
$Urea_{stab}$	100:60:0	24.7 ^{a}	63.8 ^{ab}	7.50	56.4 ^{ab}	38.8 ^{a}
$Urea_{stab}$	160:0:0	22.1 ^{a}	56.9 ^{a}	4.23	46.6 ^{a}	37.8 ^{a}
$Urea_{incorp}$ [3]	100:60:0	22.4 ^{a}	60.8 ^{ab}	6.00	57.0 ^{ab}	37.0 ^{a}
CAN + UAN_{liquid} [4]	50:50:60	23.8 ^{a}	76.0 ^{b}	7.09	76.7 ^{c}	30.9 ^{a}
Year						
2017		32.2 ^{b}	72.7 ^{b}	4.77 ^{a}	68.4 ^{b}	43.7 ^{b}
2018		22.0 ^{a}	63.5 ^{a}	8.38 ^{b}	55.1 ^{a}	34.6 ^{a}
ANOVA [5]						
N treatment (N)		***	**	n.s.	*	***
Year (Y)		***	***	*	***	***
N × Y		***	n.s. [7]	n.s.	n.s.	***

[1] Calcium ammonium nitrate, [2] Urea with nitrification inhibitor, [3] Incorporated Urea, [4] Liquid urea ammonium nitrate, [5] Analysis of variance, [6] Different letters indicate significant differences. Significant effects: $p < 0.05$ (*), $p < 0.01$ (**) and $p < 0.001$ (***), [7] Not significant, N_{SUPP} = supplied N, N_{FERT} = fertilized N, NY_{AGB} = N yield of above-ground biomass.

The NUE in the control was 34.1 kg grain kg^{-1} N_{SUPP} (= +144%, mean over both years) higher than in the N fertilized treatments, which did not differ among each other. There was a significant N fertilization × year interaction. In 2017, there were no differences in the NUE between splitting regimes of the individual fertilizers. Whereas in 2018, $Urea_{stab}$ with two and three splits had a higher NUE than with one split. The NUE was higher in all treatments, except for $Urea_{stab}$ 100:60:0 and $Urea_{incorp}$, in 2017 than in 2018 (Table 5).

The NUpE in the control was 14.4% (= +22%, mean over both years) higher than in the N fertilized treatments. It was highest in the control and CAN + UAN_{liquid} and lowest in $Urea_{stab}$ (160:0:0), with other treatments showing intermediate values. The NUpE was 14% higher in 2017 than in 2018.

The ANE did not differ between fertilizers and splitting rates: it was 76% higher in 2018 than in 2017.

The ANRE of N fertilized treatments was highest for CAN + UAN_{liquid} and lowest for $Urea_{stab}$ (160:0:0). The ANRE was 24% higher in 2017 than in 2018. It was affected by N fertilizers as follows (Table 7): The highest ANRE was found in CAN + UAN_{liquid} (50:50:60), which was 64% higher than the lowest ANRE in $Urea_{stab}$ (160:0:0). The other treatments showed intermediate values. Through splitting of CAN, Urea and $Urea_{stab}$, the ANRE increased from 58.0% (two splits) to 67.2% (three splits) (mean over fertilizers and years, $p = 0.024$).

The NUtE in the control was 35.8 kg grain kg^{-1} NY_{AGB} (= +101%, mean over both years) higher than in the N fertilized treatments. The NUtE was 26% higher in 2017 than in 2018. A highly significant N fertilization × year interaction was found for NUtE. The NUtE was highest in both years in the control and lowest with CAN + UAN_{liquid}. Splitting regimes of individual fertilizers did not affect the NUtE. The NUtE in the control, CAN 100:60:0, Urea 160:0:0 and $Urea_{stab}$ 100:60:0 was higher in 2017 than in 2018 (Table 4).

The NUE, ANE and NUtE of CAN, Urea and $Urea_{stab}$ did not significantly differ between the splitting regimes. There was a tendency of a splitting × year interaction for NUpE of CAN, Urea and $Urea_{stab}$ ($p = 0.055$, n.s.). NUpE in 2017 and 2018 for the three-splitting regime was 75.5% and 64.9% respectively, and for the two splitting regime was 65.3% and 65.1%, respectively. Splitting CAN, Urea and $Urea_{stab}$ three times (50:50:60) also tended to increase the ANE from 6.0 kg grain kg^{-1} N_{FER} to 7.9 kg grain kg^{-1} N_{FER}, thus by 1.9 kg grain kg^{-1} N_{FER} or +32% (means over the fertilizers, $p = 0.099$, n.s.). Splitting of CAN, Urea and $Urea_{stab}$ increased the ANRE from 58.0% (two splits) to 67.2% (three splits) (means over fertilizers and years, $p = 0.024$).

The N efficiency parameters are positively correlated with each other. The Pearson correlation coefficient of NUE with NUpE was 0.63 ($p < 0.001$), with ANE was 0.27 ($p = 0.46$), with ANRE was 0.54 ($p < 0.001$) and with NUtE was 0.91 ($p < 0.001$).

4. Discussion

4.1. Crop Yields, Nitrogen Concentrations and Yields

The increase of the GY with N fertilization in our experiment explained a higher grain density, which resulted from the tendency of both a higher ear density and more grains ear^{-1} with N fertilization, whereas the TKW was not affected by N. In the same location, the yield increase of oat with CAN fertilization was attributed to a higher panicle density with no differences in grains $panicle^{-1}$ or TKW [30].

The lower number of grains ear^{-1} in 2018 is explained by the dry and cold spring, whereby less productive tillers were generated. The competition of unproductive and productive tillers for water and nutrients in spring was also found by Lu et al. [31]. It is assumed that the intra-plant competition was lower in spring 2017, which allowed the plant to allocate resources into productive stems and favored a higher number of fertile florets.

Three splits compared to two splits of the fertilizers CAN, Urea and $Urea_{stab}$ (urea with nitrification inhibitor) tended to also increase the GY, whereas both the GY and also the Y_{AGB} did not differ between fertilizer types. Amount, type and time of N application is affecting the plant-available amount of N in the soil and thereby the formation of yield components [32]. Similar to our observation with a tendency of a higher number of grains ear^{-1}, N fertilization was reported to increase the spikelets of barley [33]. But, contrary to our observation that the TKW of winter wheat was not affected by N fertilization, the grain size of barley was reduced when the N dose was mainly given at early tillering [34]. N fertilization increased grain but not SY of winter wheat in both years. Contrary to that, the GY of spring-sown chickpea, pea, barley and oat was not affected by N fertilization in a two-year experiment in the same location, but the SY was increased in the more humid year but also stayed unaffected in the dry year. Similar to our study, where the GY did not differ between N fertilizer types, the SY of the four spring-sown crops was also not affected by the N fertilizer type [35].

Both GNC and SNC were increased in all N treatments compared to the control. The GNC also differed between splitting regimes of $Urea_{stab}$, where three splits caused a higher value compared to one split, whereas no differences between splits were observed for CAN and Urea. Consequently, also both the NY_{GRAIN} and NY_{STRAW} and thereby the NY_{AGB} were more than doubled with N fertilization, with little differences between N treatments, except for a higher NY_{GRAIN} and NY_{AGB} with three compared to one splits of $Urea_{stab}$, and also a tendency of higher values with more rather than less splits of CAN and Urea. A high protein content in the grain, caused by N fertilization, has a positive effect on wheat grain intended for baking bread, but negatively if the grain is to be used for malt. Higher N concentrations rather than higher yields were also the reason for higher N yields of spring-sown chickpea, pea, barley and oat in the same location in an earlier experiment, whereas in their case, higher values of both the N concentrations and the N yields during crop growth and at harvest were observed for CAN than for a depot fertilizer [36,37]. Similar to our observations, Aufhammer et al. [38] also reported a similar N uptake of maize with different N fertilizer types, attributing this to the high initial soil mineral N contents and a high mineralization from the soil. The differences of the NHI between the unfertilized control and the N treatments were considerably lower than the differences of the N yields in grain and straw, with even a higher NHI in 2017 but the lowest in 2018 for the control. Thus, the reason for the higher N yields in both grain and straw were the better uptake in both with N fertilization but not a better translocation to the grain. Splitting of N doses of Urea and Urea$_{stab}$ but not of CAN enhanced the NHI. Contrary to that, a lower translocation of macro- and micro-nutrients has been reported for oat and pea after fertilization with N [39,40].

4.2. N Surplus, N Balance, Soil Mineral Nitrate and Apparent N Loss

N fertilizer, which is not taken up by the crop during the growing season, will partly accumulate in the soil [3] and partly be lost from the plant-soil system via ammonia volatilization [41] and emissions of gases such N_2O and NO [42]. The parameters N surplus, N balance, NO_3-N content and ANL provide information if and where non-recovered fertilizer N is remaining in the system.

In our study, much more N has been removed in the fertilized treatments than in the control. Although, the N surplus and the N balance were considerably higher in fertilized treatments. The positive values of the N surplus of the fertilized treatments show that less N has been removed with the NY_{GRAIN} than has been applied with fertilization. Consequently, more N remains with fertilization after the removal of the NY_{GRAIN} in the system for the subsequent crop than initially present. Three split applications of CAN, Urea and Urea$_{stab}$ (urea with nitrification inhibitor) reduced both the N surplus and the N balance, with negative values for the N balance with three splits. At the same time, the NO_3-N content at harvest was higher with fertilization than in the control and a higher accumulation occurred with three compared to two splits in the first soil layer (0–30 cm). The combination of initial NO_3-N content, N mineralization and N_{FERT} led to a N supply exceeding the crop N demand. This has been explained as a main driving factor for N loss [28].

Both the higher values of N surplus and N balance as well as the higher NO_3-N content with three compared to two splits could be obtained as the ANL was considerably lower with three than with two splits. Already, Wallace et al. [43] have reported that deferring the N application can reduce N losses and thereby environmental damage. Enhancing the synchronization between N supply and crop demand especially with more splits is a major task for enhancing the N transfer into the crop [4,14]. But, under Pannonian climate conditions, the crop needs sufficient rainfall for the uptake of the third N dose. In case of low rainfall after application, both the grain yield and the fertilizer uptake might be reduced [43]. The best N treatment regarding highest Y_{AGB} and lowest N surplus, N balance and ANL, was CAN + UAN$_{liquid}$, whereas Urea$_{stab}$ (160:0:0) had the worst performance. Similar to that observation, Shi et al. [44] reported a lower N uptake and a higher N loss for basal compared to top-dressed of N to winter wheat.

Sieling and Kage [45] reported a poor correlation between N balances and N leaching and therefore, stated that N balance is not an appropriate indicator for the environmental impact of N fertilization in the short-term but it might if set up over a longer period. Contrary to that, we found a strong negative correlation of both N surplus and N balance with ANL. Under the semi-arid conditions of eastern Austria, N fertilizer loss through volatilization and not through leaching is the problem. For that, the N balance might be a good indicator also in the short-term.

4.3. Nitrogen Efficiency

Several authors have stated that an optimum N fertilizer management is a strategy for improving the N efficiency (e.g., References [2,15,21]). Applying N fertilizer decreased NUE, NUpE and NUtE, which has also been reported by others [3,5,46].

The N efficiency parameters differed between years. These parameters are highly impacted by water availability according to Kirda et al. [47] and Neugschwandtner et al. [46]. As the total sum of rainfall (plus irrigation in 2017) was quite similar in both years, the distribution of the rainfall or the temperature regime might have caused these differences. Crop response to applied nitrogen varies spatially and temporally [48] as environmental systems show a high degree of complexity, there is also a high variability of the nitrogen-use efficiency, depending on season, soil type, N fertilizer input and amount and distribution of rainfall [49].

The N efficiency parameters focusing on yield formation were less affected by splitting of CAN, Urea and $Urea_{stab}$ (urea with nitrification inhibitor), as also, the differences of GY between splitting regimes were small. Splitting had no effect on NUE and NUtE but tended to increase the ANE with three compared to two splits. Schulte auf´m Erley et al. [50] stated that N fertilization tends to decrease the HI and thereby is further impairing N efficiency parameters like NUtE. This effect was also absent in our study, as the HI did not differ between treatments.

Contrary to that, the N efficiency parameters focusing on N uptake and N recovery were affected by fertilization treatments as the effects of fertilization on N concentrations and thereby on N yields were higher than the effects on biomass yields. The NUpE tended to be higher in 2017 with three compared to two splits and the ANRE was significantly higher with three compared to two splits.

High grain protein concentration for high-quality wheat can be expected to have a low NUtE, which was found in our study and was confirmed by Barraclough et al. [2].

Splitting is commonly recommended for producing high-quality wheat for bread-making. This can be achieved, according to Lu et al. [31], especially by low-tillering large-spike cultivars. Schulz et al. [16] on the other hand recommended a reduction of N applications when the protein content in wheat production is not a target, i.e., in ethanol-wheat, where a single application might be sufficient. The low differences in GY in our study support this observation. Especially, as every work operation is costing time and fuel and is affecting the energy efficiency of the production system [7,8,51].

NUE is mainly controlled by NUtE and less by NUpE, as the correlation coefficients show in Section 3.4. The low correlation between NUE and ANE indicated that the two N efficiency parameters should be separately considered. The NUE components NUpE and NUtE have been typically used for characterizing newly developed cereal genotypes [2,9]. NUpE and ANRE can also show potential N losses (leaching, volatilization and denitrification) of N fertilization to the environment [52,53]. For testing the N efficiency of agronomic practices according grain yield and grain protein content, the efficiency parameters NUtE and ANE have been proposed [4,52]. Applying increased amounts of N only has a minor impact on yield above the optimal application but can give significant increases in grain N, however at the expense of NUtE [2,54]. ANE is useful for the economic assessment of the mineral N fertilization, and according to Ladha et al. [55] and Hawkesford [52], can be improved by crop management practices such as amount, timing, placement and N source.

5. Conclusions

In the Pannonian region, water is the most limiting factor for N fertilizer efficiency. In spring, it is necessary to have enough rainfall for developing plants with fertile tillers. Fertile tillers produce more grains ear^{-1} and affect the GY and N efficiency positively, with lower N losses. Splitting of N fertilizer influenced the crop yield, apparent losses and N efficiency more than the fertilizer type.

Dividing the N rate of 160 kg N ha^{-1} in three splits (50:50:60) increased the GY and the ANRE and reduced the N surplus and the N balance compared with two splits (100:60:0). Also, the ANL was lower with three compared to two splits which shows that with three splits, the N demand of the crop during the growing period could be better matched. Three splits are therefore recommended under semi-dry conditions in the Pannonian region for producing high-quality wheat for bread-making.

There are many N efficiency parameters for analyzing the fertilization systems. Each of them has a different explanatory power depending on crop yield or/and environmental perspective.

Author Contributions: Conceptualization, G.M. and H.W.; methodology, G.M. and H.W; software, G.M.; validation, R.W.N. and H.-P.K.; formal analysis, G.M.; investigation, G.M.; resources, H.W.; data curation, G.M. and R.W.N.; writing—original draft preparation, G.M. and R.W.N.; writing—review and editing, G.M.; visualization, G.M.; supervision, H.W.; project administration, H.W. and G.M. All authors have read and agreed to the published version of the manuscript.

Acknowledgments: The authors thank the technical staff of the experimental farm Groß-Enzersdorf of the University of Natural Resources and Life Sciences, Vienna (BOKU), and the students Caroline Gass, Anton Gingl, Stefan Havranek, Tobias Wittmann, Tanja Gruber, Christina Kainz, Laura Messemer, Florian Schraivogl and David

Tischler for conducting the field experiments. The authors also want to express their gratitude to Craig Jackson for proofreading.

References

1. Liu, Z.; Sun, K.; Liu, W.; Gao, T.; Li, G.; Han, H.; Li, Z.; Ning, T. Responses of soil carbon, nitrogen, and wheat and maize productivity to 10 years of decreased nitrogen fertilizer under contrasting tillage systems. *Soil Tillage Res.* **2020**, *196*. [CrossRef]
2. Barraclough, P.B.; Howarth, J.R.; Jones, J.; Lopez-Bellido, R.; Parmar, S.; Shepherd, C.E.; Hawkesford, M.J. Nitrogen efficiency of wheat: Genotypic and environmental variation and prospects for improvement. *Eur. J. Agron.* **2010**, *33*, 1–11. [CrossRef]
3. Yang, X.; Lu, Y.; Ding, Y.; Yin, X.; Raza, S.; Tong, Y. Optimising nitrogen fertilisation: A key to improving nitrogen-use efficiency and minimising nitrate leaching losses in an intensive wheat/maize rotation (2008–2014). *Field Crop. Res.* **2017**, *206*, 1–10. [CrossRef]
4. Santillano-Cázares, J.; Núñez-Ramírez, F.; Ruíz-Alvarado, C.; Cárdenas-Castañeda, M.E.; Ortiz-Monasterio, I. Assessment of fertilizer management strategies aiming to increase nitrogen use efficiency of wheat grown under conservation agriculture. *Agronomy* **2018**, *8*, 304. [CrossRef]
5. Zhang, X.; Davidson, E.A.; Mauzerall, D.L.; Searchinger, T.D.; Dumas, P.; Shen, Y. Managing nitrogen for sustainable development. *Nature* **2015**, *528*, 51–59. [CrossRef]
6. Dachler, M.; Köchl, A. Der Einfluss von Fruchtfolge, Vorfrucht, Stickstoffdüngung und Einarbeitung der Ernterückstände auf Ertrag und Rohproteingehalt von Winterweizen und nachfolgender Sommergerste. *Die Bodenkult J. Land Manag. Food Environ.* **2003**, *54*, 23–34.
7. Moitzi, G.; Neugschwandtner, R.W.; Kaul, H.-P.; Wagentristl, H. Energy efficiency of winter wheat in a long-term tillage experiment under Pannonian climate conditions. *Eur. J. Agron.* **2019**, *103*, 24–31. [CrossRef]
8. Moitzi, G.; Thünauer, G.; Robier, J.; Gronauer, A. Energieeinsatz und Energieeffizienz in der Körnermaisproduktion bei unterschiedlicher Stickstoffdüngung in der Südsteiermark. *Die Bodenkult J. Land Manag. Food Environ.* **2015**, *66*, 25–37.
9. Moll, R.H.; Kamprath, E.J.; Jackson, W.A. Analysis and interpretation of factors which contribute to efficiency of nitrogen utilization. *Agron. J.* **1982**, *74*, 562–564. [CrossRef]
10. Fischer, R.A. Optimizing the use of water and nitrogen through breeding of crops. *Plant Soil* **1981**, *58*, 249–278. [CrossRef]
11. Prey, L.; Kipp, S.; Hu, Y.; Schmidhalter, U. Nitrogen use efficiency and carbon traits of high-yielding European hybrid vs. line winter wheat cultivars: Potentials and limitations. *Front. Plant Sci.* **2019**, *9*, 1–15. [CrossRef]
12. Tian, Z.; Liu, X.; Gu, S.; Yu, J.; Zhang, L.; Zhang, W.; Jiang, D.; Cao, W.; Dai, T. Postponed and reduced basal nitrogen application improves nitrogen use efficiency and plant growth of winter wheat. *J. Integr. Agric.* **2018**, *17*, 2648–2661. [CrossRef]
13. Yin, L.; Dai, X.; He, M. Delayed sowing improves nitrogen utilization efficiency in winter wheat without impacting yield. *Field Crop. Res.* **2018**, *221*, 90–97. [CrossRef]
14. Liu, Z.; Gao, F.; Liu, Y.; Yang, J.; Zhen, X.; Li, X.; Li, Y.; Zhao, J.; Li, J.; Qian, B.; et al. Timing and splitting of nitrogen fertilizer supply to increase crop yield and efficiency of nitrogen utilization in a wheat–peanut relay intercropping system in China. *Crop J.* **2019**, *7*, 101–112. [CrossRef]
15. Chen, X.; Zhang, F.; Römheld, V.; Horlacher, D.; Schulz, R.; Böning-Zilkens, M.; Wang, P.; Claupein, W. Synchronizing N supply from soil and fertilizer and N demand of winter wheat by an improved Nmin method. *Nutr. Cycl. Agroecosyst.* **2006**, *74*, 91–98. [CrossRef]
16. Schulz, R.; Makary, T.; Hubert, S.; Hartung, K.; Gruber, S.; Donath, S.; Döhler, J.; Weiß, K.; Ehrhart, E.; Claupein, W.; et al. Is it necessary to split nitrogen fertilization for winter wheat? On-farm research on Luvisols in South-West Germany. *J. Agric. Sci.* **2015**, *153*, 575–587. [CrossRef]
17. Fischbeck, G.; Dennert, J.; Müller, R. N-Dynamik des Bodens, Ertragsbildung und Stickstoffentzug von Winterweizen bei unterschiedlicher Höhe und Verteilung der mineralischen N-Düngung. *J. Agron. Crop Sci.* **1990**, *164*, 297–311. [CrossRef]
18. Maidl, F.X.; Panse, A.; Dennert, J.; Ruser, R.; Fischbeck, G. Effect of varied N rates and N timings on yield, N uptake and fertilizer N use efficiency of a six-row and a two-row winter barley. *Eur. J. Agron.* **1996**, *5*, 247–257. [CrossRef]

19. Rossmann, A.; Pitann, B.; Mühling, K.-H. Splitting nitrogen applications improves wheat storage protein composition under low N supply. *J. Plant Nutr. Soil Sci.* **2019**, *182*, 347–355. [CrossRef]
20. Müller, S.; Ansorge, H.; Weigert, I. Einfluss der Stickstoffdüngung auf Ertrag, Ertragsstruktur und N-Verwertung von Winterweizen—Möglichkeiten und Grenzen der Bestandesführung. *Zeitschrift für Pflanzenernährung und Bodenkunde* **1991**, *154*, 115–119. [CrossRef]
21. Hartman, M.D.; Nyborg, M. Effect of early growing season moisture stress on barley utilization of broadcastincorporated and deep-banded urea. *Can. J. Soil Sci.* **1989**, *69*, 381–389. [CrossRef]
22. WRB. World reference base for soil resources. In *World Soil Resources Reports 2006*; No. 103; FAO: Rome, Italy, 2006; ISSN 0532-0488.
23. ZAMG. Klimaübersichten. Zentralanstalt für Meterologie und Geodynamik (ZAMG). ZAMG Online Database. 2019. Available online: http://www.zamg.ac.at/cms/de/klima/klimauebersichten/jahrbuch (accessed on 14 October 2019).
24. Harty, M.A.; Forrestal, P.J.; Watson, C.J.; McGeough, K.L.; Carolan, R.; Elliot, C.; Krol, D.; Laughlin, R.J.; Richards, K.G.; Lanigan, G.J. Reducing nitrous oxide emissions by changing N fertiliser use from calcium ammonium nitrate (CAN) to urea based formulations. *Sci. Total Environ.* **2016**, *563–564*, 576–586. [CrossRef] [PubMed]
25. Witzenberger, A.; Hack, H.; van den Boom, T. Erläuterungen zum BBCH-Dezimal-Code für die Entwicklungsstadien des Getreides—Mit Abbildungen. *Gesunde Pflanz.* **1989**, *41*, 384–388.
26. ÖNORM L 1091. *Chemische Bodenuntersuchungen: Bestimmung von Mineralischem Stickstoff-Nmin-Methode*; Österreichisches Normungsinstitut: Wien, Austria, 2012.
27. Winkler, R.; Botterbrodt, S.; Rabe, E.; Lindhauer, M.G. Stickstoff-/Proteinbestimmung mit der Dumas-Methode in Getreide und Getreideprodukten. *Getreide Mehl und Brot* **2000**, *54*, 86–91.
28. Hartmann, T.E.; Yue, S.; Schulz, S.; Chen, X.; Zhang, F.; Müller, T. Nitrogen dynamics, apparent mineralization and balance calculations in a maize—Wheat double cropping system of the North China Plain. *Field Crops Res.* **2014**, *160*, 22–30. [CrossRef]
29. Cui, Z.L.; Zhang, F.S.; Chen, X.P.; Miao, Y.X.; Li, J.L.; Shi, L.W.; Xu, L.W.; Ye, Y.L.; Liu, C.S.; Yang, Z.P.; et al. On-farm estimation of indigenous nitrogen supply for site specific nitrogen management in the North China Plain. *Nutr. Cycl. Agroecosyst.* **2008**, *81*, 37–47. [CrossRef]
30. Neugschwandtner, R.W.; Kaul, H.-P. Sowing ratio and N fertilization affect yield and yield components of oat and pea in intercrops. *Field Crops Res.* **2014**, *155*, 159–163. [CrossRef]
31. Lu, D.; Lu, F.; Pan, J.; Cui, Z.; Zou, C.; Chen, X.; He, M.; Wang, Z. The effects of cultivar and nitrogen management on wheat yield and nitrogen use efficiency in the North China Plain. *Field Crops Res.* **2015**, *171*, 157–164. [CrossRef]
32. Heyland, K.-U.; Triebel, U. Gezielte Stickstoffdüngung zur gesteuerten Ertragsbildung von Winterweizen unter Berücksichtigung verschiedener anbautechnischer Maßnahmen. *Die Bodenkult J. Land Manag. Food Environ.* **1986**, *37*, 133–148.
33. Aufhammer, W.; Kübler, E. Zur Leistungsfähigkeit von Gerste in Abhängigkeit von Form und Sorte sowie von Standort und Produktionstechnik—II. Die Ertragsrelevanz produktionstechnischer Maßnahmen bei verschiedenen Gerstenformen unter differenzierten Aufwuchsbedingungen. *Die Bodenkult J. Land Manag. Food Environ.* **1988**, *39*, 1–13.
34. Aufhammer, W.; Kübler, E. Zur Leistungsfähigkeit von Gerste in Abhängigkeit von Form und Sorte sowie von Standort und Produktionstechnik—III. Die Relevanz produktionstechnischer Maßnahmen für die Kornqualität verschiedenen Gerstenformen unter differenzierten Aufwuchsbedingungen. *Die Bodenkult J. Land Manag. Food Environ.* **1989**, *40*, 47–59.
35. Neugschwandtner, R.W.; Wichmann, S.; Gimplinger, D.M.; Wagentristl, H.; Kaul, H.-P. Chickpea performance compared to pea, barley and oat in Central Europe: Growth analysis and yield. *Turk. J. Field Crops* **2013**, *18*, 179–184.
36. Neugschwandtner, R.W.; Wagentristl, H.; Kaul, H.-P. Nitrogen concentrations and nitrogen yields of above-ground dry matter of chickpea during crop growth compared to pea, barley and oat in Central Europe. *Turk. J. Field Crop.* **2014**, *19*, 136–141. [CrossRef]
37. Neugschwandtner, R.W.; Wagentristl, H.; Kaul, H.-P. Nitrogen yield and nitrogen use of chickpea compared to pea, barley and oat in Central Europe. *Intern. J. Plant Prod.* **2015**, *9*, 291–304.

38. Aufhammer, W.; Kübler, E.; Kaul, H.-P. Untersuchungen zur Anpassung des Stickstoff-Angebots aus unterschiedlichen N-Quellen an den Verlauf der N-Aufnahme von Maisbeständen. *Z. Pflanz. Bodenkund.* **1996**, *159*, 471–478. [CrossRef]
39. Neugschwandtner, R.W.; Kaul, H.-P. Concentrations and uptake of macronutrients by oat and pea in intercrops in response to N fertilization and sowing ratio. *Arch. Agron. Soil Sci.* **2016**, *62*, 1236–1249. [CrossRef]
40. Neugschwandtner, R.W.; Kaul, H.-P. Concentrations and uptake of micronutrients by oat and pea in intercrops in response to N fertilization and sowing ratio. *Die Bodenkult. J. Land Manag. Food Environ.* **2016**, *67*, 1–15. [CrossRef]
41. Silva, A.G.B.; Sequeira, C.H.; Sermarini, R.A.; Otto, R. Urease inhibitor NBPT on ammonia volatilisation and crop productivity: A meta-analysis. *Agron. J.* **2017**, *109*, 1–13. [CrossRef]
42. Nayak, D.; Saetnan, E.; Cheng, K.; Wang, W.; Koslowski, F.; Cheng, Y.F.; Zhu, W.Y.; Wang, J.K.; Liu, J.X.; Moran, D.; et al. Management opportunities to mitigate greenhouse gas emissions from Chinese agriculture. *Agric. Ecosyst. Environ.* **2015**, *209*, 108–124. [CrossRef]
43. Wallace, A.J.; Armstrong, R.D.; Grace, P.R.; Scheer, C.; Partington, D.L. Nitrogen use efficiency of 15N urea applied to wheat based on fertiliser timing and use of inhibitors. *Nutr. Cycl. Agroecosyst.* **2020**, *116*, 41–56. [CrossRef]
44. Shi, Z.; Jing, Q.; Cai, J.; Jiang, D.; Cao, W.; Dai, T. The fates of ^{15}N fertilizer in relation to root distributions of winter wheat under different N splits. *Eur. J. Agron.* **2012**, *40*, 86–93. [CrossRef]
45. Sieling, K.; Kage, H. N balance as an indicator of N leaching in an oilseed rape—Winter wheat—Winter barley rotation. *Agric. Ecosyst. Environ.* **2006**, *115*, 261–269. [CrossRef]
46. Neugschwandtner, R.W.; Kaul, H.-P. Nitrogen uptake, use and utilization efficiency by oat-pea intercrops. *Field Crop. Res.* **2015**, *179*, 113–119. [CrossRef]
47. Kirda, C.; Topcu, S.; Kaman, H.; Ulger, A.C.; Yazici, A.; Cetin, M.; Derici, M.R. Grain yield response and N-fertiliser recovery of maize under deficit irrigation. *Field Crop. Res.* **2005**, *93*, 132–141. [CrossRef]
48. Robertson, G.P.; Bruulsema, T.W.; Gehl, R.J.; Kanter, D.; Mauzerall, D.L.; Rotz, C.A.; Williams, C.O. Nitrogen–climate interactions in US agriculture. *Biogeochemistry* **2013**, *114*, 41–70. [CrossRef]
49. Asseng, S.; Turner, N.C.; Keating, B.A. Analysis of water- and nitrogen-use efficiency of wheat in a Mediterranean climate. *Plant Soil* **2001**, *233*, 127–143. [CrossRef]
50. Schulte auf´m Erley, G.; Kaul, H.-P.; Kruse, M.; Aufhammer, W. Yield and nitrogen utilization efficiency of the pseudocereals amaranth, quinoa, and buckwheat under differing nitrogen fertilization. *Eur. J. Agron.* **2005**, *22*, 95–100. [CrossRef]
51. Moitzi, G.; Spiegel, H.; Sandén, T.; Vuolo, F.; Essl, L.; Neugschwandtner, R.W.; Wagentristl, H. Energieeinsatz und Energieeffizienz von Winterweizen bei unterschiedlicher mineralischer Stickstoffdüngung im Marchfeld. *Die Bodenkult J. Land Manag. Food Environ.* **2020**, *71*, 55–67. [CrossRef]
52. Hawkesford, M.J. Genetic Variation in Traits for Nitrogen Use Efficiency in Wheat. *J. Exp. Bot.* **2017**, *68*, 2627–2632. [CrossRef]
53. Raun, W.R.; Johnson, G.V. Improving nitrogen use efficiency for cereal production. *Agron. J.* **1999**, *91*, 357–363. [CrossRef]
54. Hawkesford, M.J. Reducing the reliance on nitrogen fertilizer for wheat production. *J Cereal Sci.* **2014**, *59*, 276–283. [CrossRef] [PubMed]
55. Ladha, J.K.; Pathak, H.; Krupnik, T.J.; Six, J.; van Kessel, C. Efficiency of fertilizer nitrogen in cereal production: Retrospects and prospects. *Adv. Agron.* **2005**, *87*, 85–156.

3

Potassium Supplying Capacity of Diverse Soils and K-Use Efficiency of Maize in South Asia

Saiful Islam [1,*], Jagadish Timsina [2], Muhammad Salim [3], Kaushik Majumdar [4] and Mahesh K Gathala [1]

1 International Maize and Wheat Improvement Centre, Bangladesh Office, House-10/B, Road-53, Gulshan-2, Dhaka 1212, Bangladesh; ms.islam@cgiar.org

2 Soil Research Group, Faculty of Veterinary and Agricultural Sciences, University of Melbourne, Victoria 3010, Australia; jtimsina@unimelb.edu.au

3 Department of Agronomy, Bangladesh Agricultural University, Mymensingh 2022, Bangladesh; msalimafa@yahoo.com

4 International Plant Nutrition Institute, South Asia Office, Palm Drive, B-1602, Golf Course Ext Road, Sector-66, Gurgaon 122001, Haryana, India; kmajumdar@ipni.net

* Correspondence: ms.islam@cgiar.org

Abstract: Increased nutrient withdrawal by rapidly expanding intensive cropping systems, in combination with imbalanced fertilization, is leading to potassium (K) depletion from agricultural soils in Asia. There is an urgent need to better understand the soil K-supplying capacity and K-use efficiency of crops to address this issue. Maize is increasingly being grown in rice-based systems in South Asia, particularly in Bangladesh and North East India. The high nutrient extraction, especially K, however, causes concerns for the sustainability of maize production systems in the region. The present study was designed to estimate, through a plant-based method, the magnitude, and variation in K-supplying capacity of a range of soils from the maize-growing areas and the K-use efficiency of maize in Bangladesh. Eighteen diverse soils were collected from several *upazillas* (or sub-districts) under 11 agro-ecological zones to examine their K-supplying capacity from the soil reserves and from K fertilization (100 mg K kg^{-1} soil) for successive seven maize crops grown up to V10–V12 in pots inside a net house. A validation field experiment was conducted with five levels of K (0, 40, 80, 120 and 160 kg ha^{-1}) and two fertilizer recommendations based on "Nutrient Expert for Maize-NEM" and "Maize Crop Manager-MCM" decision support tools (DSSs) in 12 farmers' fields in Rangpur, Rajshahi and Comilla districts in Bangladesh. Grain yield and yield attributes of maize responded significantly ($p < 0.001$) to K fertilizer, with grain yield increase from 18 to 79% over control in all locations. Total K uptake by plants not receiving K fertilizer, considered as potential K-supplying capacity of the soil in the pot experiment, followed the order: Modhukhali > Mithapukur > Rangpur Sadar > Dinajpur Sadar > Jhinaidah Sadar > Gangachara > Binerpota > Tarash > Gopalpur > Daudkandi > Paba > Modhupur > Nawabganj Sadar > Shibganj > Birganj > Godagari > Barura > Durgapur. Likewise, in the validation field experiment, the K-supplying capacity of soils was 83.5, 60.5 and 57.2 kg ha^{-1} in Rangpur, Rajshahi, and Comilla, respectively. Further, the order of K-supplying capacity for three sites was similar to the results from pot study confirming the applicability of results to other soils and maize-growing areas in Bangladesh and similar soils and areas across South Asia. Based on the results from pot and field experiments, we conclude that the site-specific K management using the fertilizer DSSs can be the better and more efficient K management strategy for maize.

Keywords: site-specific K management; soil K supply; maize yield response to K; maize crop manager; nutrient expert for maize

---

1. Introduction

Maize is the second most important cereal crop after rice in Asia and provides approximately 30% of the food calories to more than 4.5 billion people in 94 developing countries [1]. The world population is increasing and will continue to increase from 7.2 to 8.1 billion by 2025, reaching 9.6 billion by 2050 and 10.9 billion by 2100, with most growth occurring in the developing countries [2,3]. Maize was grown in 0.43 million hectares during 2016–2017 in Bangladesh [4].

Available soil K is deficient in many soils of Bangladesh, and crops are showing K deficiency symptoms. It is well known that the availability of K to plants does not only depend on the size of the available pool in the soil but also on the transport of K from soil solution to the root zone and from the root zone into plant roots [5]. Many plant factors (variety, root system, and antagonistic and synergistic mechanisms in ion uptake), soil factors (pH, organic matter content, texture, complementary cations, etc.) and environmental factors (rainfall, temperature, etc.) may affect these processes. However, when plant available soil K is sufficient, these factors tend to become less important. Therefore, soil K-supplying capacity is a key factor to sustain and increase crop yields.

Recent soil-test results have shown that many soils of the Indo-Gangetic Plains (IGP) of South Asia, with available K concentration of less than 0.1 cmol kg^{-1}, are becoming deficient in K despite their original high K contents [6–10]. The introduction and prevalence of high yielding varieties of rice, hybrid maize and wheat since the green revolution accelerated the removal of K from the soil than the traditional varieties did. At the same time, the application of K fertilizers, on the other hand, was limited, leading to negative input-output balance of K that depleted soil K status in most of the Asian countries [10,11]. Scientists reported 31% decline in soil K status in Bangladesh over the past 30 years, which is an alarming figure [12]. Despite this, studies in soil K received less attention than other major nutrients, because the application of K fertilizer doesn't frequently bring about a dramatic improvement in the vegetative growth of crop as is observed with nitrogen (N) fertilizer, or does not have the environmental concerns associated with its use as in N. Besides, the general perception that the South Asian soils are rich in K-bearing minerals also led to the complacency that crops may not require external K application to perform adequately. Therefore, most of the farmers, who can afford to apply fertilizers to their crops, apply only urea and phosphate fertilizers, while K application is often neglected. As a result, soils which were not deficient in K in the past have either become deficient or are likely to become deficient in the near future [13].

Of the three main macronutrients (N, phosphorus-P, and potassium-K), much work has been done in the past about N management in cropping systems [14,15]. The focus now, however in the context of maize production has been shifted to K nutrition, as the K dynamics of maize-growing soils dictate how well the K demand for high-yielding maize crops can be met. It is hypothesized that soils of maize-growing areas in Bangladesh differ in terms of mineralogy, soil K reserves, allowable drawdown, and K-supplying capacity. Thus, some soils would require more while others would require less K to grow profitable maize crops. Carefully-conducted pot and field experiments using a plant-based assay are expected to help estimate the magnitude and variation in K-supplying capacity of soils. The information generated from such experiments would help develop soil-based coefficients on allowable draw down of soil K reserves, which can be used for the determination of fertilizer K requirements of maize for Bangladesh, as well as for other maize-growing areas in South Asia.

Soil indigenous K supply, K-use efficiency, and crop yield vary spatially and temporally in the diverse irrigated maize fields in Bangladesh and South Asia. At present, however, blanket fertilizer recommendations are often applied over large areas without taking into account the wide variability in site- and season-specific crop nutrient requirements, which explains the reasons for low K-use efficiency. Further, the use of K fertilizers is often not based on crop requirements and are not balanced with other nutrients. As a result, the profitability is not optimized [6,16]. A rational and profitable K fertilizer management strategy needs to be based on better understanding of the soil K-supplying capacity. The efficiency of applied K fertilizer, in terms of agronomic efficiency of K (AE_K, kg yield increase per kg nutrient applied) and apparent K recovery efficiency (RE_K, kg K uptake per kg K applied) are commonly used as performance indicators for K management strategies [17–19]. The K-use efficiency is generally affected by yield levels, soil indigenous K-supplying capacity, amount of K fertilizer applied, and the quality of crop management operations [18]. Keeping the above points in view, the present study was undertaken to (i) determine the indigenous K-supplying capacity of major maize-growing soils in Bangladesh, and (ii) evaluate the grain yield and the K-use efficiency for maize under different K fertilizer recommendation strategies in the diverse soils of Bangladesh.

2. Materials and Methods

2.1. Pot Experiment

2.1.1. Experimental Soils

Soil samples (0–20 cm) were collected from farmers' fields of 18 *upazillas* or sub-districts (representing 18 diverse soils) located in the North, North-west, East, and South parts of Bangladesh. The 18 *upazillas* were: Birganj, Dinajpur Sadar, Gangachara, Rangpur Sadar, Mithapukur, Shibganj, Godagari, Paba, Durgapur, Tarash, Gopalpur, Modhupur, Daudkandi, and Barura (Figure 1). The soil samples were air-dried, ground and sieved through 2 mm sieve, and subject to test for their properties. Soil pH was determined in 1:1 soil/water paste by a pH meter. Exchangeable K was extracted by 1 mol L^{-1} Ammonium acetate (NH_4OAc), non-exchangeable K by boiling Nitric acid (HNO_3) method, and total K by Hydrogen fluoride (HF) digestion [20]. Particle size analysis was done by Hydrometer method [21] and the textural class was determined from Marshall's triangular co-ordinate following USDA system. The physical and chemical properties of the tested soils from the various *upazillas* used for pot study are presented in Table 1, while that for field experiment are presented in Table 2.

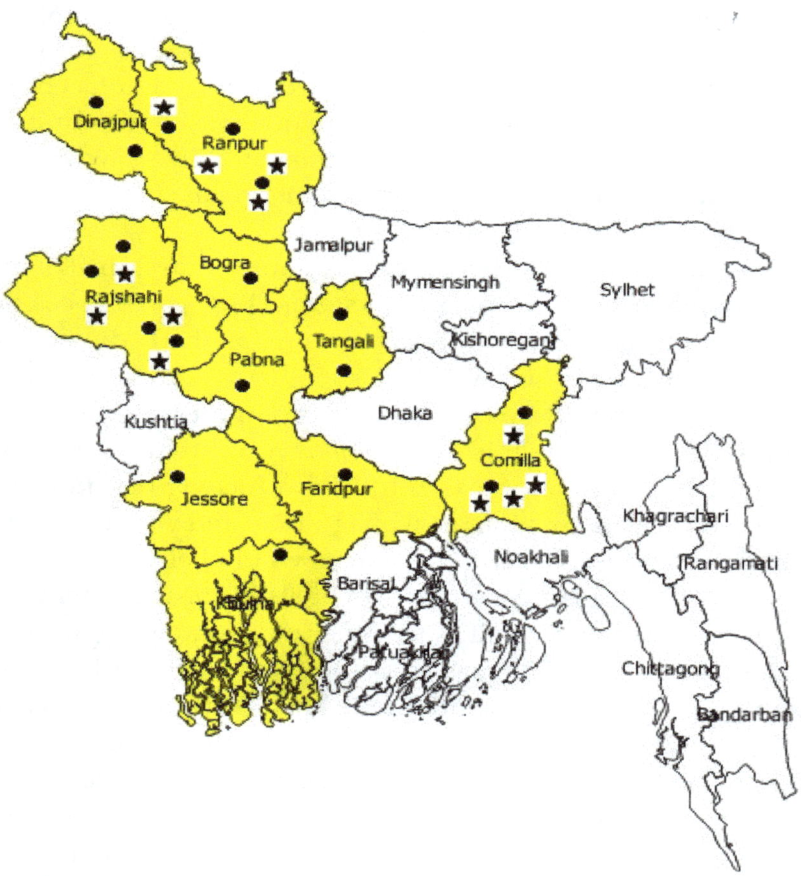

Figure 1. Dots showing the locations for 18 soil samples collected from various districts for pot experiment and stars showing the field experimental sites in Bangladesh.

Table 1. Basic chemical and physical properties of 18 soils (0–15 cm), collected from various locations, used for pot experiment with maize at Bangladesh Rice Research Institute (BRRI) net house in Gazipur, Bangladesh, 2011–2013.

Location [a]	AEZ [b]	Previous Crop	pH	OC [c] (%)	Total N [d] (%)	P [e]	K_{sl} [f]	K_{ex} [g]	K_{nex} [h]	S [i]	Soil Particles (%)			Texture
						mg kg^{-1} soil					Sand	Silt	Clay	
Birganj	1	Potato	5.85	1.29	0.088	84.46	25	57	1475	6.28	70	18	12	Sandy loam
Dinajpur Sadar	3	*Rabi* maize	5.89	0.98	0.064	12.27	40	83	1608	6.15	62	21	17	Sandy clay loam
Gangachara	2	*Rabi* maize	5.83	1.29	0.091	51.47	38	76	1544	5.91	33	50	17	Loam
Rangpur Sadar	3	Potato	6.64	0.27	0.03	19.44	45	99	2220	6.4	51	27	22	Sandy clay loam
Mithapukur	3	*Rabi* maize	5.9	1.48	0.107	77.29	60	113	3098	5.17	37	31	32	Loam
Shibganj	4	*Boro* rice	6.18	0.86	0.054	21.2	26	63	1876	6.77	41	24	35	Clay loam
Godagari	11	*Boro* rice	6.68	1.13	0.077	7.97	28	76	996	7.51	43	23	34	Clay loam
Nawabganj	11	*Boro* rice	7.15	1.4	0.100	3.82	58	83	970	6.52	47	28	25	Sandy clay loam
Paba	11	*Rabi* maize	7.25	1.76	0.131	6.06	60	85	1080	8.37	41	28	31	Clay loam
Durgapur	11	*Boro* rice	7.16	1.17	0.081	14.34	20	62	845	8.37	31	54	15	Silt Loam
Tarash	7	*Boro* rice	6.06	1.09	0.074	5.74	46	99	1345	7.75	25	36	39	Silty clay
Gopalpur	8	*Boro* rice	5.84	1.83	0.137	5.42	30	90	871	6.28	44	24	32	Clay loam
Modhupur	28	*Boro* rice	5.83	1.25	0.088	3.67	35	89	761	5.66	26	44	30	Clay loam
Modhukhali	12	*Boro* rice	7.51	1.76	0.13	1.91	105	124	2357	13.05	23	15	62	Clay
Binerpota	13	*Boro* rice	7.24	1.85	0.139	3.19	80	101	2145	13.05	24	12	64	Clay
Jhinaidah Sadar	11	*Boro* rice	6.97	1.91	0.144	3.98	75	100	2065	13.41	37	31	32	Clay loam
Daudkandi	16	*Rabi* maize	6.78	0.82	0.051	47.65	35	60	1123	12.06	43	15	42	Loam
Barura	19	*Rabi* maize	7.06	1.33	0.094	20.4	22	58	1060	7.38	47	17	36	Loam
Range			5.8–7.5	0.27–1.91	0.03–0.14	1.91–84.46	20–105	57–124	761–3098	5.17–13.41				

[a] Upazillas (Sub-districts), [b] AEZ: Agro-ecological zones, [c] OC: Organic carbon, [d] N: Nitrogen, [e] P: Available phosphorus, [f] K_{sl}: Water soluble potassium, [g] K_{ex}: Exchangeable soil potassium, [h] K_{nex}: Non-exchangeable potassium, [i] S: Available Sulphur.

Table 2. Chemical and physical properties of four soil samples for each field experimental site (Rangpur, Rajshahi and Comilla) before sowing of rabi (winter) maize, 2012–2013.

Soil Properties [a]	Methods of Determination [b]	Rangpur (n = 4)			Rajshahi (n = 4)			Comilla (n = 4)		
		Mean	Range	SD [c]	Mean	Range	SD	Mean	Range	SD
pH	By pH meter	5.78	5.46–5.96	0.22	5.94	5.18–6.66	0.80	5.23	5.10–5.38	0.15
SOC (%)	Wet digestion method	2.00	1.27–3.22	2.00	1.29	1.14–1.34	0.10	1.90	1.21–2.82	0.70
Total N (%)	Micro-Kjeldahl distillation	0.09	0.06–0.14	0.03	0.06	0.06–0.07	0.01	0.09	0.06–0.13	0.03
Available P (mg kg^{-1})	Modified Olsen's method	12.53	10.24–14.02	1.63	13.44	12.0–15.9	1.77	11.88	8.54–15.08	3.41
K_{ex} (mg kg^{-1})	1 M NH_4OAc method	41.34	37.44–43.29	2.68	44.85	43.29–47.19	1.77	38.22	28.47–54.21	11.33
K_{nex} (mg kg^{-1})	I M HNO_3 method	1269	1061–1439	197	858	827–870	21	987	857–1057	89
Available S (mg kg^{-1})	By 0.15% $CaCl_2$ extraction	14.94	12.9–17.08	2.02	15.57	13.69–18.12	2.03	13.29	11.96–15.24	1.47
Soil Textural Class	Hydrometer Methods	Silt Loam			Silt Loam			Silt Loam		
Sand (%)		43.00	36–60	11.49	39.25	26–68	19.38	39.00	34–63	15.37
Silt (%)		54.75	39–63	10.72	54.00	29–67	17.17	54.00	32–62	14.08
Clay (%)		2.75	2–5	1.50	7.00	4–10	2.45	6.00	3–9	2.83
Soil type general classification	[22]	Non-calcareous grey and brown floodplain			Calcareous dark grey and brown floodplain			Non-calcareous dark grey floodplain		
USDA classification	[22]	Typic Dystrochrept			Typic Haplaquept			Aeric Haplaquept		
Agro-ecological zone name	[23]	Tista meander floodplain			High Ganges river floodplain			Old Meghna estuarine floodplain		

[a] SOC: Soil organic carbon, K_{ex}: Soil exchangeable potassium, K_{nex}: Non-exchangeable potassium, N: Nitrogen, P: Phosphorus, S: Sulphur, [b] 1 M NH_4OAc : One molar Ammonium acetate, HNO_3: Nitric acid, $CaCl_2$: Calcium chloride, [c] SD: Standard deviation.

2.1.2. Description of Pot Experiment

A repeated pot experiment was conducted with the eighteen soils using a randomized complete block experimental design that consisted of a no-K fertilizer (K0) and a 100 mg K kg^{-1} soil (K100) treatments in four replications in a net house of BRRI, Gazipur. Soils were air dried and ground to pass a 2-mm sieve, and 7 kg of each soil was weighed into each pot. Other fertilizers such as urea, Triple Super Phosphate (TSP), gypsum, and zinc sulfate were used in every pot to supply N, P, S and Zn at the rate of 200, 50, 10, and 5 mg pot^{-1}, respectively. All nutrients except N were applied before sowing of maize while N was applied in two splits. After application of fertilizers, the soil was gently irrigated allowing smooth mixing of fertilizer materials with soils. After basal fertilizer application, five healthy seeds of hybrid maize (cv. BARI Hybrid Maize-8) were sown two cm below soil surface in each pot. Seeds were germinated after 6–11 days of sowing. Emergence was delayed in winter due to low temperature. Thinning was done 5–7 days after emergence, keeping 4 healthy plants per pot. After thinning, half N was top dressed, while the remaining 1/2 was applied at V6 stage (when 6 leaves appeared). The soils in pots were irrigated with tap water once every 3–5 days to replenish 100% soil moisture and to ensure that plants were not drought stressed. The aboveground part of the maize plant was cut at the soil surface at V10–V12 stage (55–65 days after emergence) because >90% of the total K uptake is usually accumulated by that stage [24]. The maize tissue was dried at 70 °C for 48 hours, crushed, and ground to pass a 0.5-mm sieve. A 0.5-g sub-sample was digested by an HNO_3-$HClO_{42}$ mixture at 180 to 200 °C [25] and K concentration was determined. Soil samples were collected before and after cropping for NHOAc-extractable K determination [20]. Seven maize crops were grown successively in each pot. Climate during the seven maize-growing periods together with crop duration for each period are presented in Figure 2.

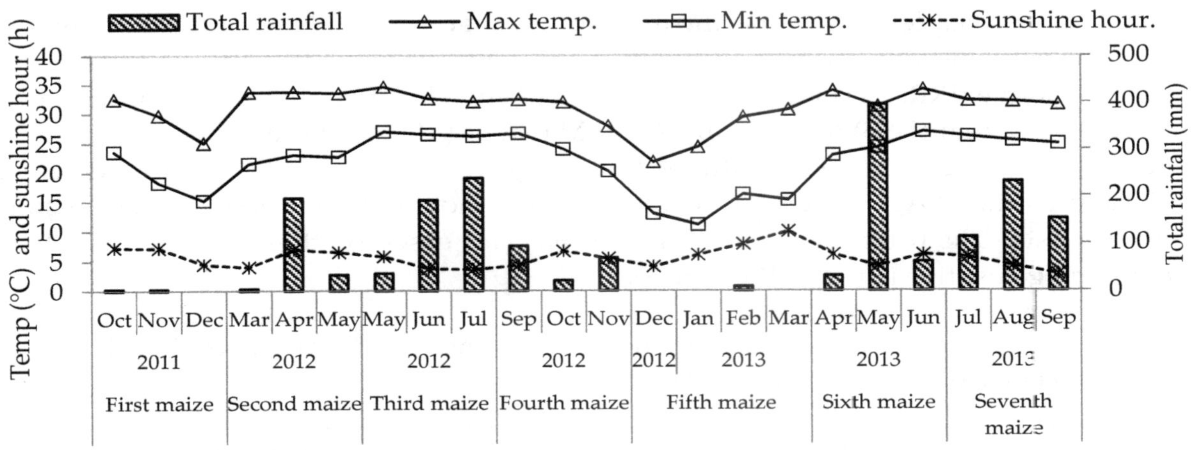

Figure 2. Mean monthly sunshine hour (h), minimum and maximum temperatures (°C), and total rainfall (mm) during successive cropping of *rabi* maize in pots from 2011 to 2013 in Gazipur, Bangladesh.

2.2. Field Experiment

2.2.1. Experimental Sites

A field experiment was conducted in 12 farmers' fields in three districts in Bangladesh, with 4 fields each in Rangpur (25.74° N, 89.28° E), Rajshahi (24.36° N, 88.62° E) and Comilla (23.46° N, 91.19° E) (Figure 1). Rajshahi is situated in the Active Ganges Floodplain (agro-ecological zone-AEZ, AEZ 11); Comilla is in the Old Meghna Estuarine Floodplain (AEZ 19); while Rangpur is spread over both the Active Tista Floodplain (AEZ 2) and Tista Meander Floodplain (AEZ 3). According to USDA Taxonomy, the soils are classified as Typic Haplaquept, Aeric Haplaquept and Typic Dystrochrept, respectively [22]. Experimental fields in Rangpur were located in AEZ 3 only.

The four farmers' fields in each district (two *upazillas* per district) were selected based on soil K (ranging from very low to high) across Bangladesh and from the literature review of K status in Bangladesh soils. The selected farmers' fields represented the diverse agro-ecosystems with variations in the cropping system, soil, and climate within each district.

2.2.2. Climatic Condition during the Experimental Period

The experimental fields were located in sub-tropical to the tropical climate, and the weather data for the three districts during the experimental period are shown in Figure 3. Most of the rainfall at all the sites occurred during the monsoon season (June–October). There was no rain from November 2012 to January 2013 in Rangpur, but Rajshahi and Comilla received ~100 mm rain in November. The total rainfall during the maize season was 532, 373, and 611 mm in Rangpur, Rajshahi, and Comilla, respectively. The mean minimum and maximum temperatures during the maize season were 9 °C and 32 °C, 9 °C and 36 °C, 11 °C and 33 °C in Rangpur, Comilla, and Rajshahi, respectively. The mean minimum and maximum temperatures were 9 °C and 25 °C respectively at the beginning of the experiment in January, with lower temperatures in Rangpur than the other two districts. The temperatures increased steadily to mean a minimum of 22 °C and mean maximum of 36 °C in April and May 2013. Mean sunshine hours across districts ranged from 5.8 to 6.8 h day^{-1}, with lowest sunshine hours recorded in Rangpur and highest in Comilla. At the beginning of the experiment, the sunshine hour was much lower in Rangpur (2.8 h day^{-1}) than the other two districts, and as the season progressed, the sunshine hour in all districts increased steadily to maximum of 10.9, 10.6, and 10.3 h day^{-1} in Rangpur, Rajshahi and Comilla, respectively (Figure 3).

2.2.3. Fertilizer Decision Support Tools

Based on the principles of site-specific nutrient management (SSNM) and experiences drawn from several years of on-farm research on maize in several Asian countries [26–31], the International Rice Research Institute (IRRI) and International Plant Nutrition Institute (IPNI) developed the fertilizer decision support system (DSS) tools, Maize Crop Manager (MCM) and Nutrient Expert for Maize (NEM), respectively, in collaboration with the International Maize and Wheat Improvement Centre (CIMMYT) and National Agricultural Research and Extension Systems (NARES) partners in South and Southeast Asia. The K fertilizer recommendation rates based on these two fertilizer DSS tools were compared against five levels of K (Table 3).

Table 3. Fertilizer recommendations for three field experimental sites in Bangladesh based on Maize Crop Manager (MCM) and Nutrient Expert for Maize (NEM).

Nutrient [a]	Rangpur (kg ha^{-1})		Rajshahi (Kg ha^{-1})		Comilla (kg ha^{-1})	
	MCM [b]	NEM [c]	MCM	NEM	MCM	NEM
N	184	164–173	184–190	162–173	115–150	141–152
P	20	22–23	15–20	19–25	15–20	19–20
K	75–100	109–125	75–100	93–105	75–100	93–103
S	7	7	7	7	7	7
Zn	3	3	3	3	3	3
B	1	1	1	1	1	1

[a] N: Nitrogen supplied from urea, P: Phosphorus supplied from Triple super phosphate (TSP), K: Potassium supplied from Muriate of potash (MoP), S: Sulphur supplied from Gypsum, Zn: Zinc supplied from Zinc sulphate, B: Boron supplied from Boric acid. [b] MCM: Maize crop manager, an online fertilizer decision tool based on SSNM, [c] NEM: Nutrient expert for maize, an offline computer-based fertilizer decisions tool based on SSNM.

2.2.4. Experimental Design and Treatments

The experiment was laid out in a Randomized Complete Block Design in 12 farmers' fields (serving as replicates) with five K levels (K_1 = 0, K_2 = 40, K_3 = 80, K_4 = 120, K_5 = 160), and two K fertilizer recommendations based on SSNM (DSS) tools (K_6 = Nutrient Expert based recommendation for maize (NEM); K_7 = Maize Crop Manager based recommendation (MCM)) in three districts. To determine the SSNM-based recommendations using two DSS tools: MCM (http://webapps.irri.org/bd/mcm/; IRRI, Philippines) and (NEM, an offline computer-based software; IPNI Offices, Delhi and Singapore), the participant farmers were asked 20 questions based on their agronomic and nutrient management practices of last (previous) year along with their field or soil characteristics. The answers to these questions were used as inputs to the MCM and NEM for generating fertilizer recommendations for each farmer (Table 3). Each treatment plot was 50 m^2 in area. Bunds of 0.5 m width were prepared between plots, and a border of 1 m width was kept around the experimental area. A medium-statured hybrid maize NK40, popularly grown in *Rabi* (winter) season was used. NK40 is tolerant of lodging and has a high yield potential of up to 20 t ha^{-1}.

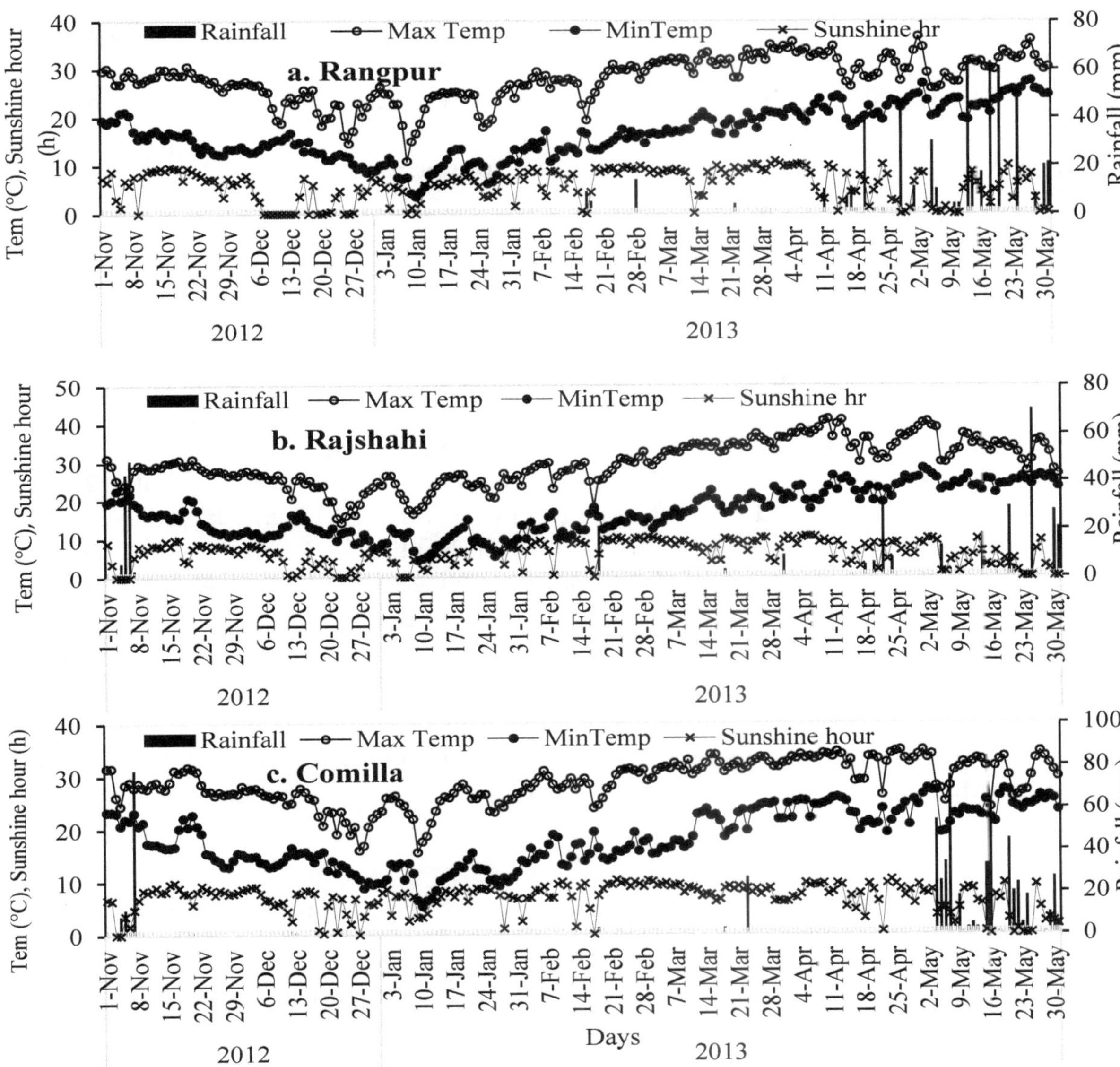

Figure 3. Daily maximum and minimum temperatures (°C), sunshine hour (h) and total rainfall (mm) during the rabi maize season from November 2012 to May 2013 in experimental sites (**a**. Rangpur, **b**. Rajshahi, **c**. Comilla) in Bangladesh.

2.2.5. Crop Management Practices

The experimental fields in all districts were irrigated with about 10 cm of water and allowed to reach proper moisture condition conducive for tillage. The fields were then prepared by 3–4 tillage with a 2-wheel operated power tiller to a depth of 8–10 cm followed by planking. Seeds in all districts were sown manually (with sowing dates ranging from 26 November 2012 to 6 January 2013) on shallow holes by dibbling and maintaining row to row and seed to seed distances of 60 and 20 cm respectively. Gap filling was done after emergence to maintain the 85,000 plant population ha^{-1}. N, P, K, S, and Zn were applied through urea, TSP, MoP, gypsum and zinc sulfate, respectively. Rates for different nutrients, including K, applied through the recommendations of MCM and NEM are presented in Table 3. N fertilizer was applied as three splits: as basal and top-dressed twice at V6 and V10; while K was applied as two splits: as basal and at V6. All other fertilizers were applied as basal.

The crops were manually weeded twice: the first weeding was done just before the first top dressing while the second was before the second top dressing, thus allowing the weeds to be removed from the fields before each top-dressing. As rainfall was not enough, four irrigations were applied in each site to avoid drought stress to the crops. First irrigation was applied at V2–V4 (2–3 leaves stage) while the second irrigation was applied after first weeding and before first top dressing at V6–V8 (6–8 leaves stage). Likewise, third irrigation was applied after second weeding and before second top-dressing at V10–V12 (10–12 leaf stage) and fourth irrigation during grain formation stage. The amount of water for each irrigation at each site was about 7.5 cm. Carbofuran 10G @ 100g per 100-meter rows was applied at planting for controlling cutworms and nematode infestation.

2.3. Data Analysis and Measurements

The crops were harvested at maturity from a 10.08 m^2 (4.2 m row length by 4 rows) area in the center of each treatment plot, excluding the two outer border rows. After harvesting, the crops were threshed with a hand thresher. The grain and stover yields and the growth and yield attributing characters (plant height, cobs plant^{-1}, cob length and girth, grains cob^{-1}, 1000-grain weight) were measured. The grain and stover samples from each plot were analyzed for total K content. For the post-harvest soil analysis for K_{ex} and K_{nex}, composite soil samples were taken from each treatment of each farmer immediately after crop harvest by using the methods as described in Table 2 [25].

Indigenous K supply (IKS) is defined as the amount of soil K that is available to maize from the soil during its growing period when other nutrients are non-limiting [31], and the IKS can be measured as the K accumulation in the above ground dry matter at harvest in the K omission plots [32]. Yield response (YR) is an effective index of soil fertility, and YR to K is defined as the yield difference between the attainable yield (measured as 85–90% of yield potential) and yield from the K omission plots [25,26]. YR to K can also be used to evaluate the soil K-supplying capacity [33]. The K concentration (%) of plant samples (stover and grain) was determined by a flame photometer [34].

The data from the K exhaustion study from the pot experiment as well as the field experiments were used to quantify the K-supplying capacity of the 18 soils to maize crops, and was calculated as follows:

$$\text{Total plant K uptake} = \text{above ground biomass} \times \text{K concentration in plant tissue} \tag{1}$$

Agronomic efficiency (AE_K; kg grain yield increase kg^{-1} applied K) was calculated using the equation:

$$AE_K = (GY_{+K} - GY_{0K}) \div F_K \tag{2}$$

where GY_{+K} is the grain yield in the treatment with K application (kg ha^{-1}), GY_{0K} is the grain yield in the treatment without K application (kg ha^{-1}), and F_K is the quantity of K applied (kg ha^{-1}).

Recovery efficiency (RE_K; kg K taken up kg^{-1} K applied) was calculated using the equation:

$$RE_K = (UK_{+K} - UK_{0K}) \div F_K \tag{3}$$

where UK_{+K} is the total plant K uptake (kg ha^{-1}) of the above-ground biomass (stover + grain) in plots that received K, UK_{0K} is total K uptake without the addition of K, and F_K is the quantity of K applied (kg ha^{-1}).

The R software [35,36] was used to analyze the means of grain and stover yields, growth and yield components, total K uptake, agronomic efficiency and recovery efficiency of K between different soils and treatments by using the least significant difference at 0.05 probability level. The Duncan's New Multiple Range Test, a mean separation technique, was applied to detect significant differences between treatment [37].

3. Results

3.1. Pot Experiment

3.1.1. Mean Shoot Dry Matter Yield and Yield Response to K Fertilizer

The mean shoot dry matter yield of maize over seven successive cropping across 18 soils varied widely from 14.52 to 39.37 g pot^{-1} in K control pots while it was from 38.83 to 47.81 g pot^{-1} in K applied pots. The lowest and highest dry matter yield with no added K soils were found in Durgapur and Mithapukur soils, respectively. While the lowest and highest dry matter yield with K fertilizer added soils were found in Modhupur and Gopalpur soils, respectively (Figure 4). The contribution of K fertilizer to the increment of maize shoot dry matter over K control was considered as the shoot dry matter response to K. The mean response of shoot dry matter over seven successive cropping ranged from 4 to 16 g pot^{-1}, with the highest response ($p \leq 0.001$) in Durgapur soil and the lowest in Modhukhali. The mean response followed the order of: Durgapur > Shibganj > Barura > Godagari > Tarash > Paba > Binerpota > Jhinaidah Sadar > Modhupur > Birganj > Nawbagnj Sadar > Gopalpur > Daudkandi > Gangachara > Rangpur Sadar > Dinajpur Sadar > Mithapukur > Modhukhali. Therefore, Durgapur and Modhukhali soil appeared as the most and least responsive to K fertilization, respectively (Figure 4). There was significant ($p \leq 0.001$) negative correlation (r^2 = 0.70, r = −0.84) between yield response to K fertilizer and soil initial K (Figure 5). The mean shoot dry matter yield in the 18 soils without K fertilization reduced drastically from 1st to 3rd crops, decreased slightly in the 4th to 5th crop due to climatic variation, and after that gradually declined up to the 7th crop. The differences in shoot dry matter yield between the K applied soils and the no added K fertilizer soils were gradually increased with the successive cropping (Figure 6).

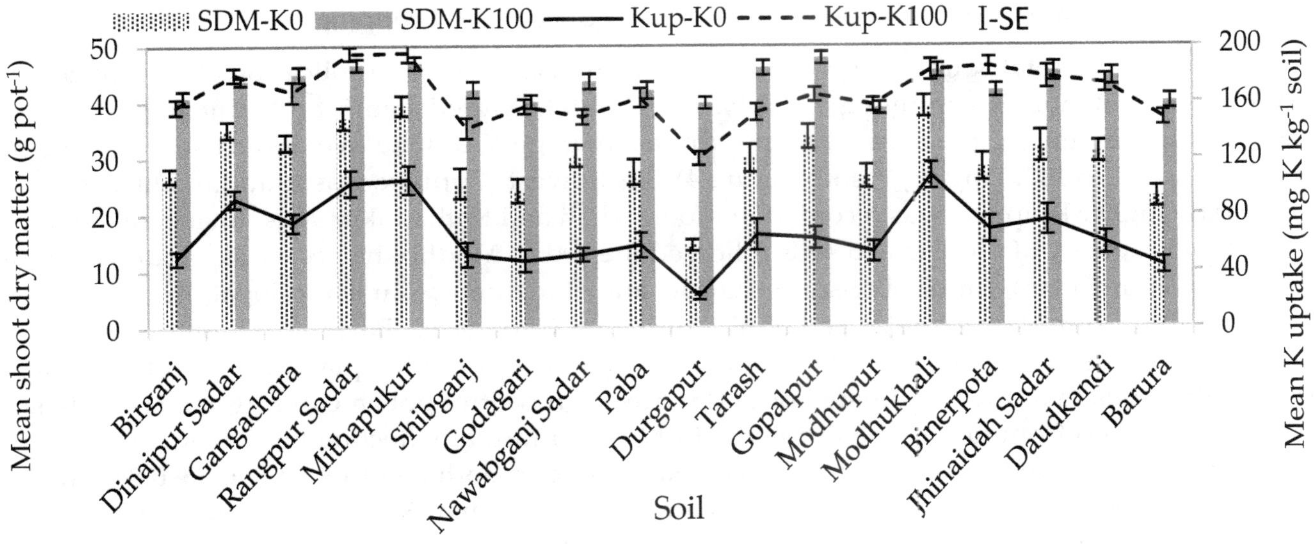

Figure 4. Trends for mean shoot dry matter (SDM) and mean K uptake over successive seven rabi maize crops in pots with K0 and K100 for 18 diverse soils of Bangladesh.

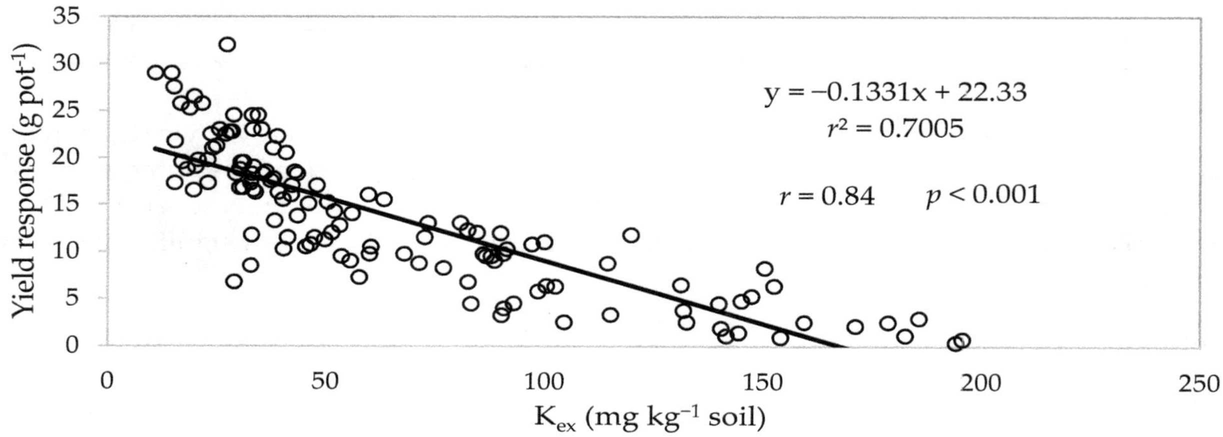

Figure 5. The relationship between initial exchangeable soil K (K_{ex}) for 18 diverse soils of Bangladesh and yield response of rabi maize grown up to V10–V12 for seven successive crops to K fertilizer in pot experiment, Gazipur, Bangladesh.

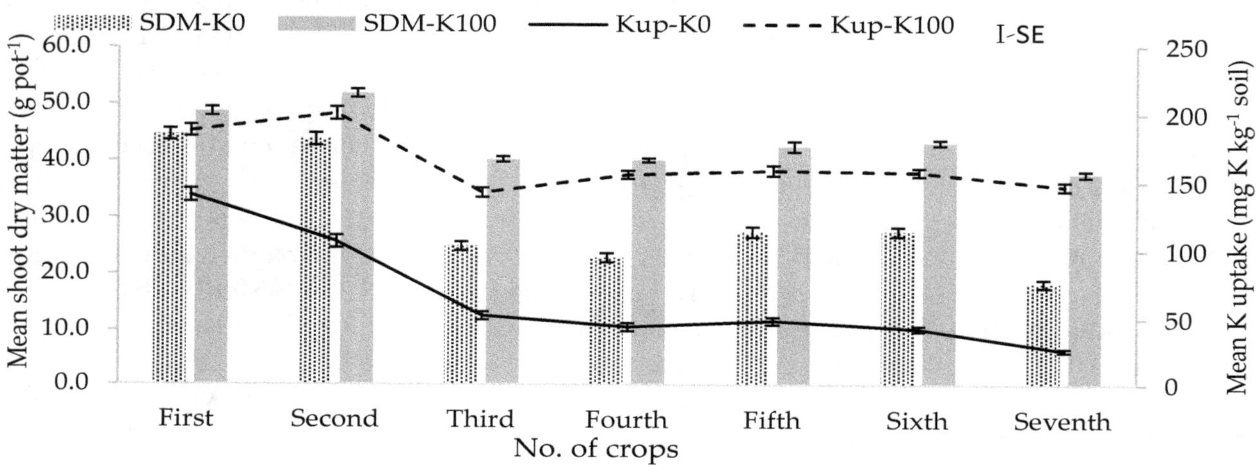

Figure 6. Trends for mean shoot dry matter (SDM) and mean K uptake over 18 diverse soils for seven successive cropping of rabi maize with K0 and K100 in pots in Gazipur, Bangladesh.

3.1.2. Soil K Supplying Capacity and K Depletion over Successive Cropping

The amount of total K uptake by plants in different soils without application of K fertilizer was defined as the potential K-supplying capacity. The K uptake varied significantly due to the variation of soil K reserves among the soils. In control pots, average K uptake over seven successive crops varied from 22.1 to 107.3 mg kg^{-1} soil (Figure 4). The lowest K uptake was found in Durgapur and significantly higher K uptake was recorded in Modhukhali and Mithapukur soils. In general, potential K-supplying capacity of these tested soils followed the order: Modhukhali > Mithapukur > Rangpur Sadar > Dinajpur Sadar > Jhinaidah Sadr > Gangachara > Binerpota > Tarash > Gopalpur > Daudkandi > Paba > Modhupur Nawabganj > Shibganj > Birganj > Godagari > Barura > Durgapur (Figure 4). Though the trends were similar for both +K and −K soils but the K uptake was 2–3 times higher for K-treated soils than the non-treated ones (Figure 4). In control pots mean K uptake over 18 soils drastically reduced from first to the third crop but it was gradually declined from fourth to seventh crops (Figure 6). There was a strong positive relationship in both soil Kex (r^2 = 0.56–0.84) and K_{nx} (r^2 = 0.48–0.63) with K uptake, but it was stronger in soil K_{ex} than K_{nx}. Moreover, the relationship became gradually stronger with increasing the number of successive cropping (Figure 7). And soil K_{ex} and K_{nex} decreased remarkably in control pots due to the growing of seven successive maize crops (Table 4).

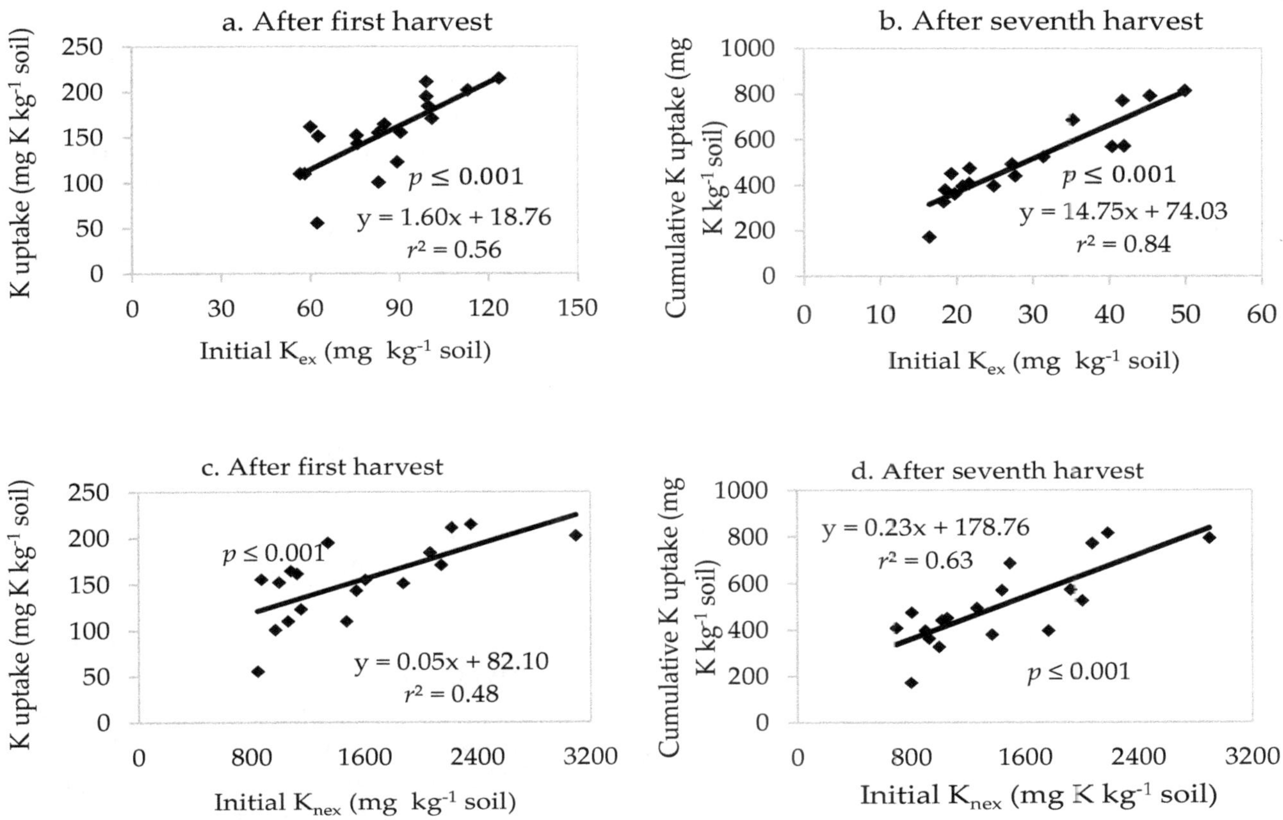

Figure 7. The relationship between exchangeable soil K (K_{ex}) and K uptake (**a.** after first harvest and **b.** after seventh harvest), and between non-exchangeable K (K_{ex}) for 18 diverse soils of Bangladesh and K uptake (**c.** after first harvest and **d.** after seventh harvest) in the above ground biomass of rabi maize grown up to V10–V12 in pot as successive crop.

Table 4. Amount of exchangeable and non-exchangeable K before the start of the experiment (initial) and after seventh harvest (grown up to V10–V12) of maize grown in pots with K_0 and K_{100} with 18 diverse soils, Gazipur, Bangladesh.

Location of Soil [a]	Pre K_{ex} [b]	Post K_{ex}		Pre K_{nex} [c]	Post K_{nex}	
		K0	K100		K0	K100
	mg kg^{-1} soil					
Birganj	57	16	39	1475	1369	1435
Dinajpur Sadar	83	37	66	1608	1498	1568
Gangachara	76	35	57	1544	1439	1502
Rangpur Sadar	99	36	76	2220	2075	2168
Mithapukur	113	39	85	3098	2900	3028
Shibganj	63	14	42	1876	1767	1849
Godagari	76	18	50	996	928	971
Nawabganj	83	20	68	970	901	944
Paba	85	23	64	1080	1018	1061
Durgapur	62	20	40	845	800	831
Tarash	99	21	67	1345	1264	1322
Gopalpur	90	20	55	871	807	846
Modhupur	89	21	67	761	698	733
Modhukhali	124	45	113	2357	2184	2360
Binerpota	101	31	83	2145	2005	2102
Jhinaidah Sadar	100	35	78	2065	1923	2016
Daudkandi	60	18	37	1123	1053	1104
Barura	58	14	37	1060	998	1045

[a] Upazilas (Sub-districts); [b] K_{ex}: Exchangeable soil K; [c] K_{nex}: Non-exchangeable K.

3.2. Field Experiment

The ANOVA for the means for all yield and yield attributing characters (except number of cobs plant^{-1}) in all the farmers' fields across the three districts showed highly significant effects between the treatments ($p \leq 0.001$) due to farmers' fields being scattered over a large area with variable soils and land types. There was no significant effect ($p \geq 0.05$) on the interaction among the same treatment of different farmers for most of the measurable variables (except yield increase over control) in the same location due to less variability of soils and maintaining same practices for all farmers.

3.2.1. Growth and Yield Components

Plant heights were 237 ± 1.35, 186 ± 5.10 and 237 ± 3.51cm in Rangpur, Rajshahi, and Comilla, respectively, while it was 217 ± 3.20 cm over all sites (Table 5). In all sites, significantly shortest plants were found in control plots, and the height increased progressively, though not significantly, with an increase of K rate up to 120 kg K ha^{-1}. In Rangpur and Rajshahi significantly taller plants were found with all K-treated plots except 40 kg K ha^{-1} plots, with slightly taller plants in MCM and NEM plots. In Comilla, however, maize plants in all K-treated plots were significantly taller than in the control plots.

Cob length varied from 15.0 to 17.8 cm overall sites, ranging from 17.0–20.1, 12.6–15.1 and 15.4–19.3 cm in Rangpur, Rajshahi, and Comilla, respectively (Table 5). Cob length increased progressively with increase of K fertilizer rate up to 120 kg ha^{-1}. In all locations, cobs were significantly longer for all levels of K and MCM- and NEM-based recommendations than in control. The longest cob was found with 120 kg K ha^{-1} plots in Rangpur and MCM-based fertilizer recommendation (88 kg K ha^{-1}) in Rajshahi, though it was statically identical with the NEM-based recommendation in all locations (Table 5).

The cob girth was 16.5 ± 0.27, 14.78 ± 0.10 and 15.6 ± 0.22 cm in Rangpur, Rajshahi, and Comilla, respectively (Table 5). Cob girth increased progressively with the increase of K rates up to 120 kg K ha^{-1}. Cob girth in control plots was lower than other treatments in all sites though it was only statistically lower to 120 and 160 kg K ha^{-1} treated plots in Rangpur. The highest cob girth for 120 kg K ha^{-1} was found in Rangpur and Rajshahi and in NE fertilizer recommended plots in Comilla, though it was statically identical to the MCM fertilizer recommended plots in all locations. In Rajshahi and Comilla, except control, all treatments were statistically identical.

The number of grains cob^{-1} was 495 ± 9.82, 367 ± 3.09 and 442 ± 17.05 in Rangpur, Rajshahi, and Comilla, respectively. The number of grains cob^{-1} increased progressively with the increase of K rate up to 120 kg ha^{-1}. It was the highest for the NEM-based recommendation in Rangpur and Comilla, 160 kg K ha^{-1} treated plots in Rajshahi, though it was statistically identical to all K levels and MCM- and NEM-based fertilizer recommendations (Table 5).

The weight of 1000-grains varied significantly among the treatments, ranging 314 ± 7.03, 3121 ± 8.61 and 307 ± 3.68 g in Rangpur, Rajshahi, and Comilla, respectively. Across the sites, it varied significantly ($p < 0.001$) from 252 g in K0 to 339 g in NEM based recommendation plot. The results showed that NEM-based fertilizer recommendation resulted in heaviest grains in all locations, followed by all levels of K and MCM-based recommendation (Table 5).

The average grain yields across all treatments were 9.18 ± 0.33, 7.67 ± 0.25 and 6.60 ± 0.26 t ha^{-1} in Rangpur, Rajshahi, and Comilla, respectively. Likewise, average biomass was 18.53 ± 0.67, 15.32 ± 0.49 and 13.18 ± 0.52 t ha^{-1}, respectively, in the three districts. The average grain and biomass yields over all districts were 7.82 ± 0.20 and 15.68 ± 0.40, respectively (Table 5). Grain and biomass yields progressively increased with the increase of K rate up to 120 kg K ha^{-1}. Grain and biomass yields in control plots were significantly lower compared to other treatments in Rajshahi, but it was similar to 40 kg K ha^{-1} plots in Rangpur and Comilla. In all districts, grain and biomass yields were significantly similar for 80, 120, 160 kg K ha^{-1}, and NEM- and MCM-based fertilizer recommendation.

Table 5. Effect of K fertilization on growth, and yield and yield attributes of *rabi* maize in three field experimental sites in Bangladesh, 2012–2013.

Treatment *	Plant Height (cm)	Cob Length (cm)	Cob Girth (cm)	Cobs Plant^{-1}	Grains cob^{-1}	1000-Grain Weight (g)	Grain Yield (t ha^{-1})	Biomass (t ha^{-1})	HI #	Yield Response (%)
					Rangpur					
K_0	218 ^{c}	17.0 ^{b}	15.4 ^{b}	0.99 ^{a}	432 ^{b}	248 ^{d}	6.35 ^{c}	12.92 ^{c}	0.49 ^{a}	
K_{40}	227 ^{b}	18.0 ^{ab}	16.2 ^{ab}	1.00 ^{a}	481 ^{ab}	303 ^{c}	7.51 ^{c}	15.19 ^{c}	0.49 ^{a}	17.72 ^{d}
K_{80}	228 ^{ab}	19.2 ^{ab}	16.4 ^{ab}	1.00 ^{a}	495 ^{ab}	312 ^{bc}	9.82 ^{ab}	19.71 ^{ab}	0.50 ^{a}	52.08^{bc}
K_{120}	231 ^{ab}	20.1 ^{a}	17.5 ^{a}	1.00 ^{a}	518 ^{a}	343 ^{a}	10.21 ^{ab}	20.70 ^{ab}	0.49 ^{a}	60.51^{a}
K_{160}	228 ^{ab}	19.5 ^{ab}	17.3 ^{a}	1.00 ^{a}	494 ^{ab}	308 ^{bc}	9.04 ^{b}	18.20 ^{b}	0.50 ^{a}	40.55^{c}
MCM	233 ^{a}	19.6 ^{ab}	16.3 ^{ab}	1.00 ^{a}	516 ^{a}	339 ^{ab}	10.38 ^{ab}	21.01 ^{ab}	0.49 ^{a}	62.76^{ab}
NEM	231 ^{ab}	19.8 ^{ab}	16.9 ^{ab}	1.01 ^{a}	527 ^{a}	345 ^{a}	10.95 ^{a}	21.98 ^{a}	0.50 ^{a}	70.47 ^{a}
Mean	228	19	16.5	1	495	314	9.18	18.53	0.5	50.68
SE (±)	1.35	0.44	0.27	0.001	9.82	7.03	0.33	0.67	0.001	4.12
					Rajshahi					
K_0	147 ^{c}	12.6 ^{e}	13.7 ^{c}	0.99 ^{a}	343 ^{c}	228 ^{d}	5.42 ^{c}	10.80 ^{d}	0.50 ^{abc}	
K_{40}	172 ^{bc}	13.7 ^{d}	14.6 ^{b}	0.99 ^{a}	368 ^{ab}	310 ^{c}	6.49 ^{b}	13.17 ^{c}	0.49 ^{bc}	22.01 ^{c}
K_{80}	184 ^{ab}	13.8 ^{cd}	14.8 ^{ab}	0.99 ^{a}	365 ^{ab}	347 ^{ab}	8.09 ^{a}	16.40 ^{ab}	0.49 ^{bc}	51.84 ^{ab}
K_{120}	203 ^{a}	14.5 ^{b}	15.2 ^{a}	0.99 ^{a}	361 ^{b}	335 ^{abc}	8.63 ^{a}	17.16 ^{a}	0.50 ^{abc}	58.75 ^{a}
K_{160}	196 ^{ab}	14.2 ^{bcd}	15.1 ^{a}	1.00 ^{a}	380 ^{a}	321 ^{bc}	8.02 ^{a}	15.51 ^{b}	0.52 ^{a}	43.49 ^{b}
MCM	204 ^{a}	15.1 ^{a}	15.1 ^{a}	0.99 ^{a}	379 ^{a}	357 ^{a}	8.41 ^{a}	17.23 ^{a}	0.49 ^{c}	59.33 ^{a}
NEM	198 ^{ab}	14.4 ^{bc}	15.0 ^{a}	1.00 ^{a}	376 ^{ab}	349 ^{a}	8.66 ^{a}	17.02 ^{a}	0.51 ^{ab}	57.49 ^{a}
Mean	186	14.01	14.78	1.00	367.4	321.1	7.67	15.32	0.50	48.82
SE (±)	5.1	0.16	0.1	0.001	3.09	8.61	0.25	0.49	0.002	3.94
					Comilla					
K_0	208 ^{b}	15.4 ^{b}	13.5 ^{b}	0.99 ^{a}	316 ^{b}	280 ^{d}	4.86 ^{b}	9.70 ^{c}	0.50 ^{a}	
K_{40}	232 ^{a}	17.4 ^{ab}	15.3 ^{a}	1.00 ^{a}	447 ^{a}	297 ^{cd}	5.65 ^{b}	11.59 ^{bc}	0.49 ^{b}	20.57 ^{b}
K_{80}	244 ^{a}	18.6 ^{a}	15.9 ^{a}	0.99 ^{a}	431 ^{a}	301 ^{bc}	6.95 ^{a}	13.70 ^{a}	0.51 ^{a}	40.79 ^{a}
K_{120}	245 ^{a}	19.3 ^{a}	16.1 ^{a}	1.00 ^{a}	470 ^{a}	310 ^{abc}	7.45 ^{a}	14.85 ^{ab}	0.50 ^{a}	54.02 ^{a}
K_{160}	244 ^{a}	19.1 ^{a}	16.1 ^{a}	0.99 ^{a}	468 ^{a}	319 ^{ab}	6.93 ^{a}	13.91 ^{ab}	0.50 ^{ab}	44.29 ^{a}
MCM	244 ^{a}	18.9 ^{a}	15.9 ^{a}	1.00 ^{a}	473 ^{a}	316 ^{abc}	7.01 ^{a}	14.03 ^{ab}	0.50 ^{a}	44.00 ^{a}
NEM	243 ^{a}	18.9 ^{a}	16.2 ^{a}	1.00 ^{a}	488 ^{a}	324 ^{a}	7.38 ^{a}	14.46 ^{a}	0.51 ^{a}	49.80 ^{a}
Mean	237	18.2	15.6	1.00	442	307	6.60	13.18	0.50	42.24
SE (±)	3.51	0.32	0.22	0.001	17.05	3.68	0.26	0.52	0.002	2.9
					All Sites (Mean)					
K_0	191 ^{c}	15.0 ^{c}	14.2 ^{c}	1.00 ^{a}	364 ^{b}	252 ^{d}	5.54 ^{d}	11.14 ^{d}	0.50 ^{ab}	
K_{40}	210 ^{b}	16.4 ^{b}	15.3 ^{b}	1.00 ^{a}	432 ^{a}	303 ^{c}	6.55 ^{c}	13.32 ^{c}	0.49 ^{c}	20.10 ^{d}
K_{80}	219 ^{ab}	17.2 ^{ab}	15.7 ^{ab}	1.01 ^{a}	430 ^{a}	320 ^{b}	8.28 ^{ab}	16.60 ^{ab}	0.50 ^{ab}	48.24 ^{bc}
K_{120}	226 ^{a}	18.0 ^{a}	16.3 ^{a}	1.00 ^{a}	450 ^{a}	329 ^{ab}	8.76 ^{ab}	17.57 ^{a}	0.50 ^{ab}	57.76 ^{a}
K_{160}	223 ^{ab}	17.6 ^{ab}	16.2 ^{ab}	1.00 ^{a}	447 ^{a}	316 ^{bc}	7.99 ^{b}	15.87 ^{b}	0.50 ^{a}	42.78 ^{c}
MCM	227 ^{a}	17.8 ^{a}	15.8 ^{ab}	1.00 ^{a}	456 ^{a}	338 ^{a}	8.60 ^{ab}	17.42 ^{ab}	0.49 ^{bc}	55.36 ^{ab}
NEM	224 ^{a}	17.7 ^{a}	16.0 ^{ab}	1.00 ^{a}	464 ^{a}	339 ^{a}	8.99 ^{a}	17.82 ^{a}	0.51 ^{a}	59.25 ^{a}
Mean	217	17.1	15.6	1.00	435	314	7.82	15.68	0.50	47.25
SE (±)	3.2	0.31	0.14	0.001	8.71	3.91	0 2	0.400	0.001	1.85

Data not sharing the same lower-case letter(s) in a column are significantly different according to Duncan's New Multiple-Range Test at 5% level of probability [37]; K_0, K_{40}, K_{80}, K_{120}, K_{160} represent control, 40, 80, 120, 160 kg K ha^{-1}, respectively; * MCM: Fertilizer recommendation ranged 75–100 kg K ha^{-1} across sites based on SSNM through "Maze Crop Manager" developed by IRRI; NEM: Fertilizer recommendations ranged 93–125 kg K ha^{-1} across sites based on SSNM through "Nutrient Expert for Maize"; SE-Standard error of mean, # HI: Harvest index.

Harvest index varied from 0.49 to 0.50, 0.49 to 0.52, and 0.49 to 0.51 in Rangpur, Rajshahi and Comilla, respectively. It varied significantly across treatments in Rajshahi and Comilla, but not in Rangpur. Yield response to K application across treatments varied from 17.72 to 70.47%, 22.01 to 59.33%, and 20.57 to 54.02% in Rangpur, Rajshahi, and Comilla, respectively, and when averaged across sites, it varied from 20.10 to 59.25% (Table 5).

3.2.2. Estimation of K Supplying Capacity

The average K concentration in maize grain over treatments were 0.439 ± 0.013, 0.439 ± 0.012 and 0.409 ± 0.014% in Rangpur, Rajshahi, and Comilla, respectively. Likewise, average K concentration in maize stover were 1.32 ± 0.05%, 1.21 ± 0.06% and 1.24 ± 0.05%, respectively, in the three districts. The average K concentration in grain and biomass yields over all districts were 0.429 ± 0.008 and 1.24 ± 0.03, respectively (Table 6). K concentration in maize grain and stover over sites varied from 0.309 to 0.520% and from 0.88 to 1.73%, respectively. K concentration in maize grain and stover progressively increased with the increase in K rate up to the maximum dose (160 kg K ha^{-1}).

K concentration across locations and treatments was lowest (0.31% in grain and 0.88% in stover) in control plots and highest in 160 kg K ha^{-1} plots (0.52% in grain and 1.73% in stover (Table 6).

The total mean K uptake by maize (grain + stover) across treatments was from 166 $\pm$ 9.66, 128 $\pm$ 7.39, and 106 $\pm$ 6.40 kg ha^{-1} in Rangpur, Rajshahi, and Comilla, respectively. Total K uptake over the sites varied significantly from 67.1 to 178.3 kg K ha^{-1}. Total K uptake progressively increased with the increase in K rate up to 160 kg K ha^{-1}. Total K uptake by maize across treatments and locations was lowest in control plots and highest in 160 kg K ha^{-1}. Across all locations, in both control and K-treated plots, the total K uptake, i.e., the K-supplying capacity of soils, was in the order: Rangpur > Rajshahi > Comilla, and this was similar to the results found in the pot study (Table 6). There was a strong positive linear relationship (r^2 = 0.73, p < 0.001) between indigenous soil K to K uptake in without K fertilized plots (Figure 8a), but no such relationship with indigenous soil K was observed in high K fertilized plots (Figure 8b).

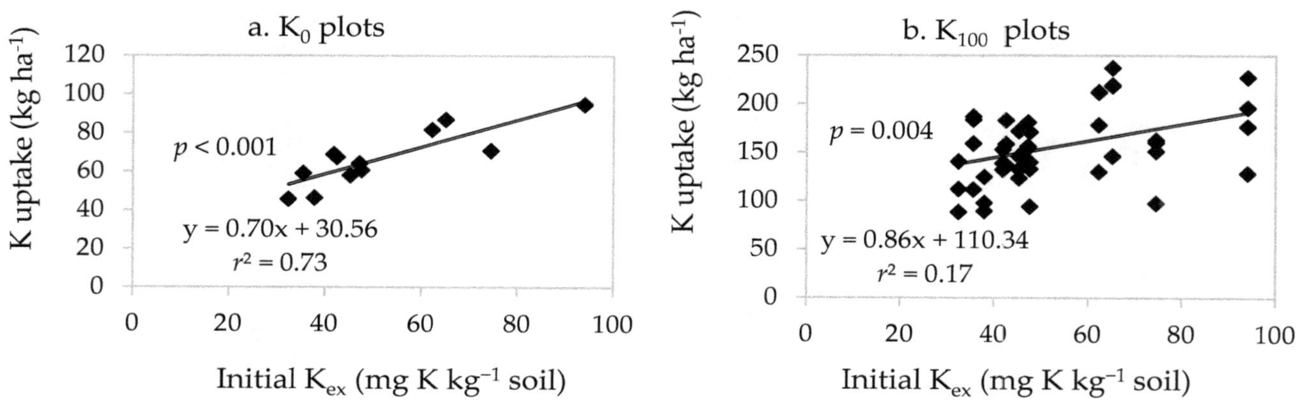

Figure 8. The relationship between initial soil K_{ex} and K uptake in (**a**) control plots, K_0 (n = 12) and (**b**) K-fertilizer applied plots, K_{100} (n = 48) in three field experimental sites in Bangladesh. 'n' indicates number.

Table 6. Effect of K fertilization on K concentration and K uptake, and agronomic and recovery efficiency of K by *rabi* maize in three field experimental sites in Bangladesh, 2012–2013.

Treatment *	K Concentration (%)		K Uptake (kg ha^{-1})			AE_K $^{\alpha}$	RE_K $^{\beta}$
	Grain	Stover	Grain	Stover	Total		
			Rangpur				
K_0	0.312 ^{d}	0.97 ^{c}	19.8 ^{c}	63.8 ^{c}	83.6 ^{c}		
K_{40}	0.404 ^{c}	1.21 ^{d}	30.4 ^{b}	93.4 ^{c}	123.8 ^{b}	29.05 ^{b}	1.01 ^{a}
K_{80}	0.447 ^{b}	1.25 ^{bc}	44.1 ^{a}	124.7 ^{b}	168.8 ^{a}	43.41 ^{a}	1.06 ^{a}
K_{120}	0.455 ^{b}	1.43 ^{b}	46.6 ^{a}	150.3 ^{ab}	196.9 ^{a}	32.19 ^{b}	0.94 ^{a}
K_{160}	0.527 ^{a}	1.73 ^{a}	47.6 ^{a}	159.6 ^{a}	207.2 ^{a}	16.80 ^{c}	0.77 ^{a}
MCM	0.452 ^{b}	1.29 ^{bc}	47.0 ^{a}	137.4 ^{ab}	184.4 ^{a}	46.10 ^{a}	1.15 ^{a}
NEM	0.476 ^{b}	1.37 ^{bc}	52.2 ^{a}	151.9 ^{ab}	204.1 ^{a}	39.57 ^{ab}	1.04 ^{a}
Mean	0.439	1.32	41.1	125.87	166.97	34.52	1.00
SE ($\pm$)	0.013	0.05	2.3	7.49	9.66	2.44	0.06
			Rajshahi				
K_0	0.317 ^{d}	0.81 ^{f}	17.2 ^{d}	43.4 ^{e}	60.5 ^{f}		
K_{40}	0.396 ^{c}	0.99 ^{e}	25.7 ^{c}	66.0 ^{d}	91.7 ^{e}	26.58 ^{a}	0.78 ^{ab}
K_{80}	0.441 ^{b}	1.11 ^{d}	35.7 ^{b}	92.2 ^{c}	127.9 ^{d}	33.30 ^{a}	0.84 ^{ab}
K_{120}	0.479 ^{b}	1.33 ^{b}	41.2 ^{a}	114.1 ^{b}	155.3 ^{b}	26.73 ^{a}	0.79 ^{ab}
K_{160}	0.519 ^{a}	1.82 ^{a}	41.4 ^{a}	135.7 ^{a}	177.2 ^{a}	16.20 ^{b}	0.73 ^{b}

Table 6. *Cont.*

Treatment *	K Concentration (%)		K Uptake (kg ha^{-1})			AE_K $^{\alpha}$	RE_K $^{\beta}$
	Grain	Stover	Grain	Stover	Total		
MCM	0.466 ^{b}	1.23 ^{bc}	39.3 ^{ab}	107.9 ^{b}	147.2 ^{bc}	34.20 ^{a}	1.00 ^{a}
NEM	0.456 ^{b}	1.19 ^{cd}	39.4 ^{ab}	99.0 ^{bc}	138.4 ^{cd}	32.37 ^{a}	0.79 ^{ab}
Mean	0.439	1.21	34.3	94.0	128.3	28.23	0.82
SE (±)	0.012	0.06	1.75	5.76	7.39	1.49	0.03
			Comilla				
K_0	0.299 ^{d}	0.88 ^{d}	14.5 ^{c}	42.7 ^{d}	57.2 ^{d}		
K_{40}	0.356 ^{cd}	1.04 ^{c}	20.1 ^{bc}	61.5 ^{cd}	81.7 ^{cd}	19.62 ^{ab}	0.61 ^{a}
K_{80}	0.419 ^{b}	1.15 ^{bc}	29.4 ^{ab}	77.3 ^{bc}	106.6 ^{bc}	26.02 ^{a}	0.62 ^{a}
K_{120}	0.447 ^{b}	1.24 ^{b}	33.6 ^{a}	91.8 ^{b}	125.4 ^{ab}	21.57 ^{ab}	0.57 ^{a}
K_{160}	0.516 ^{a}	1.65 ^{a}	35.8 ^{a}	114.8 ^{a}	150.6 ^{a}	12.90 ^{b}	0.58 ^{a}
MCM	0.430 ^{b}	1.18 ^{bc}	30.7 ^{a}	83.0 ^{b}	113.6 ^{b}	23.93 ^{a}	0.64 ^{a}
NEM	0.394 ^{bc}	1.11 ^{bc}	29.0 ^{ab}	77.9 ^{bc}	106.9 ^{bc}	25.10 ^{a}	0.50 ^{a}
Mean	0.409	1.177	27.59	78.4	106	21.52	0.59
SE (±)	0.014	0.045	1.77	4.74	6.4	1.47	0.02
			Across Sites				
K_0	0.309 ^{d}	0.88 ^{e}	17.2 ^{c}	49.9 ^{e}	67.1 ^{e}	-	-
K_{40}	0.385 ^{c}	1.08 ^{d}	25.4 ^{b}	73.7 ^{d}	99.0 ^{d}	25.09 ^{b}	0.80 ^{ab}
K_{80}	0.436 ^{b}	1.17 ^{c}	36.4 ^{a}	98.0 ^{c}	134.4 ^{c}	34.25 ^{a}	0.84 ^{ab}
K_{120}	0.460 ^{b}	1.33 ^{b}	40.5 ^{a}	118.7 ^{b}	159.2 ^{b}	26.83 ^{b}	0.77 ^{ab}
K_{160}	0.520 ^{a}	1.73 ^{a}	41.6 ^{a}	136.7 ^{a}	178.3 ^{a}	15.30 ^{c}	0.70 ^{b}
MCM	0.449 ^{b}	1.23 ^{c}	39.0 ^{a}	109.4 ^{bc}	148.4 ^{bc}	34.74 ^{a}	0.93 ^{a}
NEM	0.442 ^{b}	1.22 ^{c}	40.2 ^{a}	109.6 ^{bc}	149.8 ^{bc}	32.35 ^{a}	0.78 ^{ab}
Mean	0.429	1.24	34.3	99.4	133.8	28.1	0.8
SE (±)	0.008	0.03	1.27	4.102	5.303	1.154	0.03

Data not sharing the same lower-case letter(s) in a column are significantly different according to Duncan's New Multiple-Range Test at 5% level of probability [37]; K_0, K_{40}, K_{80}, K_{120}, K_{160} represent control, 40, 80, 120, 160 kg K ha^{-1}, respectively; * MCM: Fertilizer recommendation ranged 75–100 kg K ha^{-1} across the sites based on SSNM through "Maze Crop Manager"; NEM: Fertilizer recommendations ranged 93–125 kg K ha^{-1} across sites based on SSNM through "Nutrient Expert for Maize"; SE-Standard error of mean; $^{\alpha}$ AE_k: agronomic use efficiency of K (kg grain yield increase kg^{-1} applied K); RE_k: Recovery efficiency of K (kg K taken up kg^{-1} K applied).

3.2.3. K-Use Efficiency

There was a significant effect of K application on agronomic efficiency (AE_K) and recovery efficiency (RE_K) in each site and across all sites. The AE_K varied significantly over control from 16.8 to 46.1, 16.2 to 34.2, and 12.9 to 26.0 in Rangpur, Rajshahi, and Comilla, respectively. Likewise, the ranges of RE_K were from 0.77 to 1.15, 0.73 to 1.00, and 0.50 to 0.64, respectively in the three sites. AE_K and RE_K over the sites varied from 15.30 to 34.74 and 0.70 to 0.93, respectively, and decreased progressively with increase of K rates (Table 6). Both AE_K and RE_K were lowest for the highest K dose (160 kg ha^{-1}) compared to other rates. Though not significantly different, the MCM-based recommendation generally resulted in higher AE_K and RE_K than NEM-based recommendation (Table 6).

4. Discussion

Until a few years back, there was a general perception that agricultural soils in South and South East Asia, including those in Bangladesh, were well supplied with K, and hence there was no need to apply K fertilizer to crops. But recently, many investigations, reviews and research results have shown that the intensification of agriculture in the region with little or no K application caused gradual K mining, and crop responses to K are observed in many of those countries including Bangladesh [38–44]. Such results suggest the need for application of K fertilizers for sustaining or increasing the crop yields. Proper application and management of K require a thorough understanding of soil K dynamics and its uptake by crops at various K inputs and outputs scenarios and for different cropping systems. It is well known that the K availability to the plants does not only depend on the size of the available K pool in the soil but also K release patterns, and its transport from soil solution to the root zone for its uptake by plants [45,46]. It is hypothesized that soils K availability to plants differ in terms of mineralogy, soil K reserves, K-supplying capacity and its allowable drawdown factors [47,48]. Thus, some soils would require more while others would require less K to grow profitable maize crops [40].

Plant available K can be assessed either by plant growth analysis or by simple chemical extraction method or by a combination of both procedures with plants grown in no-K added plots. A robust relationship (r^2 = 0.82) between K uptake and grain yield was observed in no-K plots, which was considered as a good measure of soil K supply to crops [49]. When assessed by plant K uptake, it can be termed as "K-supplying capacity" whereas if assessed by extracting the soil with one or more extractions, a chemical index of available K (K-releasing capacity), can be the true index of plant available K. Consequently, many investigators reported, while "plant available K" can be equated to the K-supplying capacity of the soil, it can only be related to the K-releasing capacity. Thus, both K-releasing and K-supplying capacity can be considered the measures of the ability of a soil to supply K to plants [50–52]. The amount of total K uptake by plants from a soil depends on the potential of K-supplying capacity.

The K depletion pattern in this study was carried out in a pot experiment with the successive planting of maize for seven harvests to understand K-supplying capacity from soil reserves of major soils in Bangladesh. The 18 tested soils varied considerably in supporting K uptake over seven crops, ranging from 22.1–103.7 mg K kg^{-1} soil, and the uptake in control pots was lowest in Durgapur (Rajshahi), and highest in Mithapukur and Modhukhali (Rangpur) soils. Similar trends were observed for K-treated pots with K uptake ranging from 119.6 to 195.14 mg K kg^{-1}. The results from this study assisted us to divide the potential K-supplying capacity of 18 soils into three categories: low, medium, and high, and corresponding to average K uptake over seven successive crops of <50, 51–80 and > 80 mg K kg^{-1} soil, respectively (Figure 4). According to the classification, among the tested soils, K uptake (or soil K-supply capacity) by maize was low in Durgapur, Barura, Godagari and Birganj; medium in Nawabganj, Shibganj, Modhupur, Paba, Daudkandi, Gopalpur Tarash, Binerpota, Gangachara and Jhinaidah soils; and high in Dinajpur Sadar, Rangpur Sadar, Mithapukur and Modhukhali soils (Figure 4). In line to our study, similar observations were also recorded in Guinea grass for six successive crop harvests. In that study, the tissue K concentration and K uptake in plants, and soil K-supplying capacity varied widely in various soils and was higher for K-treated than for no K-applied soils [46,53,54].

In the multi-location field experiment, growth and yield attributing characteristics of maize, except number of cobs per plant, responded to K fertilization significantly in all locations. Yield increase over control varied 18–79%, 22–59% and 21–54% in Rangpur, Rajshahi and Comilla, respectively [55]. In all locations, significantly higher grain yield was found with NEM-based fertilizer recommendation followed respectively by MCM-based recommendation, 120, 80 and 160 kg K ha^{-1}, and lowest with no K added and 40 kg K ha^{-1} treatments. Both NEM- and MCM-based recommendations were based on the SSNM principles which considered previous crop's residues and manures and fertilizers practices, and indigenous soil fertility [28]. Previous results have also shown that the MCM and NEM have great potential to estimate K fertilizer recommendation for maize, which could help to reduce the cost of production and to increase yield and profit by reducing the over or underuse of fertilizer to the crop [27,40].

K uptake by maize was governed by K content in plant tissue and above ground dry matter. The K concentration of the plant tissue was consistent with the K availability in the soil. Therefore, K uptake by maize in K-omission plots can be a reliable measure of K-supplying capacity of soil [50–52]. In control plots, K uptake by maize varied from 57.2 to 83.6 kg ha^{-1}. The order of K-supplying capacity in the field experiment was Rangpur > Rajshahi > Comilla which was similar to the order observed in the pot study. The correlation between K uptake by maize at harvest and initial K_{ex} and K_{nex} were positive and linear. Plots of cumulative K uptake by maize from 1st to 7th harvest versus initial K_{ex} contents showed a progressively higher utilization of NH_4OAc-K by maize during the experiment (Figure 7a,b). Moreover, the highly significant r^2 values (0.56 to 0.84) revealed that 56 to 84% of the K uptake by maize was governed by the initial K_{ex} content. However, the slope (>1) of the linear regression line indicated that the initial K_{ex} was not sufficient to meet the entire uptake requirement of maize starting from the 1st crop to the 7th crop. The value of slope (1.6 to 14.8) increased with increase in crop number revealing the gradual depletion in K_{ex} pool and therefore, the uptake of K might have been complimented from other soil K pools also (Figure 7a,b). In the case of K_{nex}, the lower value of the slope and r^2 of a linear relationship for the 1st harvest indicated that the contribution of the K_{nex} pool to K uptake by maize was lower than the K_{ex} pool. But the relationship of cumulative K uptake versus K_{nex} at 7th harvest became stronger with a higher value of slope than the 1st harvest. The value of slope and r^2 increased with successive crops indicating that the K uptake dependency on the K_{nex} pools was increased due to reduced K_{ex} availability in the soil with successive exhaustive cropping in control pots (Figure 7c,d). A significant contribution of K_{nex} to crop uptake was reported in other studies also. However, K_{nex} is not measured in routine soil K test in most countries and the depletion of K_{nex} often remain unnoticed to the detriment of soil K fertility.

The amount of K uptake by maize in the K-applied plots (99.0–178.3 kg K ha^{-1}) was significantly higher than in control plots (67.1 kg K ha^{-1}) across the sites. The highest K uptake occurred with 160 kg K ha^{-1}, though MCM (75–100 kg K ha^{-1}) resulted in significantly higher agronomic and recovery efficiency of K. The K requirements vary for different crops, varieties, and locations in which they are grown [54,56]. Other studies have also shown that crop K requirement depends on the K status and K dynamics in soils, as well as efficient K use, which depends also on the rooting pattern of different crops and varieties, and their productivity [46,53]. There is thus essential need for K to be supplied at an optimum dose and maintained to augment production and ensure to improve quality crop [27,56,57]. Nevertheless, both the concomitant increase in yield and efficiency improvement with added K are important considerations for improved K management. Finally, improving nutrient use efficiency should not be the singular goal of any sustainable nutrient management program as higher efficiencies can be achieved by less and less nutrient application. In this study, the trend is yield and efficiency increase from 40 to 80 kg K, but efficiencies drop as 120 or 160 kg of K are applied, clearly suggesting that yield improvement at these rates are not enough to improve use efficiencies also. In any improved K management programs, decision support system (DSS) tools such as MCM or NEM strategies could be the better options for improved K fertilizer management, as the current study showed that, these strategies increased K application but also increased maize yield to keep the efficiencies at higher levels. SSNM strategies, such as MCM and NEM DSS tools, can take care of adequate and balanced application of all nutrients, including K, and hence their adoption in Bangladesh and South Asia would be important not just for K management but for the management of all nutrients.

5. Conclusions

In pot study, maize responded to added K in Godagari, Durgapur, and Modhupur soils out of 18 soils from the first crop and 50% soils responding from the second crop, with remaining soils responding from the third crop onwards. The mean yield response over seven successive crops across 18 soils varied from 20 to 195%, where least and most responsive soils to K fertilizer were found in Modhukhali and Durgapur soils. In control pots, K-supplying capacity over seven successive crops varied (22.1–107.3 mg K kg soil^{-1}) significantly and there was a significant ($p < 0.001$) negative correlation between yield response and indigenous K-supplying capacity. In the field validation experiment, yield and yield attributes of maize responded to K fertilizer significantly in all locations. Potassium fertilizer increased grain yield from 18 to 79%.

The current research established that the requirement of K was 111–122 kg K ha^{-1} for maize cultivation in Bangladesh. Total K uptake by the plot that did not receive K fertilizer in on-farm trials, considered as K-supplying capacity of the soils, was in the order: Rangpur > Rajshahi > Comilla, which was similar to the results of pot study. Thus, assessment of K-supplying capacity of major soils in Bangladesh will be useful for managing K fertility in soils and K nutrition for maize. The study provides evidence of the essentiality of adequate and balanced K application in maize in Bangladesh for sustainably improving or maintaining high yields. The results, however, have great implications for South Asia as a whole as maize is replacing rice and wheat in vast areas of the region because of its economic value and climate resilience. Although the results of the pot and field experiments provide sufficient information on the K-supplying capacity of diverse soils of, and the productivity maize for selected sites in, Bangladesh, which helps refine current K fertilizer recommendation rates to farmers, further study would be required to better understand K-supplying capacity, yield and K-use efficiency in more diverse soils and for robust recommendation to farmers across maize-growing areas of South Asia.

Author Contributions: S.I., J.T., M.S. and K.M conceived and designed the experiments; S.I., J.T and M.S. conducted the experiments; S.I. recorded the data; S.I., J.T and M.K.G. analyzed the data; S.I. and J.T. wrote the paper; S.I., J.T., M.S., K.M. and M.K.G. reviewed, edited, and approved the final manuscript.

Acknowledgments: Authors give special thanks to Bangladesh Rice Research Institute (BRRI) for giving the opportunity to conduct pot experiments in their net house. Authors would also like to thank Abu Saleque (BRRI) and Roland Buresh (IRRI) for their constructive suggestions during the conduct of research. We are also grateful to farmers for providing lands for the experiments and actively involved in care and maintenance of the experiments.

References

1. Shiferaw, B.; Prasanna, B.M.; Hellin, J.; Bänziger, M. Crops that feed the world 6. Past successes and future challenges to the role played by maize in global food security. *Food Secur.* **2011**, *3*, 307–327. [CrossRef]
2. Gerland, P.; Raftery, A.E.; Ševčíková, H.; Li, N.; Gu, D.; Spoorenberg, T.; Alkema, L.; Fosdick, B.K.; Chunn, J.; Lalic, N.; et al. World population stabilization unlikely this century. *Science* **2014**, *346*, 234–237. [CrossRef] [PubMed]
3. Bank, W. *World Development Indicators 2014*; World Bank: Washington, DC, USA, 2014; Volume 87, pp. 98–191.
4. DAE. *Krishi Diary*; Agricultural Information Services, Department of Agriculture Extension (DAE), Ministry of Agriculture, Khamar Bari: Dhaka, Bangladesh, 2018.
5. Barber, S.A. Potassium availability at the soil-root interface and factors influencing potassium uptake. In *Potassium in Agriculture*; Munson, R.D., Ed.; American Society of Agronomy: Madison, WI, USA, 1985; pp. 309–324.
6. Dobermann, A.; Cassman, K.G.; Mamaril, C.P.; Sheehy, J.E. Management of phosphorus, potassium, and sulfur in intensive, irrigated lowland rice. *Field Crop. Res.* **1998**, *56*, 113–138. [CrossRef]
7. Regmi, A.P.; Ladha, J.K.; Pathak, H.; Pasuquin, E.; Bueno, C.; Dawe, D.; Hobbs, P.R.; Joshy, D.; Maskey, S.L.; Pandey, S.P. Yield and Soil Fertility Trends in a 20-Year Rice-Rice-Wheat Experiment in Nepal. *Soil Sci. Soc. Am. J.* **2002**, *66*, 857. [CrossRef]
8. Srivastava, S.; Raghavareddy Rupa, T.; Swarup, A.; Singh, D. Effect of long-term fertilization and manuring on potassium release properties in a Typic Ustochrept. *J. Plant Nutr. Soil Sci.* **2002**, *165*, 352–356. [CrossRef]
9. Singh, B.; Singh, Y.; Imas, P.; Jian-chang, X. Potassium Nutrition of the Rice-Wheat Cropping System. *Adv. Agron.* **2001**, *81*, 203–259.
10. Timsina, J.; Jat, M.L.; Majumdar, K. Rice-maize systems of South Asia: Current status, future prospects and research priorities for nutrient management. *Plant Soil* **2010**, *335*, 65–82. [CrossRef]
11. Doboerman, A.; Fairhurst, T.H. *Nutrient Disorders & Nutrient Management*, 1st ed.; Potash & Phosphate Institute (PPI)-Potash & Phosphate Institute of Canada (PPIC): Norcross, GA, Canada; International Rice Institute (IRRI): Los Baños, Philippines, 2000; ISBN 9810427425.
12. Ali, M.M.; Saheed, S.M.; Kubota, D.; Masunaga, T.; Wakatsuki, T. Soil degradation during the period 1967–1995 in Bangladesh. *Soil Sci. Plant Nutr.* **1997**, *43*, 879–890. [CrossRef]

13. Saleque, M.A.; Saha, P.K.; Panaullah, G.M.; Bhuiyan, N.I. Response of wetland rice to potassium in farmers' fields of the Barind tract of Bangladesh. *J. Plant Nutr.* **1998**, *21*, 39–47. [CrossRef]
14. Majumdar, K.; Zingore, S.; Garcia, F.; Johnston, A.M. Improving nutrient management for sustainable intensifi cation of maize. In *Achieving Sustainable Cultivation of Maize*; Burleigh Dodds Science Publishing: Sawston, UK, 2017; Volume 2, pp. 1–32.
15. Panaullah, G.M.; Timsina, J.M.A.; Saleque, M.A.; Ishaque, M.; Pathan, A.B.M.B.U.; Connor, D.J.; Saha, P.K.; Quayyum, M.A.; Humphreys, E.; Meisner, C.A. Nutrient uptake and apparent balances for rice-wheat sequences. III. Potassium. *J. Plant Nutr.* **2006**, *29*, 173–187. [CrossRef]
16. Olk, D.C.; Cassman, K.G.; Simbahan, G.; Sta. Cruz, P.C.; Abdulrachman, S.; Nagarajan, R.; Tan, P.S.; Satawathananont, S. Interpreting fertilizer-use efficiency in relation to soil nutrient-supplying capacity, factor productivity, and agronomic efficiency. *Nutr. Cycl. Agro-Ecosyst.* **1999**, *1621*, 35–41. [CrossRef]
17. Cassman, K.G.; Dobermann, A.R.; Walters, D.T. Agroecosystems, Nitrogen-use Efficiency, and Nitrogen Management. *Agron. Hortic.* **2002**, *31*, 132–140. [CrossRef]
18. Dobermann, A. Nutrient use efficiency–measurement and management. In *Fertilizer Best Management Practices, Proceedings of the IFA International Workshop on Fertilizer Best Management Practices, Brussels, Belgium, 7–9 March 2007*; International Fertilizer Industry Association: Paris, France, 2007; p. 28.
19. Liu, X.; He, P.; Jin, J.; Zhou, W.; Sulewski, G.; Phillips, S. Yield gaps, indigenous nutrient supply, and nutrient use efficiency of wheat in China. *Agron. J.* **2011**, *103*, 1452–1463. [CrossRef]
20. Sparks, D.L.; Page, A.L.; Helmke, P.A.; Loeppert, R.A.; Soltanpour, P.N.; Tabatabai, M.A.; Johnston, C.T.; Sumner, M.E. Methods of soil analysis: Chemical Methods. In *Chemical Methods*; Fundación Hondureña de Investigación Agrícola (FHIA): Madison, WI, USA, 1996; p. 1390. ISBN 0891188258.
21. ASTM D422-63(2007)e2. *Report of Standard Test Method for Particle-Size Analysis of Soils*; ASTM International: West Conshohocken, PA, USA, 2007; Volume D422-63, pp. 1–8.
22. Brammer, H.; Antoine, J.; Kassam, A.H.; Van Velthuizen, H.T. *Land Resources Appraisal of Bangladesh for Agricultural Development*; FAO of United Nations: Rome, Italy, 1988; Volume II.
23. *Fertilizer Recommendation Guide 2012*; BARC (Bangladesh Agricultural Research Council): Farmgate, Dhaka, 2012; ISBN 978-984-500-000-0.
24. Ritchie, S.; Hanway, J.; Benson, G. *How a Corn Plant Develops*; Iowa State University of Science and Technology: Ames, IA, USA, 1989; pp. 1–25.
25. Jones, J.B., Jr. Laboratory Guide for Conducting Soil Tests and Plant Analysis. In *Laboratory Guide for Conducting Soil Tests and Plant Analysis*, 1st ed.; CRC (Chemical Rubber Company) Press: Boca Raton, FL, USA, 2001; p. 202. ISBN 1420025295.
26. Buresh, R.J.; Pampolino, M.F.; Witt, C. Field-specific potassium and phosphorus balances and fertilizer requirements for irrigated rice-based cropping systems. *Plant Soil* **2010**, *335*, 35–64. [CrossRef]
27. Satyanarayana, T.; Kaushik, M.; Biradar, D.P. New approaches and tools for site-specific nutrient management with reference to potassium. *Karnataka J. Agric. Sci.* **2011**, *24*, 86–90.
28. Anand, S.R.; Vishwanatha, J.; Rajkumar, R.H. Site Specific Nutrient Management (SSNM) Using " Nutrient Expert " for Hybrid Maize (*Zea mays* L.) Under Zero Tillage in Thungabhadra Project (TBP) Command Area of Karnataka. *Int. J. Curr. Microbiol. Appl. Sci.* **2017**, *6*, 3597–3605. [CrossRef]
29. Witt, C.; Found, M.G. A Site-Specific Nutrient Management Approach for Irrigated, Lowland Rice in Asia. *Better Crop. Int.* **2016**, *16*, 20–24.
30. Witt, C.; Pasuquin, J.M.C.A.; Dobermann, A. A Site-Specific Nutrient Management for Maize in Favorable Tropical Environments of Asia. In Proceedings of the 5th International Crop Science Congress, Brisbane, Australia, 26 September–1 October 2004; Better Crops Internationa: Jeju, Korea, April 2008; p. 257.
31. Xu, X.; He, P.; Qiu, S.; Pampolino, M.F.; Zhao, S.; Johnston, A.M.; Zhou, W. Estimating a new approach of fertilizer recommendation across small-holder farms in China. *Field Crop. Res.* **2014**, *163*, 10–17. [CrossRef]
32. Dobermann, A.; Witt, C.; Abdulrachman, S.; Gines, H.C.; Nagarajan, R.; Son, T.T.; Tan, P.S.; Wang, G.H.; Chien, N.V.; Thoa, V.T.K.; et al. Estimating indigenous nutrient supplies for site-specific nutrient management in irrigated rice. *Agron. J.* **2003**, *95*, 924–935. [CrossRef]
33. Xu, X.; He, P.; Pampolino, M.; Johnston, A.; Qiu, S.; Zhao, S.; Chuan, L.; Zhou, W. Fertilizer recommendation for maize in China based on yield response and agronomic efficiency. *Field Crop. Res.* **2014**, *157*, 27–34. [CrossRef]

34. Cock, J.; Yoshida, S.; Forno, D.A. Laboratory Manual for Physiological Studies of Rice. *Int. Rice Res. Inst.* **1976**, 69–72. [CrossRef]
35. Venables, W.N.; Smith, D.M.; R Development Core Team. *R Software*; R Foundation for Statistical Computing: Vienna, Austria, 2008; Volume 739, p. 409. Available online: http://www.r-project.org (accessed on 15 July 2018).
36. Chambers, J.M. *Software for Data Analysis Programming with R*; Springer Science & Business Media: Berlin/Heidelberg, Germany, 2008; Volume 15, ISBN 9780387759357.
37. Steel, R.G.D.; Torrie, J.H. *Principles and Procedures of Statistics: A Biometrical Approach*, 3rd ed.; McGraw-Hill: New York, NY, USA, 1997.
38. Dobermann, A.; Sta. CruzK, P.C.; Cassman, K.G. Fertilizer inputs, nutrient balance, and soil nutrient-supplying power in intensive, irrigated rice systems. I. Potassium uptake and K balance. *Nutr. Cycl. Agroecosyst.* **1996**, *46*, 1–10. [CrossRef]
39. Miah, M.A.M.; Saha, P.K.; Islam, A.; Hasan, M.N.; Nosov, V. Potassium fertilization in rice-rice and rice-wheat cropping system in Bangladesh. *Bangladesh J. Agric. Environ.* **2016**, *4*, 51–67.
40. Timsina, J.; Kumar Singh, V.; Majumdar, K. Potassium management in rice-maize systems in South Asia. *J. Plant Nutr. Soil Sci.* **2013**, *176*, 317–330. [CrossRef]
41. Islam, A.; Muttaleb, A. Effect of potassium fertilization on yield and potassium nutrition of Boro rice in a wetland ecosystem of Bangladesh. *Arch. Agron. Soil Sci.* **2016**, *62*, 1530–1540. [CrossRef]
42. Srinivasarao, C.; Satyanarayana, T.; Venkateswarlu, B. Potassium mining in Indian agriculture: input and output balance. *Karnataka J. Agric. Sci.* **2011**, *24*, 20–28.
43. Rijmpa, J.; Islam, F. Nutrient mining and its effect on crop production and environment. In *Seminar on Soil Health Management: DAE-SFFP Experience*; DAE-SFFP (Department of Agriculture Extension-Soil Fertility of the Paddy Fields): Dhaka, Bangladesh, 2002.
44. Grzebisz, W.; Szczepaniak, W.; Potarzycki, J.; Łukowiak, R. Sustainable Management of Soil Potassium—A Crop Rotation Oriented Concept. In *Soil Fertility*; InTech: Rijeka, Croatia, 2012.
45. Barber, S.A. *Soil Nutrient Bioavailability: A Mechanistic Approach*, 2nd ed.; John Wiley and Sons: New York, NY, USA, 1995; ISBN 978-0-471-58747-7.
46. Darunsontaya, T.; Suddhiprakarn, A.; Kheoruenromne, I.; Prakongkep, N.; Gilkes, R. The forms and availability to plants of soil potassium as related to mineralogy for upland Oxisols and Ultisols from Thailand. *Geoderma* **2012**, *170*. [CrossRef]
47. Jagadish, T.; Vinod, K.S.; Kaushik, M. Potassium management in rice–maize systems in South Asia. *J. Plant Nutr. Soil Sci.* **2013**, *176*, 317–330. [CrossRef]
48. Islam, A.; Karim, A.J.M.S.; Solaiman, A.R.M.; Islam, M.S.; Saleque, M.A. Eight-year long potassium fertilization effects on quantity/intensity relationship of soil potassium under double rice cropping. *Soil Tillage Res.* **2017**, *169*, 99–117. [CrossRef]
49. Pathak, D.S.; Aggarwal, P.K.; Rötter, R.P.; Kalra, N.; Bandyopadhaya, S.K.; Prasad, S.; van keulen, H. Modelling the quantitative evaluation of soil nutrient supply, nutrient use efficiency, and fertilizer requirements of wheat in India. *Nutr. Cycl. Agroecosyst.* **2003**, *65*, 105–113. [CrossRef]
50. Fergus, I.F.; Martin, A.E. Studies on potassium. IV. Interspecific differences in the uptake of non-exchangeable potassium. *Soil Res.* **1974**, *12*, 147–158. [CrossRef]
51. Memon, Y.M.; Fergus, I.F.; Hughes, J.D.; Page, D.W. Utilization of non-exchangeable soil potassium in relation to soil type, plant species and stage of growth. *Soil Res.* **1988**, *26*, 489–496. [CrossRef]
52. Surapaneni, A.; Tillman, R.W.; Kirkman, J.H.; Gregg, P.E.H.; Roberts, A.H.C. Potassium-supplying power of selected Pallic soils of New Zealand 1. Pot trial study. *N. Z. J. Agric. Res.* **2002**, *45*, 113–122. [CrossRef]
53. Hussain, A.; Arshad, M.; Ahmad, Z.; Ahmad, H.T.; Afzal, M.; Ahmad, M. Potassium Fertilization and Maize Physiology Potassium Fertilization Influences Growth, Physiology and Nutrients Uptake of Maize (*Zea mays* L.). *Cercet. Agron. Mold.* **2015**, *XLVIII*, 37–50. [CrossRef]
54. Askegaard, M.; Eriksen, J.; Johnson, A.E. Sustainable management of potassium. In *Managing Soil Quality: Challenges in Modern Agriculture*; Schjønning, P., Elmholt, S., Christensen, B.T., Eds.; CABI (The Centre for Agriculture and Bioscience International) Publishing: Wallingford, Oxfordshire, UK, 2004; pp. 85–102.
55. Huang, S.W.; Jin, J.Y.; Tan, D.S. Crop Response to Long-Term Potassium application as affected by Potassium Supplying Power of the Selected Soils in Northern China. *Commun. Soil Sci. Plant Anal.* **2009**, 2833–2854. [CrossRef]

56. Sadanandan, A.K.; Peter, K.V; Hamza, S. Role of Potassium Nutrition in Improving Yield and Quality of Spice Crops in India. In *Potassium for Sustainable Crop Production*; Haryana and International Potash Institute: Zug, Switzerland, 2002; pp. 445–466.
57. Singh, V.K.; Dwivedi, B.S.; Buresh, R.J.; Jat, M.L.; Majumdar, K.; Gangwar, B.; Govil, V.; Singh, S.K. Potassium fertilization in rice-wheat system across northern India: Crop performance and soil nutrients. *Agron. J.* **2013**, *105*, 471–481. [CrossRef]

4

Integrated Nutrient Management in Rice–Wheat Cropping System: An Evidence on Sustainability in the Indian Subcontinent through Meta-Analysis

Sheetal Sharma [1,*], Rajeev Padbhushan [2] and Upendra Kumar [3]

[1] International Rice Research Institute-India Office, 1st Floor, CG Block, NASC Complex, DPS Marg, Pusa, New Delhi 110012, India

[2] International Rice Research Institute-Odisha Office, Plot no. 340/C, School St., Saheed Nagar, Bhubaneswar, Odisha 751007, India; r.padbhushan@irri.org

[3] National Rice Research Institute (NRRI), Bidyadharpur, Cuttack, Odisha 753006, India; ukumarmb@gmail.com

* Correspondence: sheetal.sharma@irri.org

Abstract: Over years of intensive cultivation and imbalanced fertilizer use, the soils of the Indian subcontinent have become deficient in several nutrients and are impoverished in organic matter. Recently, this region has started emphasizing a shift from inorganic to organic farming to manage soil health. However, owing to the steadily increasing demands for food by the overgrowing populations of this region, a complete shift to an organic farming system is not possible. The rice–wheat cropping system (RWCS) is in crisis because of falling or static yields. The nations of this region have already recognized this problem and have modified farming systems toward integrated nutrient management (INM) practices. The INM concept aims to design farming systems to ensure sustainability by improving soil health, while securing food for the population by improving crop productivity. Therefore, this paper was synthesized to quantify the impact and role of INM in improving crop productivity and sustainability of the RWCS in the context of the Indian subcontinent through meta-analysis using 338 paired data during the period of 1989–2016. The meta-analysis of the whole data for rice and wheat showed a positive increase in the grain yield of both crops with the use of INM over inorganic fertilizers only (IORA), organic fertilizers only (ORA), and control (no fertilizers; CO) treatments. The increase in grain yield was significant at $p < 0.05$ for rice in INM over ORA and CO treatments. For wheat, the increase in grain yield was significant at $p < 0.05$ in INM over IORA, ORA, and CO treatments. The yield differences in the INM treatment over IORA were 0.05 and 0.13 Mg ha^{-1}, respectively, in rice and wheat crops. The percent yield increases in INM treatment over IORA, ORA, and CO treatments were 2.52, 29.2, and 90.9, respectively, in loamy soil and 0.60, 24.9, and 93.7, respectively, in clayey soil. The net returns increased by 121% (INM vs. CO) in rice, and 9.34% (INM vs. IORA) and 127% (INM vs. CO) in wheat crop. Use of integrated nutrient management had a positive effect on soil properties as compared to other nutrient management options. Overall, the yield gain and maintenance of soil health due to INM practices over other nutrient management practices in RWCS can be a viable nutrient management option in the Indian subcontinent.

Keywords: integrated nutrient management; rice; wheat; yield; net returns; soil health; sustainability

1. Introduction

Rice (*Oryza sativa* L.) and wheat (*Triticum aestivum* L.) crops are major staple foods, contributing a key portion of digestible energy and protein in human intake and occupying a premium position among all food communities [1–3]. The rice–wheat cropping system (RWCS) is one of the most prominent cropping systems prevailing on the Indian subcontinent and is considered to be of utmost importance for food security and livelihood [4–7]. The RWCS occupies about 13.5 million hectares spread over the Indian subcontinent, namely, India, Pakistan, Nepal, Bangladesh, Sri Lanka, and Bhutan, and accounting for one-fourth to one-third of total food grain production [8,9]. This cropping system covers about one-third of the total rice cultivation and two-fifths of the total wheat cultivation in the Indian subcontinent. Currently, more than nine-tenths of global rice is produced and consumed in these nations [10].

Natural resources, primarily agricultural lands, are limited globally. To meet the food demand of the ever-increasing population, agriculture must produce more food grains from limited cultivable land [11]. The crop productivity of the region is low and oscillating from 0.5 t ha^{-1} to 2.5 t ha^{-1}, with a mean of 1.5 t ha^{-1}. The increasing population and food consumption and the decline in existing arable land and other units of supply are placing exceptional pressure on the present farming system to meet the growing food demand. To counteract this problem and obtain higher yields, crop growers are shifting to fertilizer-responsive high-yielding varieties and avoiding the overuse of inputs such as synthetic fertilizers. The soils under the RWCS are now showing signs of fatigue and are no longer showing increased production with an increase in fertilizer use [12]. Even with the use of the recommended rate of fertilizer in the RWCS, a negative balance of primary nutrients has been recorded. To obtain food security in these nations, crop yields must rise considerably while ecological effects must contract significantly [13]. Figures 1 and 2 represent the trends and ratio of rice–wheat production and fertilizer consumption in the Indian subcontinent. These show that, from 1961 to 2016, rice–wheat production increased by four times, whereas fertilizer consumption increased by 67 times, which clearly indicates the slower increase in rice–wheat production even though fertilizer consumption was increasing at an exponential rate. Hence, innovative interventions are required to optimize fertilizer use and sustain the rice–wheat production system in this region.

Intensive cropping systems with continuous imbalanced use of synthetic fertilizers to feed fertilizer-responsive varieties have caused losses in soil organic carbon (SOC) [14] and soil health [15–17], often leading to unsustainability of crop production systems. The use of organic sources of nutrients in agriculture is rapidly gaining favor but, owing to the problems related to the lack of availability of a good quality and quantity of organic materials, the system may not be sufficient to achieve and sustain the production of cereal crops in the amounts required for food security [17].

Integrated nutrient management (INM) or integrated nutrient supply (INS) help to achieve efficient use of synthetic fertilizers integrated with organic sources of nutrients [18]. INM is developed with an understanding of the interactions among crops, soils, and climate, which advocates the integration of inorganic and organic sources of nutrients. This approach is based on the maintenance of plant nutrition supply to attain a certain level of crop production by enhancing the benefits from all potential sources of plant nutrition in a cohesive manner, applicable to each cropping pattern and farming scenario [19]. The inclusion of organic manures regulates the uptake of nutrients, positively affecting production, improving soil quality (physical, chemical, and biological), and producing a synergistic effect on crops [20]. INM integrates traditional and recent practices of nutrient management into an environmentally sound and cost-effective ideal farming system that uses remunerations from all probable sources of nutrition (organic, inorganic, and biological) in a careful, effective, and combined way [21,22]. It optimizes the balance between input sources and outputs with the goal of coordinating the nutritional demand of the crop and its discharge in its surroundings (Figure 3).

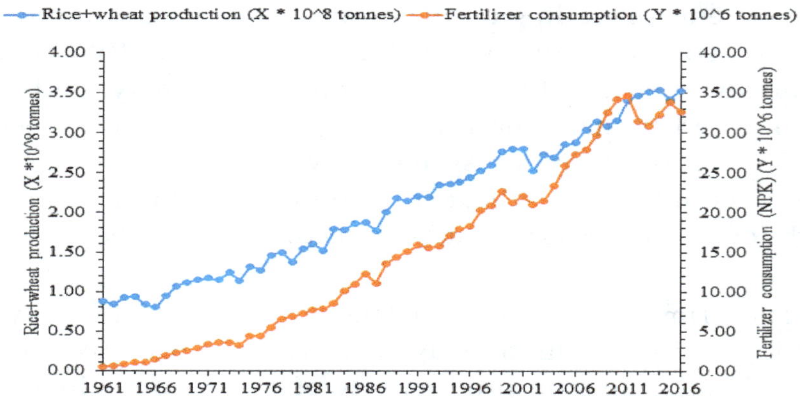

Figure 1. Trends of rice + wheat production and fertilizer consumption during the period 1961–2016 on the Indian subcontinent, Reproduced with permission from FAO.

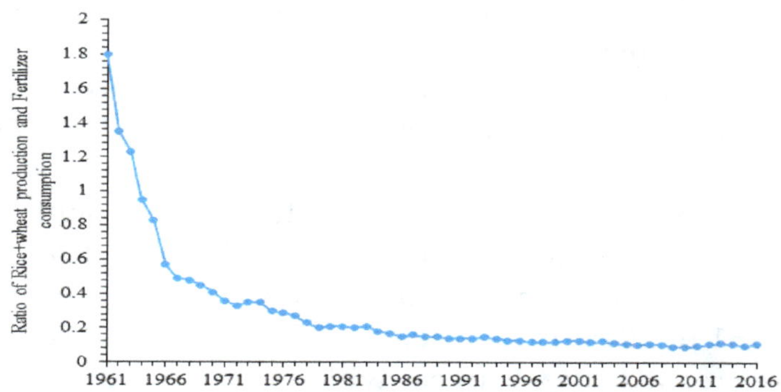

Figure 2. Ratio of rice + wheat production and fertilizer consumption during the period 1961–2016 on the Indian subcontinent, FAO.

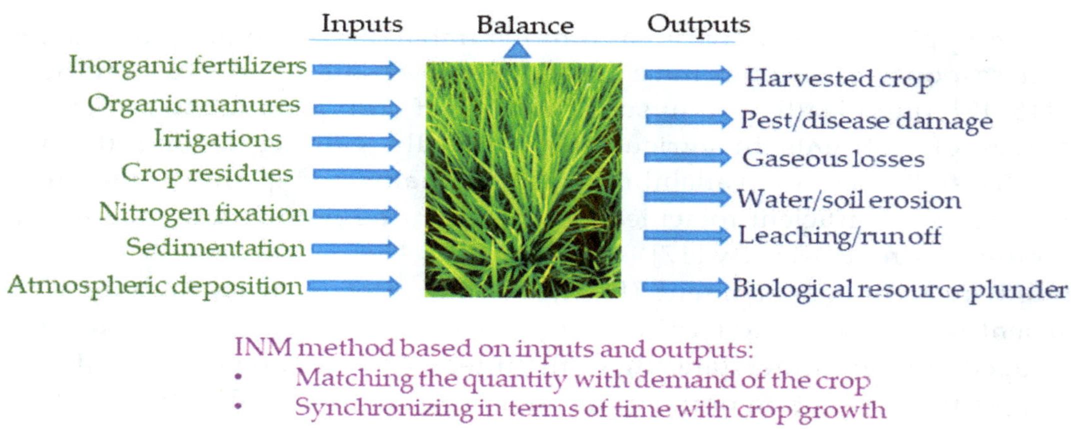

Figure 3. Mean of nutrients for inputs and outputs, and the principles of integrated nutrient management systems [22,23].

The importance of INM practices has been mentioned in several researches in the Indian subcontinent. The INM concept is now being broadened to make it more context-specific for local environmental conditions, increasing mechanization (due to the serious labor shortage as a result of migration), the increasing popularity of conservation agriculture, upcoming rain-harvesting technologies, and the increasing focus on recycling of available organic nutrient flows [22]. The incorporation of these new interventions in an INM system has developed new dimensions in the INM system and thus makes these practices synthesized for the region. An innovative approach such as INM can harness natural resources appropriately, bring about food security, and improve the livelihood of the people in the region [24]. The provision of appropriate policies, such as an incentive to adopt INM, will motivate small-holder farmers to adopt INM. Legislation on or the taxing of nutrient inputs should be enforced to minimize the use of inorganic fertilizers and encourage the integration of organic fertilizers for nutrient management.

The RWCS is a fertilizer-responsive cropping system. However, the imbalanced and excessive application of inorganic fertilizers is detrimental to the soil ecosystem and ultimately affects crop yield. Organic farming, a traditional production system relying on the use of only organic sources of nutrients, is supposed to be the production system that causes minimal damage to the ecosystem. However, the organic farming system has become a fundamental topic for discussion in recent times, with concerns about whether it can produce enough food to feed the ever-growing population of the Indian subcontinent [25,26]. On the Indian subcontinent, around a half century ago, the population was controlled and technology was not advanced. Organic manures were the only sources of nutrients. Currently, with the increase in the availability of new sources of nutrients, there is a need to recognize them and develop integrated options to increase production which are both ecologically viable and linked to economic growth.

Several researchers showed the impact of INM over other nutrient management options, such as the use of organic fertilizers only (ORA), the use of inorganic fertilizers only (IORA), or the use of no fertilizers as a control (CO), through their studies of the RWCS and undoubtedly, INM has emerged as a viable alternative nutrient management option in the RWCS for the Indian subcontinent. Researchers have compared yield and other performance parameters of nutrient management options, but an attempt to synthesize information in the RWCS, particularly in the Indian subcontinent region, is critically lacking. Such kinds of synthesis can help to prioritize research and development issues, including precise technology targeting, and articulate policy and institutional measures to facilitate large-scale adoption of nutrient management options.

Therefore, the present paper aimed to perform a comprehensive meta-analysis to understand the impacts of INM on crop performance in the RWCS of the Indian subcontinent during the period of 1989–2016. This study also outlines the sustainability of INM, noting that sole dependence on inorganic fertilizers or organic farming could not serve the purpose of the food security of the ever-growing populations in these countries. The role of INM is discussed for resolving those complications and as one of the promising strategies for addressing the current challenges of crop output and sustainability of the RWCS in these nations.

2. Materials and Methods

2.1. Data Collection

Literatures were reviewed for the period from 1989 to 2016 related to the on-station field experiments using integrated organic and inorganic nutrient sources with an aim of finding the effect of INM on crop productivity, net returns, and different soil parameters in the Indian subcontinent. After general review, they were critically analyzed, and data pertaining to INM along with other treatments (control (no fertilizer applied), organic fertilizer alone, and inorganic fertilizer alone) were selected. The treatment with 100% NPK (Nitrogen-Phosphorous-Potassium) application was selected as inorganic alone. Meta-analysis was conducted for the selected data (338 paired datasets) from several researchers to show the impact of INM over the other nutrient management options.

Only studies that met the following criteria were included:

1. Rice and wheat as study crops.
2. Nutrient management options include INM—integration of organic and inorganic sources (option 1); ORA—use of organic source of nutrient application only (option 2); or IORA—full dose of inorganic fertilizer application that mean recommended dose fertilizer application (option 3).
3. CO: Control treatments where no fertilizers were added.

4. Two soil textures: Loamy (moderately coarse to medium fine) and clayey (moderately fine to fine).

2.2. Crop and Soil Performance Parameters and Economic Analysis Used for Meta-Analysis Study

The following performance parameters were considered in the analysis: (a) Grain yield (Mg ha^{-1}), (b) soil parameters: Bulk density (BD; Mg m^{-3}), soil pH (soil water ratio:1:2), soil organic carbon (SOC; %), total nitrogen (TN; kg ha^{-1}), available phosphorus (Av. P; kg ha^{-1}), available potassium (Av. K; kg ha^{-1}), and soil microbial biomass carbon (SMBC; mg kg^{-1}), and (c) net returns (NR; $ ha^{-1}) (Net returns data from the study points was converted from INR ha^{-1} to US$ ha^{-1} for meta-analysis).

2.3. Meta-Analysis

All the variables were subjected to meta-analysis separately for rice and wheat and for soil texture. Meta-analysis has attracted considerable attention recently as a powerful statistical tool to analyze the response of treatments (i.e., nutrient management options vs. CO) from diverse individual studies to evolve to a general global trend or pattern. The meta-analysis was performed by using Meta Win 2.1 in two stages [27], in which effect size was calculated using the formula:

$$\text{Effect size} = \frac{M(\text{NM})}{M(\text{INM})} \tag{1}$$

where *M* (NM) is the mean of response variables (Grain yield, NR, soil parameters) of options 2, 3, and CO, and *M* (INM) is the mean of these variables of option 1. Since most researchers did not report the variance of the means of response variables, the effect sizes were weighted based on the number of replicates (*N*) as follows:

$$\text{Weight} = \frac{N(\text{NM}) \times N(\text{INM})}{N(\text{NM}) + N(\text{INM})} \tag{2}$$

where *N* (NM) and *N* (INM) represent the number of replications for each of the nutrient management options, in an individual study. In cases where more than one observation was included in an option, the number of observations from that research was divided by the weights. All the results are discussed as change with INM over other nutrient management options and CO. Unless stated otherwise, differences were considered significant only when p values were <0.05. The meta-analyzed value has been presented either in graph or in table to clearly show the effect of INM over the other nutrient management options.

3. Results

3.1. Impact of INM on Rice and Wheat Yield and Net Returns

The meta-analysis of the data for rice and wheat showed positive increases in the grain yield of both crops with the use of INM over IORA, ORA, and CO treatments (Figure 4). The increase in grain yield was significant at $p < 0.05$ for rice in INM over ORA and CO treatments. The data show that some of the study points in INM over IORA for rice crop were negative, but the effect was non-significant. The increases in wheat grain yield were significant in INM over IORA, ORA, and CO treatments. The respective percent yield increases in INM treatment over CO, ORA, and IORA treatments were 86.5, 28.1, and 1.2 in rice crop, and 104.6, 39.2, and 4.5 in wheat crop. On average, the yield differences in the INM treatment over IORA were 0.05 and 0.13 Mg ha^{-1} in rice and wheat crop, respectively.

The data showed a significant increase in grain yield with INM over IORA, ORA, and CO treatments under loamy soils (Figure 4). Under clayey soils, there was an increase in grain yield in INM over IORA, ORA, and CO treatments, but the effect was non-significant for IORA and ORA treatments (Figure 4). The percent yield increases in INM treatment over CO, ORA and IORA treatments were 90.9, 29.2, and 2.52, respectively, in loamy soil, and 93.7, 24.9, and 0.60, respectively, in clayey soil. The yield gains calculated through meta-analysis of various studies for INM treatment over other nutrient management options suggest that INM can be a viable nutrient management option for reversing the yield plateauing in both rice wheat cropping systems of the Indian subcontinent.

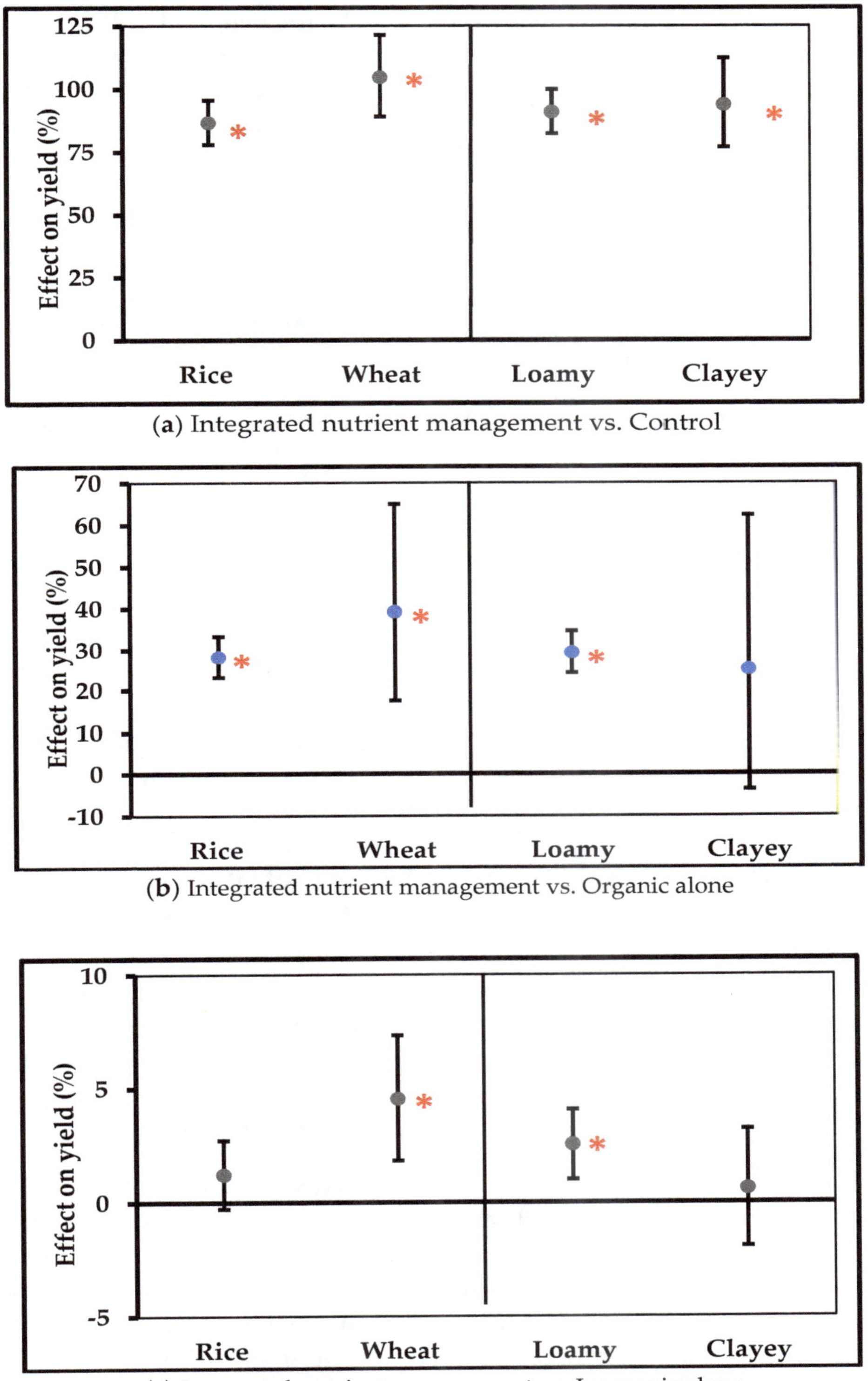

(a) Integrated nutrient management vs. Control

(b) Integrated nutrient management vs. Organic alone

(c) Integrated nutrient management vs. Inorganic alone

Figure 4. Effect of integrated nutrient management on grain yield vis-a vis control (**a**), organic alone (**b**), and inorganic alone (**c**), separated for rice and wheat crops and for soil texture (loamy and clayey). (Note: Error bars in meta-analysis indicate 95% confidence intervals (CI), where effect of yield was considered significant if 95% CI does not cover zero. * Indicates significant percent yield at $p < 0.05$).

Table 1 represents meta-analysis of the data, showing the effects of nutrient management practices on net returns in rice and wheat crops. The data for rice showed a significant positive increase in net returns with INM over the CO. The effect was significantly positive for INM over CO and IORA in wheat crop. The increases were 121 percent in NM vs. CO for rice, 9.34 and 127 per cent in INM vs. IORA and INM vs. CO, respectively, for wheat crop. The net returns increased by 0.13% in INM over ORA in wheat, but this was non-significant. Net returns decreased by 2.34% in INM vs. IORA and 0.27% in INM vs. ORA for rice crop, but these were non-significant. The negative effect in these cases was due to increased cost of cultivations in some of the studies.

Table 1. Influence on net returns in rice and wheat with integrated nutrient management over inorganic alone, organic alone, and control treatments in percent.

Nutrient Management Practices	Crops	
	Rice	Wheat
INM vs. IORA	−2.93 (−9.48 to 3.93)	9.34 (4.28 to 15.07) *
INM vs. ORA	−0.27 (−3.78 to 3.37)	0.13 (−3.85 to 4.27)
INM vs. CO	121 (101 to 142) *	127 (97 to 156) *

Mean values are given with 95% CI in parentheses. In bracket the values represent the ranges of percent net return for compared nutrient management practices. * Indicates percent net returns significant at $p < 0.05$. Where INM stands for integrated nutrient management, IORA for inorganic alone, ORA for organic alone, and CO for control (No fertilizer applied). CI used for confidence interval.

3.2. *Effect of INM on Soil Characteristics*

Crop- and texture-wise meta-analysis for the effect of INM compared to IORA, ORA, and CO treatments on various soil properties is presented in Table 2. The effect was non-significant for bulk density (BD) for both crops and soil textures. The BD increased by 0.53% (INM vs. IORA) and 1.15%(INM vs. CO) in rice crop. In wheat crop, the BD decreased by 0.98% (INM vs. IORA) and 0.55% (INM vs. CO). In loamy soil, BD decreased by 0.94% in INM vs. IORA and increased by 0.19% in INM vs. CO. In clayey soil, BD increased by 1.85% in INM vs. IORA and 1.38% in NM vs. CO.

The data for soil pH showed a positive effect of INM with an increase of 1.15% over CO in rice. However, the effect was negative by 0.55% in wheat crop for INM vs. CO. In clayey soil, pH increased by 1.85% in INM vs. IORA. The effect on soil pH was significantly negative for INM vs. CO, decreasing by 0.98% in wheat crop, and in INM vs. IORA by 0.94% in loamy soil (Table 2). The effects for all other comparisons were non-significant.

Meta-analysis of SOC showed a significant positive effect for all comparisons with sufficient data points, both crop-wise and texture-wise (Table 2). The SOC increased by 23.2% in INM vs. IORA and by 34.95% in INM vs. CO for rice. The increases were 16.2% in INM vs. IORA and 52.09% in INM vs. CO for wheat. In loamy soil, the increases were 26.5% in INM vs. IORA, and 51.21% in INM vs. CO. In clayey soils, the increases were 12.29% in INM vs. IORA and 23.4% in INM. There was no sufficient data for meta-analysis for INM vs. ORA.

For all the compared treatments, the TN, Av. P, Av. K, and SMBC showed significant positive effects for rice and wheat crops, indicating that INM treatment improved these soil properties over lone application of organic and inorganic fertilizers and CO (Table 2). Texture-wise, there was a significantly positive effect of INM over other nutrient management NM practices and CO for Av. P, Av. K, and SMBC.

Table 2. Effect of integrated nutrient management on soil properties with respect to inorganic alone, organic alone, and control treatments for rice and wheat crops and as influenced by soil texture (loamy and clayey), in percent.

Soil Parameters	Nutrient Management Practices	Crops		Texture Groups	
		Rice	Wheat	Loamy	Clayey
BD	INM vs. IORA	0.53 (−1.33 to 2.44)	−0.98 (−3.83 to 1.96)	−0.94 (−2.87 to 1.04)	1.85 (−0.79 to 4.55)
	INM vs. ORA	#	#	#	#
	INM vs. CO	1.15 (−0.74 to 3.07)	−0.55 (−3.42 to 2.42)	0.19 (−1.77 to 2.19)	1.38 (−1.24 to 4.06)
Soil pH	INM vs. IORA	0.53 (−0.13 to 1.33)	−0.98 (−1.24 to −0.68) *	−0.94 (−1.28 to −0.58) *	1.85 (0.84 to 3.02) *
	INM vs. ORA	−0.27 (−3.78 to 3.37)	0.13 (−3.85 to 4.27)	−0.07 (−2.68 to 2.60)	−0.31 (−45.34 to 81.83)
	INM vs. CO	1.15 (0.24 to 2.18) *	−0.55 (−0.99 to −0.14) *	0.19 (−1.77 to 2.19)	1.38 (−1.24 to 4.06)
SOC	INM vs. IORA	23.20 (18.55 to 27.48) *	16.20 (9.94 to 21.95) *	26.50 (21.05 to 31.19) *	12.29 (9.74 to 14.88) *
	INM vs. ORA	#	#	#	#
	INM vs. CO	34.95 (28.08 to 41.92) *	52.09 (41.79 to 66.53) *	51.21 (43.43 to 59.79) *	23.40 (17.95 to 29.38) *
TN	INM vs. IORA	49.86 (42.99 to 57.07) *	71.26 (60.11 to 83.16) *	#	#
	INM vs. ORA	29.01 (24.93 to 33.22) *	29.16 (25.08 to 33.38) *	#	#
	INM vs. CO	28.49 (25.46 to 31.59) *	29.16 (24.82 to 33.66) *	#	#
Av. P	INM vs. IORA	15.94 (12.24 to 19.76) *	4.88 (−0.45 to 10.47)	17.15 (13.18 to 21.26) *	5.30 (0.58 to 10.23) *
	INM vs. ORA	12.68 (3.97 to 22.12) *	11.28 (1.42 to 22.09) *	12.49 (5.95 to 19.45) *	7.14 (−69.39 to 275.05)
	INM vs. CO	37.18 (21.93 to 54.34) *	53.54 (31.93 to 78.68) *	54.64 (38.89 to 72.17) *	21.35 (3.40 to 42.40) *
Av. K	INM vs. IORA	15.33 (11.20 to 20.21) *	1.69 (−7.04 to 9.84)	13.36 (7.06 to 19.97) *	7.52 (4.75 to 10.69) *
	INM vs. ORA	9.10 (7.36 to 10.82) *	9.46 (7.97 to 11.02) *	9.74 (8.68 to 10.83) *	3.48 (2.84 to 4.12) *
	INM vs. CO	30.23 (24.68 to 36.12) *	18.32 (14.50 to 22.52) *	26.29 (21.85 to 31.48) *	23.97 (18.07 to 30.12) *
SMBC	INM vs. IORA	55.91 (44.09 to 68.67) *	32.92 (23.50 to 43.05) *	56.19 (46.29 to 66.76) *	28.44 (19.12 to 38.49) *
	INM vs. ORA	#	#	#	#
	INM vs. CO	97.90 (70.98 to 129.03) *	111.11 (95.15 to 128.35) *	134.03 (116.41 to 153.07) *	99.23 (79.98 to 103.16) *

Data not sufficient for meta-analysis. Mean values are given with 95% CI in parentheses. In brackets the values represent the ranges of percent net return for compared nutrient management practices. * Indicates percent net returns significant at $p < 0.05$. Where, INM stands for Integrated nutrient management, IORA for inorganic alone, ORA for organic alone, CO for control (No fertilizer applied), BD means bulk density, SOC means soil organic carbon, TN means total nitrogen, Av. P means available phosphorus, Av. K means available potassium, SMBC means soil microbial biomass carbon and CI used for confidence interval.

4. Discussion

The above results in this article for the period 1989–2016, obtained through meta-analysis, emphasize the importance of INM practices in the Indian subcontinent. The crop-wise and texture-wise analyses show positive effects of INM treatment over IORA and ORA on grain yield. The increase in yield due to the use of INM treatment puts forth that the integration of organic sources of nutrients with inorganic fertilizers could be a viable alternative nutrient management option in RWCS in the Indian subcontinent. Although the yield gains in the INM treatment over the IORA treatment were only 1.2% in rice and 4.5% in wheat, considering that the conditions of curtailing the use of inorganic fertilizers prevent the soil quality/health from ill effects of synthetic fertilizers, this would really show the benefits of the use of INM over IORA treatment. From this study, it is clear that the use of the ORA did not serve the purpose of enhancing crop productivity in both crops and hence, complete shifting towards organic farming cannot be a feasible option as far as human food security is concerned. Similar findings were also pointed out by different researchers of the Indian subcontinent. In Bangladesh, researchers reported that integrated use of *Sesbania* (dhaincha) and mung bean residues with inorganic sources of nutrients improved crop yield by 7.6% and 9.5%, respectively, over inorganic alone in *T. aman* rice [28]. The application of wheat straw or farm yard manure (FYM) along with inorganic fertilizer in the rice field recorded higher grain and straw yields of rice in comparison to the treatments with only inorganic fertilizer [29]. The grain yields of basmati rice and wheat in the RWCS when 50% of the recommended dose from chemical fertilizers was substituted with green manure (GM) significantly increased crop productivity by 100% [30]. The incorporation of green gram residues along with inorganic fertilizers resulted in a significant rice yield increase of 13% over IORA [31]. The highest yield of 4.68 Mg ha^{-1} in wheat crop was recorded when 50% N (nitrogen) of the recommended dose of fertilizer treatment was replaced with FYM in the rice crop [32].

The yield response to the soil texture group suggests that INM treatment appears to be more suitable in loamy soil in comparison to clayey soil. The reason is that loamy soils have better drainage than clayey soils, and therefore offer adequate aeration. The better performance in these soils is also because of less cracking of soil in the early state, which is critical for the rice crop. Cracking is high in clayey soil, which directly affects the plant. The plant-available water content is also higher in these soils [33].

The data show that the use of INM increased net returns in several studies over other nutrient management options. Although, increases in the cost of cultivation at a few instances resulted in negative values under meta-analysis. For wheat crop, the effect of INM treatment on net returns was positive over the other nutrient management options. In case of rice, INM vs. IORA showed declines in net returns, but the effect was not significant. This may be due to large variations in effects of treatments for net returns. Proper planning and management can decrease the cost of cultivation for INM in the instances with high cost of cultivation and can bring in a positive effect on net returns for RWCS.

The study clearly indicated the positive effect of INM on soil characteristics over other nutrient management options. The increases in SOC, TN, Av. P, Av. K, and SMBC correlated the INM with enhanced soil quality/health. Continuous use of chemical fertilizers alone affects the soil reaction by increasing hydrogen ions (H^+) in the soil during the formation of ammonia. The soil reaction in turn affects the supply of nutrients in the soil. The use of compost with inorganic fertilizers reduced soil pH significantly over the IORA treatment [34]. The reduction in soil pH is due to the formation of organic acids by the reactions between compost and the inorganic fertilizers in the presence of microorganisms. The dissolution of salts by formed acids under INM significantly improved the electrical conductivity (EC) of the soil. Thus, the use of INM regulates pH and EC of the soil due to the presence of organic sources, hence improving nutrient availability in the soil.

The integrated application of organic and inorganic fertilizers for 29 years reduced soil reactions by 0.22% over the application of inorganic fertilizers alone [32]. In a long-term fertilizer experiment under the RWCS, it was reported that INM reduced BD by 16.6%, whereas the application of IORA showed a reduction of 7.3% compared to the CO treatment [35]. The effects of INM on nutrient dynamics were recorded, and it was concluded that combining FYM with inorganic fertilizers could maintain SOC and available N and P at either equal to or greater than the initial soil nutrient levels, thus maintaining soil fertility even under continuous cultivation [36]. Incorporation of rice straw with green manure along with inorganic fertilizers increased Av. P by 12.7% and Av. K by 14.3%, as compared to treatments in which only inorganic sources of nutrients were applied [37]. Sesbania green-manuring in rice, integrated with inorganic fertilizers, increased the available N from 5.8 to 22.0 kg ha^{-1}, Av. P from 1.4 to 3.8 kg ha^{-1}, and Av. K from 2.2 to 17.9 kg ha^{-1} in comparison to IORA [38]. The use of INM in the RWCS increased SOC content by 21% [32] and 45.8% [39,40] over the application of IORA.

The study [3] analyzed apparent balances of the nutrients N, P, and K in the RWCS for an average of 28 years as influenced by a combination of organic and inorganic sources of nutrients. Apparent N balance was negative for the CO and IORA, whereas apparent N balance was positive for INM (50% recommended dose of fertilizer + 50% N straw). This shows that the nutrient N was recycled through an organic supplement. Similar positive results for apparent P balance were observed in the INM treatment over the CO. The positive balance shows more P accumulation due to an affirmative linear relationship between the available and surplus P. Apparent K balance was negative for all the treatments (INM, IORA, and CO). INM plays a key role in retaining the balance between demand and supply of nutrients in the soil–plant system. Under the current scenario, sustainable N, P, and K management strategies need to be identified for optimal production and nutrient balance. The balances deliver key information about the sustainability of the RWCS and the potential environmental impacts [41,42]. It is worth noting that one-fourth of the N and four-fifths of the K removed by the rice and wheat crop remain in the straw and are obtainable after incorporation to the soil. Application of straw residues in the field can thus reduce the recommended dose of chemical fertilizer. At present, it has been assessed that only 33% to 50% of the applied N and P fertilizers are used by crops [43,44]. The efficacy of the applied nutrients might be enhanced by the integrated use of inorganic and organic fertilizers, by increasing the availability of these nutrients to the crops [45]. Declining trends in RWCS productivity and the need for higher inputs have been observed due to the depletion of organic matter in the soil, which causes a disturbance in aggregate stability, soil productivity, and soil quality/health [46,47]. Thus, there is a need for partial substitution of inorganic fertilizers by locally available organic sources of nutrients for sustainable production [48–50].

Long-term studies conducted at several sites on the Indian subcontinent indicate that the application of required nutrients only by inorganic fertilizers affects soil health [51]. The application of unwarranted nutrients could lead to a decline in nutrient-use efficiency, making fertilizer application uneconomical and resulting in an adverse effect on the environment [52] and groundwater quality [53], causing health hazards and ill effects on the climate. Therefore, there is a need to enhance the nutrient supply system through INM that brings sustainable changes in the soil–plant system and the environment [54,55].

Soil sequesters atmospheric carbon in the soil, which improves the productivity and quality of the ecosystem [56]. The integrated use of organic manure and inorganic fertilizer is an indispensable component to manage soil in an arable crop production system. The application of INM can enhance plant available nutrients, but the amendments can also affect the soil microbial population. The use of organic manures for managing soil quality/health has been well recognized [57]. Soil biodiversity is the population of microbes that manage or enhance soil health. The microbial population enhances nutrient availability. Soil microbes and the mechanisms to control the activities are important for maintaining the long-term sustainability of farming systems [58], and are key factors in nutrient cycling and soil genesis. The activity and biomass of the microbes play a key role in enhancing soil quality/health [59], as these respond to soil and crop management practices [60]. The bacterial and fungal populations are compared in the INM-treated plot compared to the inorganic-treated plot [61]. The bacterial population increased by approximately 60%, whereas the fungal population almost doubled in the INM system in comparison with the IORA. The increase in microbial population is due to the conducive soil environment, formed by the organic addition increasing the magnitude of easily degradable carbonaceous compounds in the INM system supporting nutrition for soil inhabitants. The INM application also improves physical components such as porosity, water-holding capacity, and aggregate stability, and decreases surface crusting and soil bulk density. INM also helps in improving structure, air capacity, and water retention in the soil profile [62].

Thus, the INM system improves soil quality by regulating the soil reaction, building up SOC, improving soil physical properties, and improving nutrient solubility/mobility. By integrating organic and inorganic sources, the added SOC can aid in the sustenance of agriculture for a longer period than inorganic sources alone, especially in the tropical climate of the Indian subcontinent where the temperature remains high and organic matter decomposition is rapid. INM is the key to sustaining our soils for improving productivity and preserving soil quality and environmental sustainability. Because the agricultural production system is a combined shared effect of the soil–water–fertilizer–climate continuum (SWFCC), a sensible and methodical management of this multifaceted system is vital for improving crop productivity on a continual basis. Among the several inputs of the nutrient balance, water and nutrients are the critical inputs that contribute the most to crop yield. The use of these inputs in a better manner and interactive effect with other factors is essential for targeting crop yield potential. Practices to manage soil through tillage can improve the efficiency of these input factors. Sustainable agricultural promotion uses the efficient integration of soil, water, soil organic matter, tillage, and nutrients for achieving yield targets of the RWCS. INM aims to promote sustainable production systems by managing soil quality together with improving crop yield through the balanced integration of organic and inorganic sources [63].

The studies reported improvements in yields, net profit, and soil characteristics in the RWCS with the use of INM, and attributed the increases to improvements in the availability of nutrients, increases in resource-use efficiencies, and increases in resistance to environmental stresses with the use of INM [22,64–68]. Farmers of these nations can be benefited with the use of INM through increased productivity and profitability. This will not only reduce the excessive use of inorganic fertilizers but also improve soil quality/health and ensure food security and environmental sustainability on the Indian subcontinent.

5. Conclusions

This review paper emphasized the role and importance of an integrated nutrient management system as a management strategy that can bring sustainability to the rice–wheat cropping system of the Indian subcontinent. The meta-analysis data points of rice and wheat during the period of 1989–2016 revealed that INM treatment over inorganic alone, organic alone, and control treatments was positive on grain yield, both crop-wise as well as texture-wise. The present paper also concludes that net returns through integrated nutrient management treatment were increased by 121% and 127% in rice and wheat, respectively, compared to control. Finally, the findings of the present review suggested that INM can be one of the viable nutrient management options in the Indian subcontinent, particularly for the rice–wheat cropping system.

Author Contributions: S.S. provided overall leadership, conceived the conceptual framework and was in charge of overall direction and planning. R.P. did literature search. S.S., R.P. and U.K. wrote the manuscript. All authors discussed the results and contributed to the final manuscript.

Acknowledgments: We are grateful to all the researchers whose contributions have been cited in this paper, which have helped us to prepare this review study.

Abbreviations

RWCS	Rice-wheat cropping System
INM	Integrated nutrient management
SOC	Soil organic carbon
INS	Integrated nutrient supply
ORA	Organic only
IORA	Inorganic only
CO	Control (No fertilization)
TN	Total nitrogen
N	Nitrogen
Av. P	Soil available phosphorus
Av. K	Soil available potassium
ANB	Apparent nutrient balance
FYM	Farm yard manure
SWFCC	Soil-water-fertilizer-climate continuum

References

1. Timsina, J.; Connor, D.J. Productivity and management of rice-wheat cropping systems: Issues and challenges. *Field Crop. Res.* **2001**, *69*, 93–132. [CrossRef]
2. Singh, R.A.; Singh, J.; Yadav, D.; Singh, H.K.; Singh, J. Integrated nutrient management in rice-wheat cropping system. *Int. J. Agric. Sci.* **2012**, *8*, 523–526.
3. Das, A.; Sharma, R.P.; Chattopadhyaya, N.; Rakshit, R. Yield trends and nutrient budgeting under a long-term (28 years) nutrient management in rice-wheat cropping system under subtropical climatic condition. *Plant Soil Environ.* **2014**, *60*, 351–357. [CrossRef]
4. Gupta, R.K.; Naresh, R.K.; Hobbs, P.R.; Jiaguo, Z.; Ladha, J.K. Sustainability of post green revolution agriculture: The rice-wheat cropping systems of Indo-Gangetic plains and China. In *Improving the Productivity and Sustainability of the Rice-Wheat System: Issues and Impact*; Ladha, J.K., Hill, J.E., Duxbury, J.M., Gupta, R.K., Buresh, R.J., Eds.; ASA Special Publication No. 65; American Society Agronomy, Crop Science Society American, Soil Sciien Society American: Madison, WI, USA, 2003; pp. 1–25.
5. Mohanty, S.; Nayak, A.K.; Kumar, A.; Tripathi, R.; Shahid, M.; Bhattacharyya, P.; Raja, R.; Panda, B.B. Carbon and nitrogen mineralization kinetics in soil of rice–rice system under long term application of chemical fertilizers and farmyard manure. *Eur. J. Soil Biol.* **2013**, *58*, 113–121. [CrossRef]
6. Kalhapure, A.; Singh, V.P.; Kumar, R.; Pandey, D.S. Tillage and nutrient management in wheat with different plant geometries under rice-wheat cropping system: A review. *Basic Res. J. Agric. Sci. Rev.* **2015**, *4*, 296–303.
7. Zhao, J.T.N.; Lia, J.; Lua, Q.; Fanga, Z.; Huang, Q.; Lia, R.Z.R.; Shen, B.; Shena, Q. Effects of organic-inorganic compound fertilizer with reduced chemical fertilizer application on crop yields, soil biological activity and bacterial community structure in a rice–wheat cropping system. *Appl. Soil Ecol.* **2016**, *99*, 1–12. [CrossRef]
8. Abrol, I.P.; Bronson, K.F.; Duxbury, J.M.; Gupta, R.K. Long-term soil fertility experiments in rice-wheat cropping systems. In *Long-Term Soil Fertility Experiments with Rice-Wheat Rotations in South Asia*; Rice-Wheat Consortium Paper Series No. 1; Abrol, I.P., Ed.; Rice-Wheat Consortium for the Indo-Gangetic Plains: New Delhi, India, 1997; pp. 14–15.
9. Ladha, J.K.; Dawe, D.; Pathak, H.; Padre, A.T.; Yadav, R.L.; Singh, B.; Singh, Y.; Singh, Y.; Singh, P.; Kundu, A.L.; et al. How extensive are yield declines in long-term rice-wheat experiments in Asia? *Field Crop. Res.* **2003**, *81*, 159–180. [CrossRef]
10. Dobermann, A.; Cassman, K.G. Precision nutrient management in intensive rice systems: The need for another on-farm revolution. *Better Crop.* **1996**, *10*, 20–25.
11. Hobbs, P.R.; Sayre, K.; Gupta, R. The role of conservation agriculture in sustainable agriculture. *Philos. Trans. R. Soc. B Biol. Sci.* **2008**, *363*, 543–555. [CrossRef]
12. Benbi, D.K.; Brar, J.S.A. 25-year record of carbon sequestration and soil properties in intensive agriculture. *Agron. Sustain. Dev.* **2009**, *29*, 257–265. [CrossRef]
13. Foley, J.A.; Ramankutty, N.; Brauman, K.A.; Cassidy, E.S.; Gerber, J.S.; Johnston, M. Solutions for a cultivated planet. *Nature* **2011**, *478*, 337–342. [CrossRef] [PubMed]

14. Singh, R.P.; Mundra, M.C.; Gupta, S.C.; Agrawal, S.K. Effect of integrated nutrient management on productivity of pearl millet-wheat cropping system. *Indian J. Agron.* **1999**, *44*, 250–253.
15. Anwar, M.; Patra, D.D.; Chand, S.; Kumar, A.; Naqvi, A.A.; Khanuja, S.P.S. Effect of organic manures and inorganic fertilizer on growth, herb and oil yield, nutrient accumulation, and oil quality of French basil. *Commmun. Soil Sci. Plant Anal.* **2005**, *36*, 1737–1746. [CrossRef]
16. Kumar, U.; Shahid, M.; Tripathi, R.; Mohanty, S.; Kumar, A.; Bhattacharyya, P.; Lal, B.; Gautam, P.; Raja, R.; Panda, B.B.; et al. Variation of functional diversity of soil microbial community in sub-humid tropical rice-rice cropping system under long-term organic and inorganic fertilization. *Ecol. Indic.* **2017**, *73*, 536–543. [CrossRef]
17. Kumar, U.; Nayak, A.K.; Shahid, M.; Gupta, V.V.S.R.; Panneerselvam, P.; Mohanty, S.; Kaviraj, M.; Kumar, A.; Chatterjee, D.; Lal, B.; et al. Continuous application of inorganic and organic fertilizers over 47 years in paddy soil alters the bacterial community structure and its influence on rice production. *Agric. Ecosyst. Environ.* **2018**, *262*, 65–75. [CrossRef]
18. Mahajan, A.; Bhagat, R.M.; Gupta, R.D. Integrated nutrient management in sustainable rice-wheat cropping system for food security in India. *SAARC J. Agric.* **2008**, *6*, 1–13.
19. Mahajan, A.; Sharma, R. Integrated nutrient management (INM) system: Concept, need and future strategy. *Agrobios. Newsl.* **2005**, *4*, 29–32.
20. Yadav, D.S.; Kumar, A. Integrated nutrient management in rice-wheat cropping system under eastern Uttar Pradesh. *Indian Farm.* **2000**, *50*, 28–30.
21. Janssen, B.H. Integrated nutrient management: The use of organic and mineral fertilizers. In *The Role of Plant Nutrients for Sustainable Crop Production in Sub-Saharan Africa*; Van Reuler, H., Prins, W.H., Eds.; Ponsen and Looijen: Wageningen, The Netherlands, 1993; pp. 89–105.
22. Wu, W.; Ma, B. Integrated nutrient management (INM) for sustaining crop productivity and reducing environmental impact: A review. *Sci. Total Environ.* **2015**, *512–513*, 415–427. [CrossRef]
23. Gruhn, P.; Goletti, F.; Yudelman, M. *Integrated Nutrient Management, Soil Fertility and Sustainable Agriculture: Current Issues and Future Challenges*; International Food Policy Research Institute: Washington, DC, USA, 2000.
24. Timsina, J. Can organic sources of nutrients increase crop yields to meet global food demand? *Agronomy* **2018**, *8*, 214. [CrossRef]
25. Niggli, U.; Fliebbach, A.; Hepperly, P.; Scialabba, N. *Low Greenhouse Gas Agriculture: Mitigation and Adaptation Potential of Sustainable Farming Systems*; Food and Agriculture Organization of the United Nations: Rome, Italy, 2009.
26. Padbhushan, R.; Rakshit, R.; Das, A.; Sharma, R.P. Effects of various organic amendments on organic carbon pools and water stable aggregates under a scented rice–potato–onion cropping system. *Paddy Water Environ.* **2016**, *14*, 481–489. [CrossRef]
27. Rosenberg, M.S.; Adams, D.C.; Gurevitch, J. *MetaWin: Statistical Software for Meta-Analysis, Version 2.0*; Sinauer Associates: Sunderland, MA, USA, 2000.
28. Zaman, S.K.; Jahiruddin, M.; Panaullah, G.M.; Mian, M.H.; Islam, M.R. Integrated nutrient management for sustainable yield in rice-rice cropping system. In Proceedings of the 17th WCSS, Bangkok, Thailand, 14–21 August 2002.
29. Kumar, M.; Singh, R.P.; Rana, N.S. Effect of organic and inorganic sources of nutrition on productivity of rice. *Ind. J. Agron.* **2003**, *48*, 175–177.
30. Bhoite, S.V. Integrated nutrient management in basmati rice (*Oryza sativa*)-wheat (*Triticum aestivum*) cropping system. *Indian J. Agron.* **2005**, *50*, 98–101.
31. Surekha, K.; Rao, K.V. Direct and residual effects of organic sources on rice productivity and soil quality of vertisols. *J. Indian Soc. Soil Sci.* **2009**, *57*, 53–57.
32. Kumari, R.; Kumar, S.; Kumar, R.; Das, A.; Kumari, R.; Choudhary, C.D.; Sharma, R.P. Effect of long-term integrated nutrient management on crop yield, nutrition and soil fertility under rice-wheat system. *J. Appl. Nat. Sci.* **2017**, *9*, 1801–1807. [CrossRef]
33. Chakraborty, D.; Ladha, J.K.; Rana, D.S.; Jat, M.L.; Gathala, M.K.; Yadav, S.; Rao, A.N.; Ramesha, M.S.; Raman, A. A global analysis of alternative tillage and crop establishment practices for economically and environmentally efficient rice production. *Sci. Rep.* **2017**, *7*, 9342. [CrossRef] [PubMed]
34. Reddy, K.R.; Khaleel, R.; Overcash, M.R. Behavior and transport of microbial pathogens and indicator organism in soils treated with organic wastes. *J. Environ. Qual.* **1981**, *10*, 255–266. [CrossRef]

35. Mehdi, S.M.; Sarfraz, M.; Abbas, S.T.; Shabbir, G.; Akhtar, A. Integrated nutrient management for rice-wheat cropping system in a recently reclaimed soil. *Soil Environ.* **2011**, *30*, 36–44.
36. Chaudhary, S.; Dheri, G.S.; Brar, B.S. Long-term effects of NPK fertilizers and organic manures on carbon stabilization and management index under rice-wheat cropping system. *Soil Tillage Res.* **2017**, *166*, 59–66. [CrossRef]
37. Kharub, A.S.; Sharma, R.K.; Mongia, A.D.; Chhokar, R.S.; Tripathi, S.C.; Sharma, V.K. Effect of rice (*Oryza sativa*) straw removal, burning and incorporation on soil properties and crop productivity under rice-wheat (*Triticum aestivum*) system. *Indian J. Agric. Sci.* **2004**, *74*, 295–299.
38. Paikaray, R.K.; Mahapatra, B.S.; Sharma, G.L. Effect of organic and inorganic sources of nitrogen on productivity and soil fertility under rice (Oryza sativa)-wheat (Triticum aestivum) crop sequence. *Indian J. Agric. Sci.* **2002**, *72*, 445–448.
39. Ghuman, B.S.; Sur, H.S. Effect of manuring on soil properties and yield of rainfed wheat. *J. Indian Soc. Soil Sci.* **2006**, *54*, 6–11.
40. Bhattacharya, R.; Chandra, S.; Singh, R.D.; Kundu, S.; Srivastva, A.K.; Gupta, H.S. Long-term farmyard manure application effects on properties of a silty clay loam soil under irrigated wheat-soybean rotation. *Soil Tillage Res.* **2007**, *94*, 386–396. [CrossRef]
41. Oborn, I.; Edwards, A.C.; Witter, E.; Oenema, O.; Ivarsson, K.; Withers, P.J.A.; Nilsson, S.I.; Stinzing, A.R. Element balances as a tool for sustainable nutrient management: A critical appraisal of their merits and limitations within an agronomic and environmental context. *Eur. J. Agron.* **2003**, *20*, 211–225. [CrossRef]
42. Janssen, B.H.; de Willigen, P. Ideal and saturated soil fertility as bench marks in nutrient management. II. Interpretation of chemical soil tests in relation to ideal and saturated soil fertility. *Agric. Ecosyst. Environ.* **2006**, *116*, 147–155. [CrossRef]
43. Ladha, J.K.; Pathak, H.; Krupnik, T.J.; Six, J.; Kessel, C.V. Efficiency of fertilizer nitrogen in cereal production: Retrospect and prospects. *Adv. Agron.* **2005**, *87*, 85–156.
44. Ghosh, B.N.; Singh, R.J.; Mishra, P.K. Soil and input management options for increasing nutrient use efficiency. In *Nutrient Use Efficiency: From Basics to Advances*; Rakshit, A., Ed.; Springer: New Delhi, India, 2015.
45. Abrol, I.P.; Gill, M.S. *Regional office for Asia and Pacific (FAO) Publication*; FAO: Roma, Italy, 1994; Volume 11, pp. 172–183.
46. Modgal, S.C.; Singh, Y.; Gupta, P.C. Nutrient management in rice-wheat cropping system. *Fert. News* **1995**, *40*, 49–54.
47. Patro, H.; Dash, D.; Parida, D.; Panda, P.K.; Kumar, A.; Tiwari, R.C.; Shahid, M. Effect of organic and inorganic sources of nitrogen on yield attributes, grain yield and straw yield of rice (Oryza sativa). *Int. J. Pharma Biol. Sci.* **2011**, *2*, 1–8.
48. Acharya, D.; Mandal, S.S. Effect of integrated nutrient management on the growth, productivity and quality of crops in rice (Oryza sativa L.)—Cabbage (Brassica oleracea)—Greengram (Vigna radiata) cropping system. *Ind. J. Agron.* **2010**, *55*, 1–5.
49. Brahmachari, K.; Choudhury, S.R.; Karmakar, S.; Dutta, S.; Ghosh, P. Sustainable nutrient management in rice (Oryza sativa)—Paira chickling pea (Lathyrus sativus)—Green gram (Vigna radiata) sequence to improve total productivity of land under coastal zone of West Bengal. *Rajshahi Univ. J. Environ. Sci.* **2011**, *1*, 51–61.
50. Yadav, G.S.; Datta, M.; Basu, S.; Debnath, C.; Sarkar, P.K. Growth and productivity of lowland rice (Oryza sativa) as influenced by substitution of nitrogenous fertilizer by organic sources. *Indian J. Agric. Sci.* **2013**, *83*, 1038–1042.
51. Jaga, P.K. Effect of integrated nutrient management on wheat—A review. *Innovare J. Agric. Sci.* **2013**, *1*, 185–191.
52. Aulakh, M.S.; Adhya, T.K. Impact of agricultural activities on emission of greenhouse gases—Indian perspective. In *Proceedings of the International Conference on Soil, Water and Environmental Quality—Issues and Strategies*; Indian Society of Soil Science: New Delhi, India, 2005; pp. 319–335.
53. Aulakh, M.S.; Khurana, M.P.S.; Singh, D. Water pollution related to agricultural, industrial and urban activities, and its effects on food chain: Case studies from Punjab. *J. New Seeds* **2009**, *10*, 112–137. [CrossRef]
54. Prasad, B.; Prasad, J.; Prasad, R. Nutrient management for sustainable rice and wheat production in calcareous soil amended with green manures, organic manures and zinc. *Fert. News.* **1995**, *40*, 39–45.

55. Aulakh, M.S. Integrated nutrient management for sustainable crop production, improving crop quality and soil health, and minimizing environmental pollution. In Proceedings of the 19th World Congress of Soil Science, Soil Solutions for a Changing World, Brisbane, Australia, 1–6 August 2010.
56. Kundu, S.; Bhattacharyya, R.; Prakash, V.; Ghosh, B.N.; Gupta, H.S. Carbon sequestration and relationship between carbon addition and storage under rainfed soybean–wheat rotation in a sandy loam soil of the Indian Himalayas. *Soil Tillage Res.* **2007**, *92*, 87–95. [CrossRef]
57. Chander, K.; Goyal, S.; Mundra, M.C.; Kapoor, K.K. Organic matter, microbial biomass and enzyme activity of soils under different crop rotations in the tropics. *Biol. Fert. Soils* **1977**, *24*, 306–310. [CrossRef]
58. Wardle, D.A.; Yeates, G.W.; Nicholson, K.S.; Bonner, K.I.; Watson, R.N. Response of soil microbial biomass dynamics, activity and plant litter decomposition to agricultural intensification over a seven-year period. *Soil Biol. Biochem.* **1999**, *31*, 1707–1720. [CrossRef]
59. Schloter, M.; Dilly, O.; Munch, J.K. Indicators for evaluating soil quality. *Agric. Ecosyst. Environ.* **2003**, *98*, 255–262. [CrossRef]
60. Livia, B.; Uwe, L.; Frank, B. Microbial biomass, enzyme activities and microbial community structure in two European long-term field experiments. *Agric. Ecosyst. Environ.* **2005**, *109*, 141–152.
61. Kuttimani, R.; Somasundaram, E.; Velayudham, K. Effect of integrated nutrient management on soil microorganisms under irrigated banana. *Int. J. Curr. Microbiol. Appl. Sci.* **2017**, *6*, 2342–2350. [CrossRef]
62. Pernes-Debuyser, A.; Tessier, D. Soil physical properties affected by long-term fertilization. *Eur. J. Soil Sci.* **2004**, *55*, 505–512. [CrossRef]
63. Singh, G.; Singh, O.P.; Singh, S.; Prasad, K. Weed management in late sown wheat (Triticum aestivum) after rice (*Oryza sativa*) in rice–wheat system in rainfed lowland. *Indian J. Agron.* **2010**, *55*, 83–88.
64. Prasad, P.; Satyanarayana, V.; Murthy, V.; Boote, K.J. Maximizing yields in rice- groundnut cropping sequence through integrated nutrient management. *Field Crop. Res.* **2002**, *75*, 9–21. [CrossRef]
65. Parkinson, R. System based integrated nutrient management. *Soil Use Manag.* **2013**, *29*, 608. [CrossRef]
66. Zhang, F.; Cui, Z.; Chen, X.; Ju, X.; Shen, J.; Chen, Q.; Liu, X.; Zhang, W.; Mi, G.; Fan, M.; et al. Integrated nutrient management for food security and environmental quality in China. *Adv. Agron.* **2012**, *116*, 1–40.
67. Das, B.; Chakraborty, D.; Singh, V.; Aggarwal, P.; Singh, R.; Dwivedi, B.; Mishra, R. Effect of integrated nutrient management practice on soil aggregate properties, its stability and aggregate-associated carbon content in an intensive rice-wheat system. *Soil Tillage Res.* **2014**, *136*, 9–18. [CrossRef]
68. Khan, M.U.; Qasim, M.; Sarhad, I.U.K. Effect of integrated nutrient management on crop yields in rice-wheat cropping system. *J. Agric.* **2007**, *23*, 4.

5

Phosphorus and Potassium Fertilizer Application Strategies in Corn–Soybean Rotations

Timothy J. Boring [1], Kurt D. Thelen [2,*], James E. Board [3], Jason L. De Bruin [4], Chad D. Lee [5], Seth L. Naeve [6], William J. Ross [7], Wade A. Kent [8] and Landon L. Ries [9]

1 Michigan Agri-Business Association, East Lansing, MI 48824, USA; boringti@msu.edu
2 Plant, Soil, and Microbial Sciences, Michigan State University, East Lansing, MI 48824, USA
3 Emiritus, Department of Plant, Environmental, and Soil Science, Louisiana State University, Baton Rouge, LA 70803, USA; jboard@agcenter.lsu.edu
4 DuPont Pioneer, Macomb, IL 61455, USA; jason.debruin@pioneer.com
5 Department of Plant and Soil Sciences, University of Kentucky, Lexington, KY 40546, USA; cdlee2@uky.edu
6 Department of Agronomy and Plant Genetics, University of Minnesota, St. Paul, MN 55108, USA; naeve002@umn.edu
7 Department of Crop, Soil, and Environmental Science, University of Arkansas, Little Rock, AR 72204, USA; jross@uaex.edu
8 Becks Hybrids, Urbandale, IA 55032, USA; wkent@beckshybrids.com
9 DuPont Pioneer, Algona, IA 50511, USA; Landon.l.ries@gmail.com
* Correspondence: thelenk3@msu.edu

Abstract: To determine if current university fertilizer rate and timing recommendations pose a limitation to high-yield corn (*Zea mays* subsp. *mays*) and soybean (*Glycine max*) production, this study compared annual Phosphorous (P) and Potassium (K) fertilizer applications to biennial fertilizer applications, applied at 1× and 2× recommended rates in corn–soybean rotations located in Minnesota (MN), Iowa (IA), Michigan (MI), Arkansas (AR), and Louisiana (LA). At locations with either soil test P or K in the sub-optimal range, corn grain yield was significantly increased with fertilizer application at five of sixteen site years, while soybean seed yield was significantly increased with fertilizer application at one of sixteen site years. At locations with both soil test P and K at optimal or greater levels, corn grain yield was significantly increased at three of thirteen site years and soybean seed yield significantly increased at one of fourteen site years when fertilizer was applied. Site soil test values were generally inversely related to the likelihood of a yield response from fertilizer application, which is consistent with yield response frequencies outlined in state fertilizer recommendations. Soybean yields were similar regardless if fertilizer was applied in the year of crop production or before the preceding corn crop. Based on the results of this work across the US and various yield potentials, it was confirmed that the practice of applying P and K fertilizers at recommended rates biennially prior to first year corn production in a corn–soybean rotation does not appear to be a yield limiting factor in modern, high management production systems.

Keywords: phosphorous; potassium; corn–soybean rotation; management; production system

1. Introduction

Phosphorus (P) and potassium (K) are essential nutrients for corn and soybean, comprising a significant proportion of total fertilizer expenditures, and can be yield limiting in many major crop production areas in the United States. Determining optimum application rates and timings for these fertilizers has been an ongoing research focus for decades and efforts continue to refine recommendations. Perceptions by producers of stagnant crop yield increases, particularly in soybeans, have spurred interest in revisiting the role of production inputs in an attempt to determine yield-limiting factors. Previous studies have investigated optimizing factors including planting date, seed treatment, weed management, variety selection, and tillage in a high management scheme, although no studies have explored the role of fertilizer recommendation schemes on a multi-regional scale.

A central tenet of crop fertilizer recommendations is the identification of the soil test critical level, the point below which crop growth and yield will be limited by nutritional deficiency. Specific soil critical test values differ by geographic region, soil characteristics, crop, and environment and emphasize the need for regional specific recommendations. Fertilizer recommendations for soils with nutrient values below the critical level include fertilizer to meet intended crop needs and to elevate soil test levels above the critical point. Terminology differs by state for soils testing above the critical level, with soils classified as either optimum or medium. Soils in this range are expected to have a good probability of not responding in yield increases with the addition of fertilizer. As such, this range is considered the most economical category in which to maintain soil test values.

While the specific numerical delineations of soil test ranges vary by state, the premise of these ranges serving as a guide to quantify the likelihood of economic yield response to fertilizer application is a consistent theme. Fertility experiments conducted in the time following the development of these recommendations have generally validated existing fertility standards. As expected, corn and soybean grain yield increases to applications of P and K have generally not been found in soils testing at or above medium ranges [1–5].

In a corn–soybean crop rotation, a biennial fertilizer application of P and K preceding soybean has become a common management practice. While fertilizer recommendations have been specifically developed to meet a single year of crop production following a soil sample, multiple year fertilization for corn and soybeans produced on soils testing in an optimum or higher range is a reasonable practice when combining the recommended rates for each crop in one application [6,7]. The lack of documented yield response to fertilizer applications on high testing soils supports fertility management that consists of periodic fertilizer application to maintain optimal soil test levels. On high testing soils, extended intervals between maintenance fertilizer applications may be possible. Dodd and Mallarino (2005) [8] found eight to nine years of non-fertilizer corn and soybean production could be conducted on high P testing soils before yield responses could be seen from fertilizer application. Buah et al. (2000) [2] documented conflicting soybean yield results comparing annual and biennial P and K applications at eight site years on Iowa farmer fields, but recorded increased soybean yield from annual P applications compared to biennial in two of three years on research station plots testing optimum for P. McCallister et al. (1987) [9] measured greater extractable P when applying P fertilizer annually rather than equivalent total applications made every two, three, or six years. Corn grain yields were not responsive to application frequency, but the authors suggested that smaller, more frequent fertilizer applications lead to increase plant available phosphorous.

The objective of this study was to determine if current fertilizer rate and timing recommendations constitute a yield limitation in modern, high yielding corn and soybean rotations across a range of production regions in the United States.

2. Materials and Methods

2.1. Site Description

Research trials were established at four sites in Minnesota, three sites each in Iowa, Arkansas and Louisiana, and two sites in Michigan in 2009 (Table 1). Experiments were established on sites with a history of corn–soybean rotations in a randomized complete block design, blocked by crop rotation, with four replications. Corn was established in 2009 for rotation to soybeans in 2010. A second rotation cycle was initiated in 2010 adjacent to existing plots; corn was planted in 2010 for rotation to soybean in 2011. Corn and soybean production was conducted following local cultural practices for tillage, row spacing, population and variety.

Table 1. Description of soil characteristics at research sites used in this study in the years 2009 to 2011.

Site	Soil Type	Soil Series	pH	Initial P [a]	Initial K
				Mg kg^{-1}	
Minnesota					
Delavan	Fostoria loam	Aquic Hapludolls	6.8	9 M [b]	141 H
Lamberton	Ves loam	Calcic Hapludolls	5.5	24 VH	114 M
Morris	Tara silt loam	Aquic Hapludolls	7.9	6 L	150 H
St. Charles	Seaton silt loam	Typic Hapludolls	6.2	14 M	77 L
Iowa					
Lewis	Marshall silty clay loam	Typic Hapludolls	6.8	9 L	160 H
Sutherland	Sac silty clay loam	Oxyaquic Hapludolls	5.9	20 O	198 VH
Ames 2009–2010	Clarion loam	Typic Hapludolls	7.2	9 L	128 L
Ames 2010–2011	Canisteo silty clay loam	Typic Endoaquolls	7.2	11 O	82 VL
Michigan					
Ingham	Capac loam	Aquic Glossudalfs	6.3	37 AO	149 AO
Branch	Matherton loam	Typic Argiaquolls	6.5	25 O	136 AO
Arkansas					
Colt	Calhoun silt loam	Typic Glossaqualfs	6.3	15 L	100 M
Keiser	Sharkey silty clay	Chromic Epiaquerts	6.8	45 O	201 AO
Rohwer	Henery silt loam	Typic Fragiaqualfs	7.3	26 M	58 VL
Louisiana					
Baton Rouge	Commerce silt loam	Fluvaquentic Endoaquepts	6.6	29 M	136 L
St. Joseph	Commerce silt loam	Fluvaquentic Endoaquepts	6.6	51 H	239 H
Winnsboro	Gigger silt loam	Typic Fragiudalfs	6.3	31 M	115 L

[a] Minnesota soil test P determined by the Olson method at Delavan and Morris and with the Bray-P1 method at Lamberton and St Charles. Iowa and Michigan soil test P determined with the Bray-P1 method. Kentucky, Arkansas and Louisiana soil test P determined with the Mehlich 3 method. [b] Soil test ranges determined by state soil test labs. AO, Above optimum; H, high; L, low; M, medium; O, optimum; VH, very high; VL, very low.

2.2. *Treatment Plans*

Fertility treatments consisted of annual applications of P and K prior to corn in 2009 and soybeans in 2010 or biennial applications of P and K for both corn and soybeans applied preceding the corn crop. Fertilizer rates were determined from state specific fertilizer recommendations according to soil test values and applied at 1× and 2× rates in both annual and biennial practices [7,10–12]. Specific fertilizer rates, which varied by site, can be found at Boring, 2013 [13]. Soils testing below optimal or medium were fertilized with specific state recommended rates; crop removal rate combined with additional fertilizer to bring test levels to optimal ranges. Sites testing above the optimal or medium level were fertilized at crop removal rates. These treatments were compared to a control receiving no P or K fertilizer. Fertilizer treatments were broadcast and incorporated in the spring with mono-ammonium phosphate (10-52-0) and potassium chloride (0-0-62). Tillage, row spacing, variety, plant population, planting date and harvest date can be referenced at Boring, 2013 [13] in the study done at Michigan State University, East Lansing, Michigan, USA. Nitrogen management was conducted in accordance with local practices.

Fertility rates were determined for each site from the composite of 10–15 cores sampled to a 15 cm depth. All samples were analyzed in the originating state's university soil testing lab following standard soil testing procedures. Phosphorus analysis was determined in Minnesota with the Olson method [14] at sites with pH levels of 6.8 and greater, the Bray-P1 method [14] at sites with pH below 6.8, in Iowa and Michigan with the Bray-P1 method, and Arkansas and Louisiana with the Mehlich 3 method [15]. Potassium analysis was determined by ammonium acetate extraction [16] in Minnesota, Iowa and Michigan and by Mehlich 3 in Arkansas and Louisiana. Soils were categorized as to relative P and K test levels according to state fertilizer recommendations. Categorization of soil test ranges varied by state, both the number and names of categories and the degree to which additional soil information is utilized. Soil test classification included texture, cation exchange capacity, subsoil nutrient concentrations and irrigation supplementation. A commonality among all state classifications is the delineation between sub-optimal, optimal, and above optimal test levels. Sub-optimal ranges include very low and low in Minnesota, Iowa and Louisiana; below optimal in Michigan; and very low, low, and medium in Arkansas. Optimal ranges include medium in Minnesota and Louisiana and optimal in Iowa, Michigan and Arkansas. Above optimal soil test ranges include high and very high in Minnesota and Iowa, above optimal in Michigan and Arkansas, and high in Louisiana. These three ranges, sub-optimal, optimal, and above optimal, are used for delineation of individual sites in this study (Table 2). Plot width varied between 3.9 and 7.8 m and ranged from 9.1 and 12.2 m in length. Soybean seed yields were obtained by machine from plot centers and adjusted to 130 g kg^{-1} H_2O. Corn grain yields were obtained by either hand or machine harvest and adjusted to 150 g kg^{-1} H_2O.

Table 2. Description of site soil test levels used in the study in the years 2009 to 2011.

Site	P Range	K Range
Ames 2009–2010, Colt, Rohwer	Sub-optimal	Sub-optimal
Lewis, Morris	Sub-optimal	Above optimal
Ames 2010–2011, Baton Rouge, St. Charles, Winnsboro	Optimal	Sub-optimal
Branch, Delavan, Keiser, Sutherland	Optimal	Above optimal
Lamberton, St. Joseph	Above optimal	Optimal
Ingham	Above optimal	Above optimal

2.3. Statistical Analysis

Data were analyzed with SAS statistical software (SAS Institute, Inc. 2014. SAS v. 9.3. SAS Institute, Cary, NC, USA). Analysis of variance was performed using PROC MIXED. Coefficient of variance was determined using PROC GLM. Treatment means were considered statistically different at the $p < 0.10$ level. The 90% confidence interval was chosen because it was considered conservative enough to meet the objective of evaluating fertilizer application practices across the broad study area [17,18].

3. Results

3.1. Sites Testing in the Sub-Optimal Range for Soil P and K

At 10 total site-years testing sub-optimal for both soil P and K, fertilizer applications significantly increased corn yields in two of five comparisons and soybean yields in one of five. In all cases where a yield response was documented, annual fertilizer applications at the 1× rate maximized yield of corn. At Ames, corn grain yield was significantly increased in 2009 by all fertilizer treatments (Table 3) and the annual and biennial 1× rates resulted in significantly greater yield than the 2× biennial treatment. Corn grain yield in 2010 was significantly increased at Colt by 1× annual and 2× biennial fertilizer treatments compared to the control and 1× biennial treatments. At Colt, the corn yield in 2009 was not influenced by fertilizer treatments. At Rowher, the 2× biennial fertilizer treatment significantly reduced 2010 corn grain yield compared to the control. Yield for all other fertilizer treatments were similar to the control. No differences in corn grain yield were observed at Rohwer in 2009. Soybean grain yield was significantly increased by annual 2× and biennial 1× fertilizer applications compared to the biennial 2× and control treatments at Ames in 2009. Soybean seed yield was not affected by fertilizer applications in 2010 or 2011 at Colt and Rohwer locations. At Colt following the 2010–2011 rotation, soil test P and K were significantly increased with biennial 2× application compared to all other treatments (Table 3). Following the 2010–2011 rotation at Rowher, all fertilizer treatments significantly increased soil test K compared to the control. Fertilizer rates at the 2× rate significantly increased soil test K compared to 1× rates.

Table 3. Corn and soybean seed yield and soil test results affected by fertilizer treatments at sites with sub-optimal P and sub-optimal K.

Site	Crop Rotation	Control	Annual		Biennial		P > F	CV
			1×	2×	1×	2×		
				Grain yield (Mg ha^{-1})				
Ames, IA	2009 Corn	10.39 c *	12.89 a	12.29 a,b	12.91 a	11.54 b	0.0026	6.37
	2010 Soybean	3.62 b	4.00 a,b	4.03 a	4.02 a	3.80 b	0.0132	4.11
Colt, AK	2009 Corn	7.32	7.98	7.51	7.53	6.90	0.9090	21.44
	2010 Soybean	2.62	2.56	2.49	2.26	2.43	0.6031	13.16
	2010 Corn	5.32 b	6.29 a	5.80 a,b	5.28 b	6.04 a	0.0856	10.51
	2011 Soybean	3.80	3.83	3.78	4.10	4.03	0.2233	5.77
Rohwer, AK	2009 Corn	12.73	11.79	11.87	12.27	11.81	0.7851	11.45
	2010 Soybean	2.95	3.05	3.31	3.11	3.02	0.4121	8.66
	2010 Corn	9.38 a	9.10 a	8.82 a,b	9.36 a	8.50 b	0.0293	4.55
	2011 Soybean	2.28	2.38	2.21	2.11	2.28	0.3412	8.10
				Soil test results (Mg ha^{-1})				
Colt, AK	2010–2011 P	29 b	32 b	33 b	31 b	39 a	0.0255	11.48
	2010–2011 K	94 b	96 b	100 b	94 b	106 a	0.0277	5.18
Rohwer, AK	2010–2011 P	37	42	49	41	48	0.1546	14.90
	2010–2011 K	75 c	103 b	116 a	94 b	117 a	0.0004	8.74

* Letters following numbers within a row represent differences between treatments at the $p \leq 0.10$ level.

3.2. Sites Testing in the Sub-Optimal Range for P and Above Optimal Range for K

There were eight site-years with suboptimal P and optimal soil test K. All fertilizer treatments significantly increased corn grain yield compared to the control at Lewis in 2009 (Table 4). No significant differences in corn grain yield were observed at Lewis in 2010 or at Morris in 2009 or 2010. Soybean seed yield was not affected by fertilizer treatment at either site. No significant differences in soil test P or K levels were noted at either site following 2009–2010 or 2010–2011 crop rotations (Table 4).

Table 4. Corn and soybean grain yield and soil test results affected by fertilizer treatments at sites with sub-optimal P and above optimal K.

Site	Crop Rotation	Control	Annual		Biennial		P > F	CV
			1×	2×	1×	2×		
				Grain yield (Mg ha^{-1})				
Lewis, IA	2009 Corn	11.88 b *	13.78 a	13.98 a	14.06 a	14.52 a	0.0014	5.02
	2010 Soybean	4.07	4.19	4.03	4.27	4.18	0.4448	4.61
	2010 Corn	13.30	13.19	13.52	12.98	12.91	0.7904	5.76
	2011 Soybean	4.32	4.16	4.41	4.46	4.32	0.1176	3.10
Morris, MN	2009 Corn	9.73	10.41	10.83	10.51	11.26	0.4766	11.38
	2010 Soybean	3.01	3.16	2.98	3.14	3.06	0.7899	7.74
	2010 Corn	14.77	14.91	14.97	14.81	15.14	0.7201	2.69
	2011 Soybean	2.34	3.21	2.68	2.92	2.82	0.3405	20.48
				Soil test results (Mg ha^{-1})				
Lewis, IA	2010–2011 P	28	39	40	32	27	0.5467	45.58
	2010–2011 K	196	252	213	240	195	0.1098	14.47
Morris, MN	2009–2010 P	7	9	6	5	4	0.4408	58.29
	2009–2010 K	156	157	152	152	144	0.9076	14.56
	2010–2011 P	9	11	10	9	9	0.8148	27.63
	2010–2011 K	165	170	164	169	166	0.8985	6.65

* Letters following numbers within a row represent differences between treatments at the $p \leq 0.10$ level.

3.3. Sites Testing in the Optimal Range for P and Sub-Optimal Range for K

There were 14 comparisons with optimal soil test P and sub-optimal soil test K. A fertilizer treatment significantly increased yields in only one site-year for each corn and soybean. At Baton Rouge in 2009, all fertilizer treatments resulted in significantly higher corn yield compared to the control (Table 5). Yields were significantly greater for biennial 1× and 2× treatments compared to the annual treatment. No significant differences in corn grain yield were observed at other sites testing in the optimal range for P and sub-optimal range for K. In 2010 at Baton Rouge, soybean seed yields were significantly lower with annual and biennial 2× treatments compared to the control. Soybean seed yield did not significantly differ between fertilizer treatments at other sites. At Baton Rouge, soil test P levels following the 2010–2011 crop rotation were highest with the biennial 2× treatment compared to all other treatments (Table 5). Additionally, soil test P levels were significantly greater with annual 1×, annual 2×, and biennial 1× treatments compared to the control. Soil test P and K following the 2009–2010 rotation and soil test K following the 2010–2011 rotation did not differ by fertilizer treatment at Baton Rouge. At Ames 2010–2011 and St. Charles, no differences in soil test P and K were observed.

Table 5. Corn and soybean grain yield and soil test results affected by fertilizer treatments at sites with optimal P and sub-optimal K.

Site	Crop Rotation	Control	Annual		Biennial		P > F	CV
			1×	2×	1×	2×		
				Grain yield (Mg ha^{-1})				
Ames, IA	2010 Corn	12.31	12.38	13.17	13.13	13.73	0.1405	6.94
	2011 Soybean	3.48	3.83	3.62	3.73	3.86	0.9278	20.26
Baton Rouge, LA	2009 Corn	3.56 c *	4.73 a,b	4.45 b	5.12 a	5.20 a	0.0068	11.67
	2010 Soybean	4.32 a,b	3.94 b,c	3.76 c	4.36 a	3.71 c	0.0752	9.68
	2010 Corn	6.33	8.57	8.62	7.92	8.12	0.1693	12.46
	2011 Soybean	2.74	2.58	2.88	2.68	2.58	0.7907	14.47
St. Charles, MN	2009 Corn	12.10	13.29	12.52	13.02	13.11	0.1372	4.61
	2010 Soybean	3.36	3.48	3.61	3.16	3.51	0.4408	10.07
	2010 Corn	12.76	12.52	13.19	12.80	12.58	0.8802	7.70
	2011 Soybean	3.32	3.28	3.47	3.34	3.26	0.6177	5.78
Winnsboro, LA	2009 Corn	11.23	11.26	11.44	11.48	10.86	0.9181	9.29
	2010 Soybean	4.10	4.03	4.43	4.57	4.63	0.2777	10.36
	2010 Corn	10.76	10.41	10.95	11.13	10.68	0.4588	5.70
	2011 Soybean	2.06	1.23	2.15	1.87	1.66	0.1494	28.56
				Soil test results (Mg ha^{-1})				
Ames, IA	2010–2011 P	16	20	27	21	34	0.2086	49.28
	2010–2011 K	143	58	176	167	199	0.1097	17.28
Baton Rouge, LA	2009–2010 P	54	56	53	48	54	0.2606	9.62
	2009–2010 K	230	244	222	215	228	0.1250	6.59
	2010–2011 P	47 c	52 b	54 b	53 b	59 a	0.0033	6.39
	2010–2011 K	229	230	230	216	234	0.6141	7.73
St. Charles, MN	2009–2010 P	6	6	6	6	7	0.9577	41.25
	2009–2010 K	72	79	77	73	71	0.5816	10.71
	2010–2011 P	5	6	5	5	6	0.4382	29.68
	2010–2011 K	78	71	70	79	74	0.4112	10.17

* Letters following numbers within a row represent differences between treatments at the $p \leq 0.10$ level.

3.4. *Sites Testing for the Optimal Range for P and Above Optimal Range in K*

There were 15 comparisons with optimal soil test P and above optimal soil test K. A fertilizer treatment significantly increased corn grain yields in three of seven comparisons and soybean seed yield in one of eight site-years. Corn grain yield was significantly increased by any fertilizer application at Branch 2009, Delavan in 2010 and Sutherland in 2010 (Table 6). All fertilizer applications at Branch in 2009 and Delavan in 2010 resulted in significantly higher corn grain yield compared to the untreated control. At Sutherland in 2010, fertilizer application of any kind significantly increased corn yield above that of the control, but yield was significantly lower for the biennial 1× treatment compared to other fertilizer treatments. Corn grain yield at Delavan in 2009, Keiser in 2009 and 2010, and Sutherland in 2009 were not influenced by fertilizer application. At Branch following the 2009–2010 rotation, all treatments except the biennial 1× treatment significantly increased soil test K compared to the control (Table 6). Soil test K was significantly increased at Branch for 2× rate treatments compared to 1× treatments. Soil test K following the 2010–2011 rotation and soil test P following the 2009–2010 and 2010–2011 crop rotations did not differ by fertilizer treatment. At Delavan, soil test P and K following 2009–2010 and 2010–2011 rotations were not affected by fertilizer treatment. No significant differences in soil test P or K were observed following the 2010–2011 crop rotation at Keiser. Following the 2010–2011 crop rotation at Sutherland, all fertilizer treatments increased soil test P compared to the control. All fertilizer treatments except the biennial 1× treatment increased soil test K compared to the control. Soil test P and K was significantly greater for 2× rate treatments compared to 1× rate treatments.

Table 6. Corn and soybean grain yield and soil test results affected by fertilizer treatments at sites with optimal P and above optimal K.

Site	Crop Rotation	Control	Annual		Biennial		P > F	CV
			1×	2×	1×	2×		
					Grain yield (Mg ha^{-1})			
Branch, MI	2009 Corn	5.36 b *	6.86 a	6.99 a	7.01 a	7.21 a	0.0013	7.46
	2010 Soybean	2.70	2.78	2.86	2.83	2.89	0.9775	16.26
	2011 Soybean	0.94	0.86	1.06	1.04	0.95	0.6076	20.43
Delavan, MN	2009 Corn	12.50	13.59	13.45	14.05	13.66	0.3965	8.11
	2010 Soybean	2.72	2.73	2.97	3.00	2.94	0.5453	10.68
	2010 Corn	9.79 b	12.16 a	12.82 a	11.82 a	12.34 a	0.0019	6.91
	2011 Soybean	2.39 b	3.04 a	3.11 a	3.05 a	3.12 a	0.0159	9.69
Keiser, AK	2009 Corn	10.04	10.72	9.98	10.13	10.16	0.6352	7.46
	2010 Soybean	3.80	3.76	3.71	3.79	3.76	0.9932	7.79
	2010 Corn	10.73	10.71	10.43	11.00	10.63	0.4529	3.91
	2011 Soybean	4.31	4.32	4.40	4.37	4.51	0.4131	3.53
Sutherland, IA	2009 Corn	13.60	14.27	14.3	14.46	14.41	0.3242	4.51
	2010 Soybean	4.65	4.60	4.66	4.82	4.63	0.5157	3.67
	2010 Corn	12.61 c	13.72 a	13.90 a	13.20 b	13.46 a	0.0032	2.79
	2011 Soybean	3.74	3.94	3.93	4.04	3.96	0.1382	3.90
					Soil test results (Mg ha^{-1})			
Branch, MI	2009–2010 P	69	68	70	69	69	0.1238	0.85
	2009–2010 K	16 c	28 b	42 a	21 c	40 a	<0.0001	18.30
	2010–2011 P	42	33	54	36	24	0.2943	48.33
	2010–2011 K	117	89	126	115	77	0.3497	36.15
Delavan, MN	2009–2010 P	7	9	10	9	11	0.9223	67.03
	2009–2010 K	141	149	148	138	131	0.4673	10.68
	2010–2011 P	4	4	8	5	5	0.3789	63.46
	2010–2011 K	135	136	133	124	130	0.7199	10.28
Keiser, AK	2010–2011 P	48	56	54	56	57	0.4226	13.19
	2010–2011 K	337	341	338	341	341	0.9981	6.77
Sutherland, IA	2010–2011 P	17 d	28 b,c	43 a	23 c	34 a,b	<0.0001	16.27
	2010–2011 K	188 d	218 b,c	241 a	200 c,d	222 b	0.0019	7.15

* Letters following numbers within a row represent differences between treatments at the $p \leq 0.10$ level.

3.5. *Sites Testing in the Optimal Range for Soil P and K*

There were eight comparisons with optimal soil test P and optimal soil test K, though fertilizer application had no effect on corn or soybean seed yield (Table 7). At Lamberton, no differences in soil test P or K levels following 2009–2010 and 2010–2011 crop rotations were observed. At St. Joseph following the 2010–2011 crop rotation, the annual 2× fertilizer application resulted in significantly greater soil test with both P and K compared to the control (Table 7).

Table 7. Corn and soybean grain yield and soil test results affected by fertilizer treatments at sites with above optimal P and optimal K.

Site	Crop Rotation	Control	Annual		Biennial		P > F	CV
			1×	2×	1×	2×		
					Grain yield (Mg ha^{-1})			
Lamberton, MN	2009 Corn	8.22	8.84	10.31	10.33	9.23	0.3951	20.14
	2010 Soybean	3.45	3.64	3.73	3.63	3.48	0.3221	5.74
	2010 Corn	11.82	13.00	12.81	12.33	12.99	0.2422	6.40
	2011 Soybean	2.89	3.30	3.20	3.15	3.18	0.1348	6.68
St. Joseph, LA	2009 Corn	11.49	11.35	11.06	11.16	11.41	0.6394	3.67
	2010 Soybean	3.58	3.89	3.84	3.80	3.79	0.6005	3.63
	2010 Corn	10.29	10.19	10.71	10.53	10.39	0.4447	3.90
	2011 Soybean	3.74	3.63	3.52	3.67	3.76	0.1003	3.38
					Soil test results (Mg ha^{-1})			
Lamberton, MN	2009–2010 P	12	12	11	10	12	0.9423	31.29
	2009–2010 K	111	107	111	115	121	0.7398	14.10
	2010–2011 P	6	8	7	7	6	0.2872	15.99
	2010–2011 K	104	111	104	106	110	0.3129	5.40
St. Joseph, LA	2010–2011 P	43b	49b	61a	47b	49b	0.0067	10.57
	2010–2011 K	172b	189b	211a	174b	180b	0.0446	9.18

Letters following numbers within a row represent differences between treatments at the $p \leq 0.10$ level.

3.6. Sites Testing in the Above Optimal Range for Both P and K

One site had above optimal soil test levels for both P and K, for a total of four site year comparisons. No differences were measured in corn or soybean seed yield (Table 8). Annual fertilizer application at the 2× rate significantly increased soil test P following the 2010–2011 rotation compared to all other treatments. No differences were observed in soil test P following the 2009–2010 crop rotation or soil test K following the 2009–2010 and 2010–2011 crop rotations.

Table 8. Corn and soybean grain yield and soil test results affected by fertilizer treatments at sites with above optimal P and above optimal K.

Site	Crop rotation	Control	Annual		Biennial		P > F	CV
			1×	2×	1×	2×		
				Grain yield (Mg ha^{-1})				
Ingham, MI	2009 Corn	9.58	9.6	9.68	9.61	10.23	0.9249	12.86
	2010 Soybean	2.92	2.9	3.1	3.01	3.04	0.6311	7.05
	2010 Corn	9.78	9.65	9.07	9.63	9.88	0.2297	5.01
	2011 Soybean	2.81	2.66	2.66	2.7	2.73	0.1842	3.47
				Soil test results (Mg ha^{-1})				
Ingham, MI	2009–2010 P	67	67	67	66	67	0.7257	1.71
	2009–2010 K	64	69	90	78	89	0.1411	21.42
	2010–2011 P	63 b	69 b	80 a	59 b	66 b	0.0415	12.54
	2010–2011 K	209	222	238	207	236	0.2243	10.17

Letters following numbers within a row represent differences between treatments at the $p \leq 0.10$ level.

4. Discussion

4.1. Fertilizer Recommendations

Fertilizer recommendations are built upon the principle that soil test values exceeding the critical soil test level are expected to supply adequate nutrients to support optimal economic growth [10]. However, variability in soil test values and crop responses mean yield responses can be observed from fertilizer applications at all soil test levels, although with decreasing frequency as soil test levels increase. Iowa fertilizer recommendations cite the probability of yield responses as 80% on very low testing sites, 65% on low testing, 25% on optimal soils, 5% for soils in the high range and less than 1% in the very high range [7]. Bruulsema (2004) [19] noted that, in Ontario, Canada, corn and soybean have been documented to have a 59% and 49%, respectively, probability of response to fertilizer applications on soils testing in the medium soil test level. Results from this multi-state study generally agree with these response frequencies, both in terms of the greater probability of a corn yield response compared to soybean and the overall likelihood of yield response. Of the 14 site years with soil test P and K values in the optimal or higher range, corn yield was significantly increased at three, or 21% of sites, while soybean yield was significantly increased at one, or 7% of sites. Mallarino et al. (2011) [20] observed corn grain responses to broadcast P and K fertilizer treatments at five sites with soil test P and K testing at optimal levels or below. The only site they found to be unresponsive to preplant broadcast fertilizer application tested in the very high range for soil P and medium for soil K. In our study, sites with soil test levels in the sub-optimal range for at least one nutrient demonstrated a significant response on corn grain yield at five of sixteen site years. Bordoli and Mallarino (1998) [21] observed frequent corn grain yield responses to P fertilizer on low testing sites in Iowa, but not every low testing site responded to fertilizer application. In our study, soybean seed yield was significantly increased with fertilizer application at two of sixteen site years. These responses are less than those predicted in the Iowa Fertilizer recommendations, but follow the trend illustrated by Bruulsema (2004) [19] of an increased frequency of fertilizer response in corn compared to soybean.

4.2. Corn Yield after Treatment

Corn yield increases in response to fertilizer applications, when observed, did not follow trends with rate or timing. At Branch in 2009, Delavan in 2010, and Lewis in 2009, all fertilizer treatments resulted in a similar increase in yield compared to the control. At Baton Rouge in 2009, yield generally increased with increasing fertilizer rate. This was contrasted by high fertilizer rates detrimentally impacting yield at Ames in 2009 and Rowher in 2010. At these locations, the biennial 2× treatment had significantly lower yields than other fertilizer treatments. At Rowher, yield from the untreated control exceeded the biennial 2× treatment, indicating a yield reducing effect of fertilizer application rather than a simple lack of response. Fertilizer applications have been previously associated with reductions in crop growth and yield, although generally these responses have been attributed to proximity of fertilizer and seed. Anghinoni and Barber (1980) [22] measured decreasing corn root length with increasing P rate in pots. Heckman and Kamprath (1992) [23] observed decreased early season corn growth and K accumulation with increased broadcast K rates, a phenomenon they attributed to high salt concentrations. Numerous site years of phosphorus and potassium rate studies in Arkansas have generally observed a lack of response on high testing soils, response to rate on low testing sites, and agreement with established fertilizer recommendations. Muir and Hedge (2001) [24] measured corn yield reductions when increasing the fertilizer rate from 101 kg K_2O ha^{-1}, the recommended rate, to 202 kg K_2O ha^{-1}, on both a low and a high K testing site.

4.3. Soil Test Levels

Yield responses at several sites failed to follow patterns that could be directly correlated to fertilizer rate. At Colt in 2010 and Sutherland in 2010, yield among the fertilizer treatments was lowest for the 1× biennial treatment compared to other fertilizer treatments. Fertilizer rates for this treatment were neither the lowest nor highest, precluding high fertilizer rate injury as an explanation. The occurrence of this trend at three sites in different production regions suggests more than a statistical anomaly or protocol errors. Soil test values following soybean did not differ between fertilizer treatments at Branch. At Colt, soil test P and K were increased by biennial 2× applications compared to all other treatments. While fertilizer applications at 2× rates would be expected to increase soil test values compared to 1× rates, this effect would be expected to be more pronounced when fertilizer applications were made closer in time to soil sampling. The lack of expected increases in soil test P and K levels, particularly the lack of differences in the annual 1× from the annual 2×, may provide an explanation for differences in yield between these two treatments. At Sutherland in 2010, corn grain yield was maximized with annual 1×, annual 2×, and biennial 2× fertilizer treatments. Grain yield for the biennial 1× treatment was greater than the control, but lagged behind these other treatments. Increases in both soil test P and K following the 2010–2011 crop rotation generally followed the same pattern of increase as corn grain yield. Initial soil test K levels were in the high range and tested in this same range in 2011. Soil test P values were initially in the low range, but fertilizer treatments with the highest corn yields in 2010 had soil test P levels in the high range when tested following the two-year rotation, medium test values for the 1× biennial treatment that was associated with lower corn yield in 2010, and low soil test P values for the control. These trends in soil test results seem to explain 2010 corn grain results, but the low yielding biennial 1× treatment received medium amounts of P and K fertilizer in comparison to other treatments. Examples of decreased yield at medium fertilizer rates have been observed by other researchers as well. Muir and Hedge (2002) [25] noted reduced corn grain yield with a 78 kg P_2O_5 ha^{-1} application compared to 39 kg P_2O_5 ha^{-1}, 157 kg P_2O_5 ha^{-1} and an untreated control at one site. Mozaffari et al. (2012) [1] observed significant corn yield increases with 224 kg K_2O ha^{-1} compared to the control at a low K testing site, though a 179 kg K_2O ha^{-1} rate decreased yield compared to the control. At two medium K testing sites, yield responses were not statistically significant, but a similar trend of decreased yield with a 179 kg K_2O ha^{-1} application was apparent.

4.4. Soybean Yield

Soybean yield responses to fertilizer applications were observed at three sites: Ames in 2010 with sub-optimal P and K, Baton Rouge in 2010 with optimal P and sub-optimal K, and Delavan in 2011 with optimal P and above optimal K. All fertilizer applications resulted in similar yields at Delavan, but fertilizer applications at Ames 2009–2010 and Baton Rouge did not always result in yield increases. At Ames 2009–2010, yield was significantly increased by annual 2× and biennial 1× treatments compared to the control, but yield from annual 1× and biennial 2× treatments were similar to the control. At Baton Rouge, annual 2× and biennial 2× fertilizer treatments resulted in significant yield reductions compared to the control. Instances of yield reduction from high rates of broadcast fertilizer are not commonly observed, but have been noted to occur from both P and K applications. Ebelhar and Varsa (2000) [26] documented greater soybean yield at 56 kg K ha^{-1} than at higher K fertilizer rates, suggesting yield sensitivity to salt concentrations. Farmaha et al. (2011) [27] saw decreasing soybean yields with increasing K application in no-till systems but were unable to identify a definitive cause. They pointed to Ebelhar and Varsa's postulation of salt injury as a likely explanation. In Arkansas, Slaton et al. (2008) [28] observed a slight decrease in soybean yield with 179 kg K_2O ha^{-1} compared to 90 kg K_2O ha^{-1}. Slaton et al. (2001) [29] observed decreased soybean yield with 134 kg P_2O_5 ha^{-1} compared to 90 kg P_2O_5 ha^{-1} applied annually in a soybean–rice rotation. While maximum P and K rates at Baton Rouge were less than those applied at other sites, salt injury from high rates of broadcast applied fertilizer appear to have negatively impacted yield.

Soybean yield was influenced by the timing of fertilizer application in the rotation at two of thirty-two site years of soybean. At Baton Rouge in 2010, the biennial 1× treatment resulted in significantly greater yield than the 1× annual treatment. Both annual and biennial 2× fertilizer treatments resulted in significantly lower yield compared to the control. It is possible that the difference in yield between the 1× annual and 1× biennial occurred due to a greater time interval between fertilizer application and soybean production in the biennial fertilizer treatment. At Ames in 2010, no differences were observed when comparing annual 1× and biennial 1× application, but annual 2× applications resulted in higher yield compared to biennial 2×.

4.5. Comparisons of Soil Test Levels

Soil test levels following each crop rotation did not consistently differ between treatments. When responses in soil test P or K were observed, treatments did not tend to have consistent effects. At Branch following the 2009–2010 rotation, Sutherland following the 2010–2011 rotation, and Rowher following the 2010–2011 rotation, soil test K levels were increased with 2× application rates compared to 1× rates. At all three sites, corn yield increases were documented in the first year of the rotation. Soil test K levels in control treatments at both Branch and Rowher were in the sub-optimal range, suggesting K levels could pose a yield limiting condition. While 2× application rates resulted in higher soil test K levels than 1× rates, these increased levels were still below the critical level at both sites. Soil test K levels at Sutherland were initially characterized as very high and continued to test in this range following the 2010–2011 crop rotation.

The effects of fertilizer application timing were not consistent when increases in soil test P or K were observed. Soil test P levels at Ingham following the 2010–2011 rotation, soil test K levels at Sutherland following the 2010–2011 rotation, and soil test P and K levels at St Joseph following the 2010–2011 rotation were increased by the annual 2× treatment compared to all other treatments. In contrast, biennial 2× fertilizer applications resulted in the highest soil test P and K at Colt following the 2010–2011 rotation and soil test P following the 2010–2011 rotation at Baton Rouge. Treatment effects on corn grain yield were only observed at the Sutherland and Colt sites. At both sites, the biennial 1× application resulted in the lowest yield of all fertilizer treatments. Soil test value responses to fertilizer applications were not consistently increased, but, when observed, increases from annual applications tended to be more associated with greater corn grain yield than when soil test values were increased with biennial applications.

5. Conclusions

Results of this study conducted at several locations showed that present fertilizer recommendations meet nutrient needs of corn and soybean. Corn and soybean yield increases in response to fertilizer applications were observed at sites with adequate fertility levels, but these responses were infrequent and in line with expected response frequencies. Corn yields were more responsive to fertilizer timing and rate than soybean. High fertilizer rates in excess of university fertilizer recommendations were observed to, at best, result in no significant increase in corn or soybean seed yield, and, at worst, result in significant yield reductions. These responses serve to reinforce the current widespread practice of biennial fertilizer applications preceding corn, applied at recommended rates, to supply both corn and soybean fertility needs in a high yield environment.

Author Contributions: Conceptualization, methodology and validation were undertaken by J.E.B., C.D.L., S.L.N., W.J.R. and K.D.T. Formal analysis and investigation were undertaken by T.J.B., L.L.R., W.A.K. and J.L.D.B. The original draft was written by T.J.B. and K.D.T. Review and editing were done by C.D.L., S.L.N., W.J.R. Visualization, supervision, project administration and funding acquisition were undertaken by C.D.L., S.L.N., W.J.R. and K.D.T.

Acknowledgments: The authors wish to thank the field personnel at Michigan State University, University of Arkansas, University of Kentucky, University of Minnesota, Iowa State University, and Louisiana State University, for their technical assistance and support in conducting this research. The authors would also like to thank the United Soybean Board for funding this research.

References

1. Mozaffari, M.; Slaton, N.A.; Hayes, S.; Griffin, B. Corn response to soil applied phosphorus and potassium fertilization in Arkansas. In *Sabbe Arkansas Soil Fertility Studies 2011*; Wayne, E., Slaton, N.A., Eds.; Research Series; Arkansas Agricultural Experiment Station, University of Arkansas: Fayetteville, AK, USA, 2012; p. 599.
2. Buah, S.S.J.; Polito, T.A.; Killorn, R. No-tillage soybean response to banded and broadcast and direct and residual fertilizer phosphorus and potassium applications. *Agron. J.* **2000**, *92*, 657–662. [CrossRef]
3. Webb, J.R.; Mallarino, A.P.; Blackmer, A.M. Effects of residual and annual applied phosphorus on soil test vales and yields of corn and soybean. *J. Prod. Agric.* **1992**, *5*, 148–152. [CrossRef]
4. Mallarino, A.P.; Webb, J.R.; Blackmer, A.M. Corn and soybean yields during 11 years of phosphorus and potassium fertilization on a high-testing soil. *J. Prod. Agric.* **1991**, *4*, 312–331. [CrossRef]
5. DeMoody, C.J.; Young, J.L.; Kaap, J.D. Comparative response of soybeans and corn to phosphorus and potassium. *Agron. J.* **1973**, *65*, 851–855. [CrossRef]
6. Murdock, L.; Schwab, G. *2010–2011 Lime and Nutrient Recommendations*; University of Kentucky College of Agriculture: Lexington, KY, USA, 2010.
7. Sawyer, J.E.; Mallarino, A.P.; Killorn, R.; Barnhart, S.K. *A General Guide for Crop Nutrient and Limestone Recommendations in Iowa*; Publ. PM 1688 (Rev.); Iowa State Univercity Extension and Outreach: Ames, IA, USA, 2011.
8. Dodd, J.R.; Mallarino, A.P. Soil-test phosphorus and crop grain yield response to long-term phosphorus fertilization for corn-soybean rotations. *Soil Sci. Soc. Am. J.* **2005**, *69*, 1118–1128. [CrossRef]
9. McCallister, D.L.; Shapiro, C.A.; Raun, W.R.; Anderson, F.N.; Rehm, G.W.; Engelstad, O.P.; Russelle, M.P.; Olson, R.A. Rate of phosphorus and potassium buildup/decline with fertilization for corn and wheat on Nebraska Mollisols. *Soil Sci. Soc. Am. J.* **1987**, *51*, 1646–1652. [CrossRef]
10. Vitosh, M.L.; Johnson, J.W.; Mengel, D.B. *Tri-State Fertilizer Recommendations for Corn, Soybean, Wheat and Alfalfa*; Extension Bulletin E-2267; Michigan State University: East Lansing, MI, USA; Ohio State University: Columbus, OH, USA; Purdue University: West Lafayette, IN, USA, 1995.
11. Espinoza, L.; Slaton, N.; Mozaffari, M. *Understanding the Numbers on Your Soil Test Report*; Cooperative Extension Service FAS2118; University of Arkansas: Fayetteville, AR, USA, 2012.
12. Kaiser, D.E.; Lamb, J.A.; Eiason, R. *Fertilizer Guidelines for Agronomic Crops in Minnesota*; University of Minnesota: Minneapolis, MN, USA, 2011.

13. Boring, T.J. Field Investigations of Foliar Fertilizer Strategies of Soybean Mn deficiency in Michigan and Phosphorus and Potassium Fertilizer Application Strategies in Corn-Soybean Rotations in the United States. Ph.D. Thesis, Michigan State University, East Lansing, MI, USA, 2013.
14. Frank, K.; Beegle, D.; Denning, J. *Phosphorus: Recommended Chemical Soil Test Procedures for the North Central Region*; North Central Regional Publ. 221 (Rev.); University of Missouri: Columbia, MO, USA, 1998; pp. 21–29.
15. Mehlich, A. Mehlich 3 soil extractant: A modification of Mehlich 2 extractant. Commun. *Soil Sci. Plant Anal.* **1984**, *15*, 1409–1416. [CrossRef]
16. Warncke, D.; Brown, J.R. *Potassium and Other Basic Cations: Recommended Chemical Soil Test Procedures for the North Central Region*; North Central Regional Publ. 221 (Rev.); University of Missouri: Columbia, MO, USA, 1998; pp. 31–33.
17. Young, F.J.; Hammer, R.D.; Maatta, J.M. Confidence intervals for soil properties based on differing statistical assumptions. *Conf. Appl. Stat. Agric.* **1992**. [CrossRef]
18. Loescher, H.; Ayres, E.; Duffy, P.; Luo, H.; Brunke, M. Spatial variation in soil properties among North American Ecosystems and guidelines for sampling designs. *PLoS ONE* **2014**, *9*, e83216. [CrossRef] [PubMed]
19. Bruulsema, T.W. Understanding the science behind fertilizer recommendations. *Better Crops Plant Food* **2004**, *88*, 16–19.
20. Mallarino, A.P.; Bergmann, N.; Kaiser, D.E. Corn responses to in-furrow phosphorus and potassium starter fertilizer applications. *Agron. J.* **2011**, *103*, 685–694. [CrossRef]
21. Bordoli, J.M.; Mallarino, A.P. Deep and shallow banding of phosphorus and potassium as alternatives to broadcast fertilization for no-till corn. *Agron. J.* **1998**, *90*, 27–33. [CrossRef]
22. Anghinoni, I.; Barber, S.A. Phosphorus application rate and distribution in the soil and phosphorus uptake by corn. *Soil Sci. Soc. Am. J.* **1980**, *44*, 1041–1044. [CrossRef]
23. Heckman, J.R.; Kamprath, E.J. Potassium accumulation and corn yield related to potassium fertilization rate and placement. *Soil Sci. Soc. Am. J.* **1992**, *56*, 141–148. [CrossRef]
24. Muir, J.H.; Hedge, J.A. Corn response to phosphorus and potassium fertilization at different soil test levels. In *Arkansas Soil Fertility Studies 2000*; Norman, R.J., Chapman, S.L., Eds.; Research Series 480; Arkansas Agricultural Experiment Station, University of Arkansas: Fayetteville, AK, USA, 2001.
25. Muir, J.H.; Hedge, J.A. Corn response to phosphorus and potassium fertilization at different soil test levels. In *Sabbe Arkansas Soil Fertility Studies 2001*; Wayne, E., Slaton, N.A., Eds.; Research Series 490; Arkansas Agricultural Experiment Station, University of Arkansas: Fayetteville, AK, USA, 2002.
26. Ebelhar, S.A.; Varsa, E.C. Tillage and potassium placement effects on potassium utilization by corn and soybean. *Commun. Soil Sci. Plant Anal.* **2000**, *31*, 2367–2377. [CrossRef]
27. Farmaha, B.S.; Ferández, F.G.; Nafziger, E.D. No-till and strip-till soybean production with surface and subsurface phosphorus and potassium fertilization. *Agron. J.* **2011**, *103*, 1862–1869. [CrossRef]
28. Slaton, N.A.; DeLong, R.E.; Mozaffari, M.; Shafer, J.; Branson, J. Soybean response to phosphorus and potassium fertilization rate. In *Sabbe Arkansas Soil Fertility Studies 2007*; Wayne, E., Slaton, N.A., Eds.; Research Series; Arkansas Agricultural Experiment Station, University of Arkansas: Fayetteville, AK, USA, 2008; p. 558.
29. Slaton, N.A.; DeLong, R.E.; Ntamatungiro, S.; Clark, S.D.; Boothe, D.L. Phosphorus fertilizer rate and application time effect on soybean yield. In *Arkansas Soil Fertility Studies 2000*; Norman, R.J., Chapman, S.L., Eds.; Research Series; Arkansas Agricultural Experiment Station, University of Arkansas: Fayetteville, AK, USA, 2001; p. 480.

6

Bacillus Pumilus Strain TUAT-1 and Nitrogen Application in Nursery Phase Promote Growth of Rice Plants under Field Conditions

Khin Thuzar Win [1,†], Aung Zaw Oo [1,‡], Naoko Ohkama-Ohtsu [2] and Tadashi Yokoyama [2,*]

1. Faculty of Agriculture, Tokyo University of Agriculture and Technology, Saiwaicho 3-5-8, Fuchu, Tokyo 183-8509, Japan; khinthuzarwin@gmail.com (K.T.W.); aungzawo@gmail.com (A.Z.O.)
2. Institute of Agriculture, Tokyo University of Agriculture and Technology, Saiwaicho 3-5-8, Fuchu, Tokyo 183-8509, Japan; nohtsu@cc.tuat.ac.jp

* Correspondence: tadashiy@cc.tuat.ac.jp

† Current Address (K.T.W.): Central Region Agricultural Research Center, National Agriculture and Food Research Organization (NARO), Tsukuba, Kanondai 2-1-18, Ibaraki 305-8666, Japan.

‡ Current Address (A.Z.O.): Natural Science Research Unit, Tokyo Gakugei University, Koganei, Tokyo 184-8501, Japan.

Abstract: The aims of this study were to boost growth attributes, yield, and nutrient uptake of rice in paddy fields using a combination of *Bacillus pumilus* strain TUAT-1 biofertilizer and different nitrogen (N) application rates in nursery boxes. *Bacillus pumilus* strain TUAT-1 was applied as an inoculant biofertilizer in conjunction with different rates of N fertilizer to rice seedlings in a nursery. Plant growth and yield parameters were evaluated at two stages: in 21-day-old nursery seedlings and in mature rice plants growing in a paddy field. Inoculation with TUAT-1 significantly increased the seedling growth and root morphology of 21-day-old nursery seedlings. There was a marked increase in chlorophyll content, plant height, number of tillers, and tiller biomass of rice plants with the use of TUAT-1 and N fertilizers alone, and their combinations, at the maximum tillering stage in the field. The combination of TUAT-1 and 100% N (farmer recommended rate of N) resulted in the greatest tiller number and biomass at the maximum tillering stage, and positively affected other growth attributes and yield. The growth and yield were similar in the TUAT-1 + 50% N and 100% N (uninoculated) treatments, because TUAT-1 promoted root development, which increased nutrient uptake from the soil. These results suggest that the *B. pumilus* strain TUAT-1 has a potential to enhance the nutritional uptake of rice by promoting the growth and development of roots.

Keywords: biofertilizer; *Bacillus pumilus*; growth promotion; N fertilizer; rice; yield

1. Introduction

Rice (*Oryza sativa* L.) is considered as one of the world's most important staple foods and is the key to food security, especially under the threats of climate change in the coming decades [1]. The global rice cultivation area in 2015/2016 approached 158.8 million hectares, and total global rice production amounted to 711.24 million metric tons [2]. Nitrogen (N) fertilizers are used extensively in rice cultivation to meet the growth demands of the crop. However, excessive use of chemical fertilizers, in recent decades, has led to soil toxicity by contamination with toxic heavy metals, which adversely affect the health of rice plants [3]. Inoculating rice plants with plant growth promoting rhizobacteria (PGPR) can significantly enhance rice production, thus reducing the need for N fertilizers and contributing to sustainable rice production and reduced environmental problems [4]. Therefore, the use of biological fertilizers for reducing chemical fertilizers is one of the most effective steps towards sustainable agriculture [5].

Several root-colonizing *Bacillus* species have been shown to enhance plant growth [6]. It is likely that the growth-promoting effects of various PGPR are due to bacterial production of plant growth regulators, such as auxins, gibberellins, and cytokinins [7,8]. *Bacillus pumilus* strain TUAT-1 (hereafter referred to as TUAT-1) has been shown to increase shoot and root growth in rice, mustard, radish, and komatsuna [9,10], mainly due to its effects to promote nutrient uptake by plant roots. Inoculation of plants with TUAT-1 biofertilizer at sowing and transplanting resulted in significant changes in plant biomass, nutrient uptake, tissue N content, tiller number, root length, and number of roots in the forage rice "leaf star" [10]. Despite the large body of experimental evidence on the growth-promoting effects of TUAT-1, our knowledge of the conditions required for a consistent positive interaction between the bacteria and the plant (i.e., increased grain yield), in field conditions, is limited. Generally, bacterial inoculation improves plant growth and rice yield, but not uniformly. The yield response to inoculation is more pronounced in the presence of moderate levels of N fertilizer [11]. Fertilization management can also affect the community structures of plant-associated bacteria [12]. There is an ongoing debate regarding the impacts of fertilization strategies on the effects of PGPR [13].

Several studies have focused on interactions between PGPR and N fertilizers. However, the optimization of inoculant biofertilizer in conjunction with chemical N fertilizer for rice from the nursery tray to field stage has not been reported. We hypothesized that the combination of TUAT-1 biofertilizer and nitrogen fertilizer (N) applied to rice seedlings in nursery trays will promote plant growth and ensure a consistent positive interaction between the bacterium and the plant in terms of grain yield under field conditions. The specific objectives of this study were as follows: (1) to determine the effect of nursery application of TUAT-1 biofertilizer and different N fertilizer rates on growth and root morphology of rice seedlings; and (2) to evaluate the effect these treatments on the growth, yield, and N content of the mature rice plants after transplanting into field conditions.

2. Materials and Methods

The TUAT-1 and N fertilizer treatments were applied to rice seedlings growing in nursery trays in a greenhouse. The treated rice plants were transplanted into a paddy field and grown to maturity.

2.1. Soil Preparation and Chemical Analysis

The experimental field site was the Fuchu Honmachi paddy field (35°41′ N, 139°29′ E, 59 MSL) of Tokyo University of Agriculture and Technology, Tokyo, Japan. Soil samples were collected from four points in each plot at a depth of 0–15 cm, before the seedlings were transplanted. The samples were air dried, ground, and passed through a 2-mm sieve. The soil physicochemical properties were measured using conventional methods [14,15], and are summarized in Table 1.

Table 1. Physiochemical properties of paddy soil (0–15 cm) in experimental plots.

Soil Physiochemical Property	Value
pH (H_2O)	6.25
Total carbon content (%)	5.25
Total nitrogen content (%)	0.33
Cation-exchange capacity (cmolc/kg)	20.55
Sand (%)	40.12
Silt (%)	32.27
Clay (%)	27.61

2.2. Nursery Preparation

One of the main cultivars of paddy rice (*Oryza sativa* L.) in Japan, Koshihikari, was used in this study. Seeds (100 g nursery tray^{-1}) were surface-sterilized by immersion in 1% sodium hypochlorite solution for 2–3 min, and in 80% ethanol for 3–4 min, before being washed thoroughly with distilled water. The seeds were then allowed to imbibe in tap water for 72 h, and incubated for 12 h to hasten germination. Pre-germinated seeds were uniformly broadcast in each nursery tray (30 cm × 60 cm) on 22 April 2015. Each nursery tray contained 3.2 kg commercial rice nursery soil (Shinano Soil, Shinano Baiyoudo Co., Ltd., Nagano, Japan). Before sowing the pre-germinated seeds, biofertilizer with or without TUAT-1 was mixed thoroughly into the soil at a rate of 5 g granular biofertilizer per 100 g soil (160 g in 3.2 kg soil). The density of TUAT-1 *Bacillus* cells in biofertilizer was approximately 1.2×10^7 colony forming units (cfu) g^{-1}. The experiment included eight different treatments combining N fertilizer at various application rates with or without TUAT1; (1) 0% N, (2) TUAT-1 + 0% N, (3) 50% N, (4) TUAT-1 + 50% N, (5) 100% N, (6) TUAT-1 + 100% N, (7) 150% N, and (8) TUAT-1 + 150% N. The N fertilizer was solid form of ammonium sulfate, $(NH_4)_2SO_4$. The 100% N rate consisted of 2.64 g $(NH_4)_2SO_4$ per nursery tray, which is the application rate recommended for rice farmers in Japan.

The TUAT-1 isolate was grown as a liquid culture in trypticase soy broth (TSB) (Becton Dickinson, Sparks, MD, USA). When the culture density reached 10^7 cfu mL^{-1}, overnight cultures were then centrifuged at 10,000 rpm for 10 min at 6 °C, and were then washed twice in sterile Milli-Q water and diluted to an optical density of 0.4 at 600 nm, corresponding to approximately 10^7 cells mL^{-1}. Five hundred milliliters were applied to each nursery seedling, once per week for 3 successive weeks. The treatments were arranged in a completely randomized design (CRD) and were replicated three times. At 21 days after sowing (DAS), 30 seedlings were randomly selected from three positions in each nursery tray, and the following data were recorded: chlorophyll content (*SPAD* value), shoot length, shoot biomass, total root length, root number, root surface area, root biomass, and total biomass.

Upon harvest, root systems were removed gently from the nursery tray and stored in 70% ethanol until root parameters were measured. At the time of measurement, the roots were washed gently with deionized water and the root surface area and total root length were measured with an image analyzer (Win-Rhizo REG V 2004 b; Regent Inc., Quebec, Canada).

2.3. Transplanting of Rice Plants into the Paddy Field

The field experiment was conducted at the experimental paddy field of the Tokyo University of Agriculture and Technology in Fuchu Honmachi, Tokyo, Japan. The experiment was conducted in a split-plot design with four replications. The N, phosphorus, and potassium fertilizers were applied as basal dressings at the rate of 80 ($(NH_4)_2SO_4$), 100 (P_2O_5), and 30 (K_2O) kg ha^{-1}, respectively. The roots of 21-day-old seedlings were reinoculated in a bacterial suspension (10^7 cfu mL^{-1}) or tap water (control) for 1 night before transplanting. The main plot treatment factor was presence/absence of TUAT-1, and the subplot treatment factor was N fertilization rate (0%, 50%, 100%, and 150% N). Each plot was about 52 m^2 (7.2 m × 7.2 m).

The 21-day-old seedlings were manually transplanted on 13 May 2015 into the paddy field at a planting density of 22.2 hills m^2. Each hill contained three seedlings with 30 cm row spacing and 15 cm intra-row spacing. The experiment was conducted in irrigated paddy conditions. At 45 days after transplanting (DAT) in the field, chlorophyll content (*SPAD* value), plant height, tiller number, and tiller biomass were measured at the maximum tillering stage.

Upon harvest, 120 DAT, the following data were recorded: number of panicles, panicle length, panicle weight, straw yield, aboveground biomass, and grain yield. Straw yield and grain yield were recorded from 44 hills (2 m^2) of each plot, which is reasonable for sampling in some area of Japan [16,17]. Grain yield was adjusted to 14% moisture content while straw yield was recorded on an oven-dry basis (80 °C). Nitrogen contents in the grain and straw samples were estimated colorimetrically after H_2SO_4–H_2O_2 wet digestion, as described by Mizuno and Minami [18].

2.4. Data Analysis

Analysis of variance was performed for all measurements with the CROP-STAT version 7.2 software (International Rice Research Institute, IRRI, Philippines). The statistical model included replication, TUAT-1, N rate, and the interaction between TUAT-1 treatment and N rate. The results were subjected to a two-way analysis of variance, and mean values were then compared by 5% level by Fisher (LSD) test ($p < 0.05$) when the F probability value was significant, using XLSTAT Version 2017 (Addinsoft, Paris, France). Data were reported as means ± the standard deviation (SD).

3. Results

3.1. Seedling Growth at Nursery Stage

Root morphology was significantly changed with TUAT-1 inoculation in each N treatment when compared with its uninoculated treatment (Table 2). Rice seedlings receiving TUAT-1 + 150% N showed the greatest root surface area and total length than that of these root parameters of uninoculated plants and other inoculated plants receiving 100% N, 50% N, and 0% N (Table 2). Treatments of TUAT-1 + 150% N and TUAT-1 + 100% N were on par with each other with respect to root surface area and it did not reach significant level between these two treatments. Root parameters increased by either of inoculation or N fertilizer application alone. The TUAT-1 inoculation and N fertilizer interactions were significant with respect to total root length.

Table 2. Effects of nitrogen (N) levels and TUAT-1 on chlorophyll content (*SPAD* value), shoot length plant-1 (SL), shoot biomass plant-1 (SB), root biomass plant-1 (RB), Total root length plant-1 (TRL) and root surface area plant-1 (RSA) (g) of 21 days old seedlings. Means in columns followed by the different letters are significantly different according to Least Significant difference (LSD) test ($p < 0.05$). p values indicate that the differences are statistically significant (2-way ANOVA, $p < 0.05$), ns = non-significant.

Treatments	*SPAD*	SL (cm)	SB (g)	RB (g)	TRL (cm)	RSA (cm^2)
0% N	18.6 ^{c}	12.1 ^{cd}	16.0 ^{b}	15.3 ^{d}	12.7 ^{f}	240.3 ^{d}
0% N + TUAT-1	19.6 ^{c}	13.6 ^{c}	21.6 ^{b}	20.6 ^{c}	19.7 ^{de}	335.7 ^{c}
50% N	18.1 ^{bc}	15.2 ^{ab}	27.0 ^{a}	18.3 ^{c}	17.1 ^{ef}	268.7 ^{cd}
50% N + TUAT-1	24.8 ^{cd}	15.0 ^{bc}	29.1 ^{a}	25.0 ^{b}	29.2 ^{bc}	466.9 ^{b}
100% N	21.4 ^{bc}	15.0 ^{ab}	25.3 ^{a}	19.6 ^{c}	18.7 ^{de}	315.7 ^{c}
100% N + TUAT-1	26.8 ^{bc}	15.2 ^{ab}	25.9 ^{a}	25.9 ^{b}	31.9 ^{b}	520.9 ^{ab}
150% N	25.4 ^{ab}	16.3 ^{a}	26.9 ^{a}	23.8 ^{b}	26.6 ^{cd}	405.3 ^{b}
150% N + TUAT-1	32.1 ^{a}	16.2 ^{a}	28.1 ^{a}	29.2 ^{a}	39.2 ^{a}	682.0 ^{a}
Analysis of variance				p value		
Nitrogen (N)	0.001	0.0001	0.0001	0.0001	0.0001	0.0001
TUAT-1 (T)	0.006	0.957	0.127	0.0001	0.0001	0.0001
N × T	0.439	0.512	0.387	0.76	0.074	0.127

TUAT-1 inoculation significantly enhanced seedling growth. Rice nursery seedlings receiving TUAT-1 + 150% N showed the highest chlorophyll content, shoot length, and shoot biomass (Table 2). The lowest growth parameters were recorded in uninoculated plants receiving 0% N. Chlorophyll content was not significantly different among the treatments with and without TUAT-1 inoculation (Figure 1A). At each N treatment, TUAT-1 inoculation showed a significant greater root biomass, as compared to those of their respective uninoculated plants. TUAT-1 inoculation did not enhance the shoot length and shoot biomass significantly over its respective uninoculated plants.

3.2. Growth at Maximum Tillering Stage

The effect of TUAT-1 on growth at maximum tillering stage was significant at an F probability level $p < 0.002$. Increase of growth parameters measured at maximum tillering stage were also significant, due to N fertilization, with an F probability level of $p < 0.004$. The interaction effect between TUAT-1 and N fertilizer was also significant for chlorophyll content, plant height, and tiller biomass (Table 3).

Table 3. Effects of N levels and TUAT-1 on chlorophyll content (*SPAD* value), plant height, tiller number (TN), and tiller biomass plant^{-1} (TB) at 45 days after transplanting (DAT). The significance (*) derived from two ways analysis of p value *, **, and ***, significant at $p \leq 0.05$, $p \leq 0.01$, and $p \leq 0.001$, respectively. Means in columns, followed by the different letters, are significantly different according to LSD test ($p < 0.05$). p values indicate that the differences are statistically significant (2-way ANOVA, $p < 0.05$), ns = non-significant.

Treatments	*SPAD*	Plant Height (cm)	TN	TB (g)
0% N	39.6 ^{d}	82.2 ^{c}	8.7 ^{c}	9.5 ^{d}
0% N + TUAT-1	40.2 ^{d}	86.7 ^{ab}	13.6 ^{ab}	17.0 ^{c}
50% N	40.0 ^{d}	82.8 ^{c}	11.2 ^{bc}	10.7 ^{d}
50% N + TUAT-1	41.4 ^{c}	86.8 ^{ab}	13.4 ^{ab}	18.4 ^{bc}
100% N	40.5 ^{c}	84.4 ^{bc}	12.5 ^{ab}	16.2 ^{bc}
100% N + TUAT-1	41.7 ^{b}	88.4 ^{a}	15.7 ^{a}	20.9 ^{a}
150% N	42.1 ^{a}	87.6 ^{a}	13.7 ^{ab}	18.6 ^{ab}
150% N + TUAT-1	42.3 ^{ab}	89.1 ^{a}	16.4 ^{a}	21.8 ^{a}
Analysis of variance		p value		
Nitrogen (N)	0.002	0.0001	0.004	0.0001
TUAT-1 (T)	0.001	0.001	0.002	0.0001
N × T	0.232	0.023	0.536	0.029

Under the field condition, chlorophyll content at maximum tillering stage was significant difference between the treatments with and without TUAT-1 inoculation at 50% N and 100% N. TUAT-1 inoculation also promoted plant height at 0%, 50%, and 100% N (Table 3). Besides plant height, tiller number and tiller biomass are also important parameters in terms of the vegetative growth of the plants, and these were also found to be increased in response to nursery application of TUAT-1 inoculation. The best combination of nursery seedling treatment with TUAT-1 and different N rates on rice tiller number was observed in TUAT-1 + 100% N. The combined application of TUAT-1 + 100% N also gave the best effect on dry matter accumulation (tiller biomass) by plants at maximum tillering stage followed by TUAT-1 + 50% N. The nursery application of TUAT-1 + 150% N did not further enhance the above plant growth parameters significantly over application of TUAT-1 + 100% N.

3.3. Yield and Yield Component Parameters

The yield and yield components showed similar trends to those of other growth parameters at the maximum tillering stage. Application of N to nursery seedlings, alone or in combination with TUAT-1, led to significant increases in the number of panicles, panicle length, panicle weight, straw yield, aboveground biomass, and grain yield, compared with uninoculated plants with 0% N (Figure 1). The values of yield and yield components were significantly higher in TUAT-1-inoculated plants than in uninoculated plants, with F probability levels range of p-0.002 to p-0.048. N fertilization also increased yield and yield components with F probability values of $p < 0.001$. Interaction between TUAT-1 and N levels was not observed for yield and yield component characters.

Nursery treatments of TUAT-1 + 100% N gave the best values on panicle length, number, and weight (Figure 1A–C). No further significant increase in the above parameters was found in the treatment of TUAT-1 + 150% N. In the 0% N treatments, panicle length and number were significantly greater in TUAT-1-inoculated plants than in uninoculated plants (Figure 1A,B). In addition, TUAT-1 led to significant increases in the panicle weight compared with uninoculated plants with 50% N treatment (Figure 1C).

The maximum estimated straw yield was in the TUAT-1 + 150% N treatment (11.16 ton ha^{-1}), followed by the 150% N treatment (9.72 ton ha^{-1}), the TUAT-1 + 100% N treatment (9.69 ton ha^{-1}), and then the TUAT-1 + 50% N treatment (9.18 ton ha^{-1}) (Figure 1D). In the 50% and 150% N treatments, the estimated straw yield was significantly higher in TUAT-1-inoculated plants than in the uninoculated plants. The aboveground biomass was also significantly higher in TUAT-1-inoculated plants with the 50% N treatment (Figure 1E).

The highest estimated grain yield was in the TUAT-1 + 100% N treatment (4.89 ton ha^{-1}) and the TUAT-1 + 150% N treatment (5.05 ton ha^{-1}) (Figure 1F). The grain yield of plants in the TUAT-1 + 50% N treatment (4.4 ton ha^{-1}) was statistically insignificant to that of plants in the 100% N treatment (4.6 ton ha^{-1}) or TUAT-1 + 100 % N.

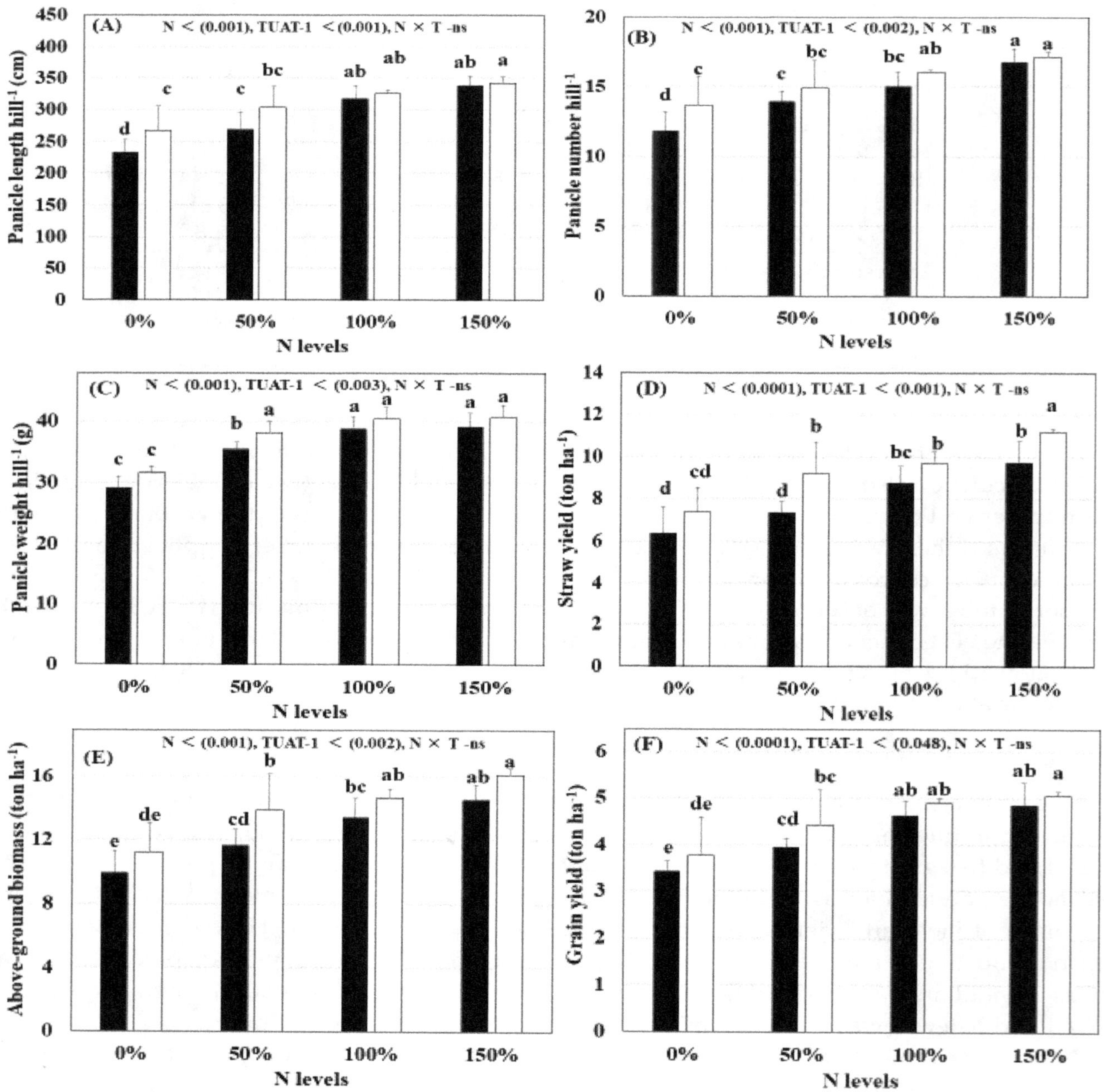

Figure 1. Effect of TUAT-1 biofertilizer and nitrogen fertilizer treatments at the seedling stage on yield and yield components of mature rice plants. (**A**) Panicle length hill^{-1}, (**B**) panicle number hill^{-1}, (**C**) panicle weight hill^{-1}, (**D**) straw yield, (**E**) aboveground biomass, and (**F**) grain yield. Black and white bars represent uninoculated and inoculated with TUAT-1. Error bars indicate standard deviation. Different letters indicate significant difference at 5% level by Fisher (LSD) test ($p < 0.05$). p values are shown in brackets. p values indicate that the differences are statistically significant (2-way ANOVA, $p < 0.05$), ns = non-significant.

3.4. Nitrogen Content (%) in Grain and Straw

The patterns of N content in rice grain and straw were affected by TUAT-1 inoculation and N application levels in the nursery (Figure 2A,B). In the 50% N treatment, grain N content was significantly higher in TUAT-1-inoculated plants than in uninoculated plants. However, TUAT-1 inoculation did not significantly increase grain N content in the other treatments. Nitrogen fertilization alone did not significantly promote N content in rice grain. TUAT-1 significantly increased the N content in straw in the 0% N treatments (Figure 2B).

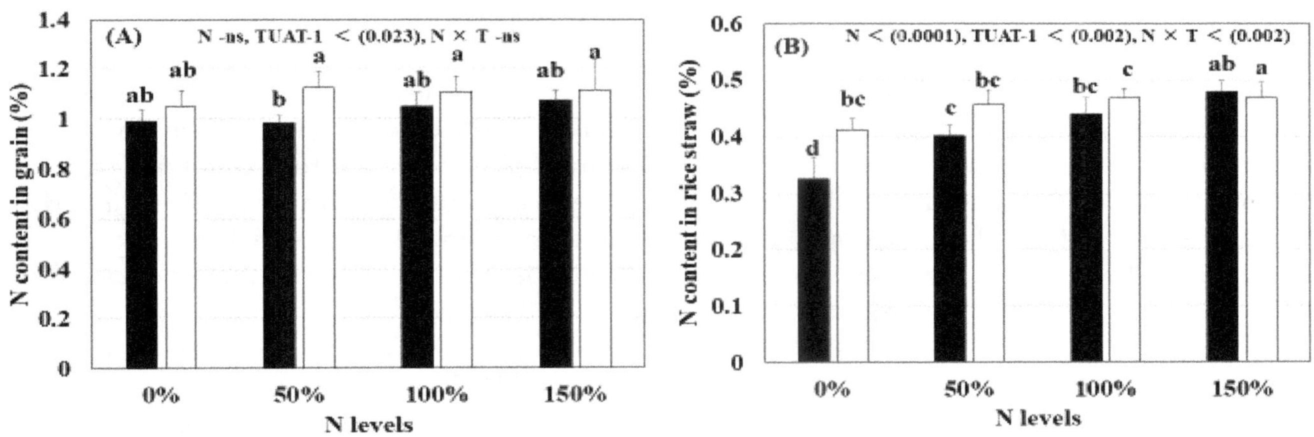

Figure 2. Effect of TUAT-1 biofertilizer and nitrogen fertilizer treatments at the seedling stage on N content (%) (**A**) in grain and (**B**) in straw of mature rice plants. Black and white bars represent uninoculated and inoculated with TUAT-1, respectively. Error bars indicate standard deviation. Different letters indicate significant difference at 5% level by Fisher (LSD) test ($p < 0.05$). p values are shown in brackets. p values indicate that the differences are statistically significant (2-way ANOVA, $p < 0.05$), ns = non-significant.

4. Discussion

Seedling growth is the most important growth stage of any crop, as it determines the amount of biomass generated. In rice, it is also important for tiller development [19]. In our study, nursery seedling growth was increased by TUAT-1 inoculation (Table 2). The values of root parameters (total root length and root surface area) were significantly higher for TUAT-1-inoculated plants than for uninoculated plants. A larger and stronger root system can increase seedling vigor, plant growth, and micronutrient status [20]. An increase in rice root biomass in response to PGPR inoculation was also reported by Souza et al. [4]. The root system plays an important role in plant productivity because roots take up essential nutrients from the soil [21]. Various PGPR have been shown to enhance root hair proliferation and deformation, increase root branching, promote seedling emergence, and increase leaf surface area, vigor, biomass, endogenous plant hormone levels, and uptake of minerals and water [22,23].

After 21 days in the nursery, in each N treatment, rice seedlings inoculated with TUAT-1 produced approximately 17% greater total biomass than did uninoculated seedlings (Supplementary Figure S1). Like other PGPR, TUAT-1 produces growth-promoting and growth-regulating substances to support root growth, allowing inoculated rice plants to absorb more nutrients to enhance their growth. A wide range of nutrients and signaling molecules are exuded from roots, directing plant–microbe interactions [24]. The associated microbes may affect root morphogenesis [25], and many studies suggest that the acceleration of plant growth by PGPR involves phytohormone modulation. Numerous studies have reported on Plant growth-promoting bacteria (PGBs), particularly on *Bacillus* spp. exerting a number of characteristics enabling mobilization of soil nutrients and synthesis of phytohormones, leading to plant growth promotion [26–28]. Further research is required to determine the exact mechanism by which TUAT-1 promotes rice growth.

The combination of TUAT-1 and different rates of N application in the nursery also have a significant effect on chlorophyll content, plant height, and tiller biomass at 45 DAT (Figure 2A). Promoting crop growth in the early stage is important for early and rapid tiller production. Herein, we suggested that TUAT-1 inoculation had the higher tiller biomass production as a result of vigorous early growth and early leaf expansion at tillering stage. Similar responses to fertilizer addition at the nursery stage were reported by Ros et al. [29] and Panda et al. [30]. Increased tiller number and tiller biomass have been reported for rice plants treated with PGPR [31].

Furthermore, in the other section of this study, the results of the present study also showed that the seedling growth parameters of root, such as total root surface area, total root length, and root biomass, increased significantly due to TUAT-1 inoculation, which can improve uptake of water and nutrition from soil. Wang et al. [32] also recently reported vigorous root growth and long and dense root hairs ensured efficient acquisition of macro- and micronutrients during early growth, and a high root length to shoot dry matter ratio favored high macronutrient concentrations in the shoots, which is assumed to be important for later plant development.

Herein, inoculating rice with TUAT-1 was also reported to improve the water and nutrition absorption capacity of the root system, and promote the absorption of N from soil during the early growth period [10]. Matsumura et al. [33] suggested that increased N absorption may promote tillering, increase the number of ears, and maintain photosynthetic activity during the growth period of the plant, so that crop yield is boosted. The increase in tiller number and tiller biomass by TUAT-1 may be due to the effects of *Bacillus* species to improve tolerance of plants to the drought and salinity stresses [34,35] and their water and nutrient uptake ability [10]. Furthermore, compared with untreated seedlings, the seedlings treated with TUAT-1 and N fertilizer with vigorous dense root systems and strong stems may have coped better with transplanting shock by rapidly developing new roots. It is likely that the vigorous dense root systems and strong stems contributed to the higher tiller production (biomass) in TUAT-1-inoculated plants than in uninoculated plants.

Nursery application of TUAT-1 with N fertilizer led to higher tiller number at the maximum tillering stage, which resulted in increased panicle number, panicle length, panicle weight, and yield (Figure 1). The higher panicle number resulted from increased number of effective tillers. Hence, increased panicle number resulted in increased panicle weight, leading to increased grain number per panicle and ultimately increased yield. The beneficial effects of PGPR on rice yield have been observed in both greenhouse and field conditions [36,37]. In our study, the combination of TUAT-1 and N fertilizer applied to rice plants at the nursery stage clearly increased the values of most growth parameters.

The greatest yield was shown in the TUAT-1 + 150% N and the TUAT-1 + 100% N treatments. The grain yield of plants in the TUAT-1 + 50% N treatment (4.4 ton ha^{-1}) was statistically insignificant to that of uninoculated plants in the 100% N treatment (4.6 ton ha^{-1}); that is, inoculation with TUAT-1 has a potential of reducing fertilizer use at the seedling stage by 50% without negatively affecting yield. This result is consistent with earlier studies by Okon and Labandera-Gonzalez [38] and Dobbelaere et al. [39], and beneficial effects of PGPB on crop yield were mainly observed under intermediate levels of fertilizer, rather than maximum or minimum fertilization. The similar rice grain yield in TUAT-1 + 50% N and 100% N may be ascribed to the significantly increased root to shoot growth at the seedling stage (Table 2), which can improve dry matter accumulation (tiller biomass) at maximum tillering stage (Table 3) by supplying a sufficient amount of nutrients, water, and phytohormones to shoots and, subsequently, ensure an increase in rice productivity [10]. We suggested that the nursery applications of N fertilizer at appropriate levels with TUAT-1 biofertilizer in this field study may have masked our ability to observe significant growth impacts on rice grain yields. Previous investigations suggested that strategies combining both reduced rates of agriculture fertilizers and biofertilizers can benefit plant development and nutrient uptake [40,41].

Hence, on the basis of our results, we suggested that the inoculation of TUAT-1 resulted in improving the growth of the rice plants and grain yield, and its inoculation may be applied to spare the use of chemical N fertilization. However, biofertilizer performance may be specific to each situation, as its effectiveness depends on factors like plant species, soil type, and environmental conditions [42]. Since the results reported here for enhancing of growth and yield of rice in the field experiment, as influenced by the application of a combination of TUAT-1 and N fertilizer at the nursery stage, was conducted only for a one year experiment, we suggest that further studies should evaluate the performance of TUAT-1 in different field conditions/locations with different rice cultivars and environmental field conditions of specific issues for a given year. Such studies will determine whether TUAT-1 can substantially decrease the use of chemical N fertilizers in rice cultivation, which is an important economic and environmental goal.

The increases in N contents in grain and straw as a result of N fertilization were consistent with the results of a previous study [43]. Growth promotion by inoculation with the TUAT-1 was accompanied by increased N levels in all plant tissues tested in forage rice [10]. TUAT-1 inoculation significantly promoted straw N content (%) at 0% N and grain N content (%) at 50% N than those of their respective uninoculated plants (Figure 2). It has been suggested that PGPR improves mineral nutrition, especially under low-N input conditions [19]. Hence, it is also postulated that growth promotion effects of TUAT-1 with a moderate amount of N application yielded the result, at least partly, of an enhanced N uptake efficiency. In the other studies, the enhancement of N uptake by plants inoculated with the PGPR strains might be through associative N fixation and phosphorus solubilization [44]. However, such evidence is lacking for TUAT-1. These observations could indicate that growth promotion mechanisms other than nitrogen fixation, such as phytohormone production, improved nutrient uptake balance. In addition, the higher nutrient uptake might be attributed to morphological changes in rice roots, such as increased root number, length, and thickness [45].

5. Conclusions

The increase in grain yield resulting from application of TUAT-1 combined with N fertilizer to rice seedlings at the nursery stage could be related to the increased size of the root system at the early growth stage, which increased nutrient uptake to promote tiller growth (biomass), and yield. As observed the straw and yield were similar in the TUAT-1 + 50% N and 100% N (un-inoculated) treatments, we conclude that TUAT-1 biofertilizer should be used with N fertilizer at appropriate levels to maximize benefits in terms of saving fertilizer and improving yield.

Author Contributions: The individual contribution and responsibilities of the authors were as follows: K.T.W.: experiment, sample analysis, data analysis and article writing. T.Y.: research idea and design, and supervision of data collection and analysis. N.O.-O.: research idea and design and A.Z.O.: conceptualization, research idea and data analysis.

Acknowledgments: This study was supported by MAFF Science and Technology Research Promotion Project for Agriculture, forestry and fisheries and food industries (26073C).

References

1. Ali, J.; Aslam, U.M.; Tariq, R.; Murugaiyan, V.; Schnable, P.S.; Li, D.; Marfori-Nazarea, M.C.; Hernandez, J.E.; Arif, M.; Xu, J.; et al. Exploiting the Genomic Diversity of Rice (*Oryza sativa* L.): SNP-Typing in 11 Early-Backcross Introgression-Breeding Populations. *Front. Plant Sci.* **2018**, *9*, 849. [CrossRef] [PubMed]
2. World Rice Acreage from 2008/2009 to 2015/2016 (In Million Hectares): Statista; 2017. Available online: https://www.statista.com/statistics/271969/world-rice-acreage-since-2008/ (accessed on 21 August 2017).
3. Habibah, J.; Le, P.T.; Khairiah, J.; Ahmad, M.R.; Fouzi, B.A.; Ismail, B.S. Speciation of heavy metals in paddy soils from selected areas in Kedah and Penang, Malaysia. *Afri. J. Biotechnol.* **2011**, *4*, 13505–13513. [CrossRef]
4. Souza, R.; Beneduzi, A.; Ambrosini, A.; Costa, P.B.; Meyer, J.; Vargas, L.K.; Schoenfeld, R.; Passaglia, L.M.P. The effect of plant growth-promoting rhizobacteria on the growth of rice (*Oryza sativa* L.) cropped in southern Brazilian fields. *Plant Soil.* **2013**, *366*, 585–603. [CrossRef]
5. Shariatmadari, Z.; Riahi, H.; Seyed-Hashtroudi, M.; Ghassempour, A.; Aghashariatmadary, Z. Plant growth promoting cyanobacteria and their distribution in terrestrial habitats of Iran. *Soil Sci. Plant Nutri.* **2013**, *59*, 535–547. [CrossRef]
6. Idris, E.E.S.; Iglesias, D.J.; Talon, M.; Borriss, R. Tryptophan-dependent production of Indole 3-Acetic Acid (IAA) affects level of plant growth promotion by *Bacillus amyloliquefaciens* FZB42. *Mol. Plant Microbe Interact.* **2007**, *20*, 619–626. [CrossRef] [PubMed]
7. Bottini, R.; Cassan, F.; Picolli, P. Gibberellin production by bacteria and its involvement in plant growth promotion and yield increase. *Appl. Microbiol. Biotechnol.* **2004**, *65*, 497–503. [CrossRef] [PubMed]
8. Bloemberg, G.V.; Lugtenberg, B.F.J. Molecular Basis of plant growth promotion and biocontrol by rhizobacteria. *Curr. Opin. Plant Biol.* **2001**, *4*, 343–350. [CrossRef]
9. Aung, H.P.; Salem, D.; Oo, A.Z.; Aye, Y.S.; Yokoyama, T.; Suzuki, S.; Sekimoto, H.; Bellingrath-Kimura, S.D. Growth and ^{137}Cs uptake of four Brassica species influenced by inoculation with a plant growth-promoting rhizobacterium *Bacillus pumilus* in three contaminated farmlands in Fukushima prefecture, Japan. *Sci. Total Environ.* **2015**, *521*, 261–269. [CrossRef] [PubMed]
10. Torii, A. Analysis of field factors resulting frutuations of yield and nutritional uptakes of forage rice Leaf Star with inoculation of an endophytic nitrogen fixing bacteria TUAT-1. Master Thesis, Tokyo University of Agriculture and Technology, Tokyo, Japan, March 2012.
11. Rao, V.R.; Jena, P.K.; Adhya, T.K. Inoculation of rice with nitrogen-fixing bacteria problems and perspectives. *Biol. Fertil. Soils.* **1987**, *4*, 21–26.
12. Ikeda, S.; Okubo, T.; Kaneko, T.; Inaba, S.; Maekawa, T.; Eda, S.; Sato, S.; Tabata, S.; Mitsui, H.; Minamisawa, K. Community shifts of soybean stem-associated bacteria responding to different nodulation phenotypes and N levels. *ISME J.* **2010**, *4*, 315–326. [CrossRef] [PubMed]
13. Sasaki, K.; Ikeda, S.; Eda, S.; Mitsui, H.; Hanzawa, E.; Kisara, C.; Kazama, Y.; Kushida, A.; Shinano, T.; Minamisawa, K.; et al. Impact of plant genotype and nitrogen level on rice growth response to inoculation with *Azospirillum* sp. Strain B510 under paddy field conditions. *Soil Sci. Plant Nutri.* **2010**, *56*, 636–644. [CrossRef]
14. Blakemore, L.C.; Searle, P.L.; Daly, B.K. *Science Report: Methods for Chemical Analysis of Soils*; New Zealand Soil Bureau: Australia, 1977.
15. Klute, A. Methods of soil analysis, Part 1. In *Physical and Mineralogical Methods*, 2nd ed.; Soil Science Society of America, Inc.: Madison, WI, USA, 1990.
16. Yoshinaga, S.; Heinai, H.; Ohsumi, A.; Furuhata, M.; Ishimaru, T. Characteristics of growth and quality, and factors contributing to high yield in newly developed rice variety 'Akidawara'. *Plant Prod. Sci.* **2018**, *21*, 186–192. [CrossRef]
17. Saito, K.; Fukuta, Y.; Yanagihara, S.; Ahouanton, K.; Sokei, Y. Beyond NERICA: Identifying high-yielding rice varieties adapted to rainfed upland conditions in Benin and their plant Characteristics. *Jpn. J. Trop. Agric.* **2014**, *58*, 51–57.
18. Mizuno, N.; Minami, M. Rapid decomposition method for determination of N, K, Mg, Ca, Fe and Mn in agricultural plants with sulfuric acid and hydrogen peroxide. *Jpn. Soil Sci. Plant Nutr.* **1980**, *51*, 418–420. (In Japanese)
19. Sharma, A.; Shankhdhar, D.; Sharma, A.; Shankhdhar, S.C. Growth promotion of the rice genotypes by PGPRs isolated from rice rhizosphere. *J. Soil Sci. Plant Nutri.* **2014**, *14*, 505–517. [CrossRef]

20. Lynch, J.P.; Lynch, A.F.; Jonathan, P. Roots of the second green revolution. *Aust. J. Bot.* **2007**, *55*, 493–512. [CrossRef]
21. Shaharoona, B.; Naveed, M.; Arshad, M.; Zahir, Z.A. Fertilizer-dependent efficiency of Pseudomonads for improving growth, yield, and nutrient use efficiency of wheat (*Triticum aestivum* L.). *Appl. Microbiol. Biotechnol.* **2008**, *79*, 147–155. [CrossRef] [PubMed]
22. Spaepen, S.; Vanderleyden, J.; Remans, R. Indole-3-acetic acid in microbial and microorganism-plant signaling. *FEMS Microbiol Rev.* **2007**, *31*, 425–448. [CrossRef] [PubMed]
23. Podile, A.R.; Kishore, G.K. Plant growth promoting rhizobacteria. In *Plant Associated Bacteria*; Gnanamanickam, S.S., Ed.; Springer: Amsterdam, The Netherlands, 2006; pp. 195–230. [CrossRef]
24. Badri, D.V.; Vivanco, J.M. Regulation and function of root exudates. *Plant Cell Environ.* **2009**, *32*, 666–681. [CrossRef] [PubMed]
25. Persello-Cartieaux, F.; Naussaume, L.; Robaglia, C. Tales from the underground: Molecular plant-rhizobacteria interactions. *Plant Cell Environ.* **2003**, *26*, 189–199. [CrossRef]
26. Niu, D.D.; Liu, H.X.; Jiang, C.H.; Wang, Y.P.; Wang, Q.Y.; Jin, H.L.; Guo, J.H. The plant growth-promoting rhizobacterium *Bacillus cereus* AR156 induces systemic resistance in *Arabidopsis thaliana* by simultaneously activating salicylate-and jasmonate/ethylene-dependent signaling pathways. *Mol. Plant Microbe.* **2011**, *24*, 533–542. [CrossRef] [PubMed]
27. Wahyudi, A.T.; Astuti, R.P.; Widyawati, A.; Meryandini, A.; Nawangsih, A.A. Characterization of *Bacillus* spp. strains isolated from rhizosphere of soybean plants for their use as potential plant growth for promoting Rhizobacteria. *J. Microbiol. Antimicrob.* **2011**, *2*, 406–417.
28. Ortíz-Castro, R.; Valencia-Cantero, E.; López-Bucio, J. Plant growth promotion by *Bacillus megaterium* involves cytokinin signaling. *Plant Signal Behav.* **2008**, *3*, 263–265. [CrossRef] [PubMed]
29. Ros, C.; White, P.F.; Bell, R.W. Nursery Fertilizer Application Increases Rice Growth and Yield in Rainfed Lowlands with or without Post-Transplanting Crop Stress. *Am. J. Plant Sci.* **2015**, *6*, 2878–2892. [CrossRef]
30. Panda, M.M.; Reddy, M.D.; Sharma, A.R. Yield performance of rainfed lowland rice as affected by nursery fertilizer under conditions of intermediate deepwater (15–50 cm) and flash floods. *Plant Soil.* **1991**, *132*, 65–71. [CrossRef]
31. Kumar, K.V.K.; Yellareddygari, S.K.; Reddy, M.S.; Kloepper, J.W.; Lawrence, K.S.; Zhou, X.G.; Groth, D.E.; Krishnam Raju, S.K.; Miller, M.E. Efficacy of *Bacillus subtilis* MBI 600 against sheath blight caused by *Rhizoctonia solani* and on growth and yield of rice. *Rice Sci.* **2012**, *9*, 55–63. [CrossRef]
32. Wang, Y.; Thorup-Kristensen, K.; Jensen, L.S.; Magid, J. Vigorous Root Growth Is a Better Indicator of Early Nutrient Uptake than Root Hair Traits in Spring Wheat Grown under Low Fertility. *Front. Plant Sci.* **2016**, *7*, 865. [CrossRef] [PubMed]
33. Matsumura, S.; Ban, T.; Kanda, S.; Win, A.T.; Toyota, K. Biomass production and nutrient cycling. In *Research Approaches to Sustainable Biomass Systems*; Tojo, S., Hirasawa, T., Eds.; Elsevier: Oxford, UK, 2013; pp. 290–294.
34. Khan, A.; Sirajuddin, M.; Zhao, X.Q.; Javed, T.; Khan, K.S.; Bano, A.; Shen, R.F.; Masood, S. *Bacillus pumilus* enhances tolerance in rice (*Oryza sativa* L.) to combined stresses of NaCl and high boron due to limited uptake of Na^+. *Environ. Experi. Bot.* **2016**, *124*, 120–129. [CrossRef]
35. Armada, E.; Azcón, R.; López-Castillo, O.M.; Calvo-Polanco, M.; Ruiz-Lozano, J.M. Autochthonous arbuscular mycorrhizal fungi and *Bacillus thuringiensis* from a degraded Mediterranean area can be used to improve physiological traits and performance of a plant of agronomic interest under drought conditions. *Plant Physiol. Biochem.* **2015**, *90*, 64–74. [CrossRef] [PubMed]
36. Yadav, J.; Verma, J.P.; Jaiswal, D.K.; Kumar, A. Evaluation of PGPR and different concentration of phosphorus level on plant growth, yield and nutrient content of rice (*Oryza sativa*). *Ecol. Eng.* **2014**, *62*, 123–128.
37. Mishra, R.P.N.; Singh, R.K.; Jaiswal, H.K.; Kumar, V.; Maurya, S. Rhizobium-mediated induction of phenolics and plant growth promotion in rice (*Oryza sativa* L.). *Curr. Microbiol.* **2006**, *52*, 383–389. [CrossRef] [PubMed]
38. Okon, Y.; Labandera-Gonzalez, C.A. Agronomic applications of Azospirillum—An evaluation of 20 years worldwide field inoculation. *Soil Biol. Biochem.* **1994**, *26*, 1591–1601. [CrossRef]
39. Dobbelaere, S.; Croonenborghs, A.; Thys, A.; Ptacek, D.; Vanderleyden, J.; Dutto, P.; Labandera-Gonzalez, C.; Caballero-Mellado, J.; Aguirre, J.F.; Kapulnik, Y.; et al. Responses of agronomically important crops to inoculation with Azospirillum. *Aust. J. Plant Physiol.* **2001**, *28*, 871–879. [CrossRef]
40. Adesemoye, A.O.; Torbert, H.A.; Kloepper, J.W. Increased plant uptake of nitrogen from ^{15}N-depleted fertilizer using plant growth promoting rhizobacteria. *Appl. Soil Ecol.* **2010**, *46*, 54–58. [CrossRef]

41. Duarah, I.; Deka, M.; Saikia, N.; Deka, B.H.P. Phosphate solubilizers enhance NPK fertilizer use efficiency in rice and legume cultivation. *3 Biotech* **2011**, *1*, 227–238. [CrossRef] [PubMed]
42. Adesemoye, A.O.; Egamberdieva, D. Beneficial effects of plant growth-promoting rhizobacteria on improved crop production: Prospects for developing economies. In *Bacteria in Agrobiology: Crop Productivity*; Maheshwari, D.K., Saraf, M., Aeron, A., Eds.; Springer: Berlin, Germany, 2013; pp. 45–63. [CrossRef]
43. Liu, X.; Wang, H.; Zhou, J.; Hu, F.; Zhu, D.; Chen, Z.; Liu, Y. Effect of N fertilization pattern on rice yield, N use efficiency and fertilizer-N fate in the Yangtze river Basin, China. *PLoS ONE* **2016**, *11*, e0166002. [CrossRef] [PubMed]
44. Çakmakçi, R.; Dönmez, F.; Aydın, A.; Şahin, F. Growth promotion of plants by plant growth-promoting rhizobacteria under greenhouse and two different field soil conditions. *Soil Biol. Biochem.* **2006**, *38*, 1482–1487. [CrossRef]
45. Biswas, J.C.; Ladha, J.K.; Dazzo, F.B.; Yanni, Y.G.; Rolfe, B.G. Rhizobial inoculation influences seedling vigor and yield of rice. *Agron. J.* **2000**, *92*, 880–886. [CrossRef]

7

Soil Properties for Predicting Soil Mineral Nitrogen Dynamics Throughout a Wheat Growing Cycle in Calcareous Soils

Marta Aranguren [1,*], Ana Aizpurua [1], Ander Castellón [1], Gerardo Besga [1] and Nerea Villar [2]

1 NEIKER-Basque Institute for Agricultural Research and Development, Berreaga, 1, 48160 Derio, Biscay, Spain; aizpurua@neiker.eus (A.A.); acastellon@neiker.eus (A.C.); gbesga@neiker.eus (G.B.)

2 Department of Plant Biology and Ecology, The University of the Basque Country, Barrio Sarriena s/n, 48940 Leioa, Biscay, Spain; nereavillar01@gmail.com

* Correspondence: maranguren@neiker.eus

Abstract: A better understanding of the capacity of soils to supply nitrogen (N) to wheat can enhance fertilizer recommendations. The aim of this study was to assess the soil mineral N (N_{min}) dynamics throughout the wheat growing season in crucial stages for the plant yield and grain protein content (GPC). To this aim, we evaluated the utility of different soil properties analyzed before sowing: (i) commonly used soil physicochemical properties, (ii) potentially mineralizable N or N_o (aerobic incubation), and (iii) different extraction methods for estimating N_o. A greenhouse experiment was established using samples from 16 field soils from northern Spain. Wheat N uptake and soil N_{min} concentrations were determined at following growing stages (GS): sowing, GS30, GS37, GS60, harvest, post-harvest, and pre-sowing. Pearson's correlation analysis of the soil properties, aerobic incubations and chemical extractions with the soil N_{min} dynamics and N uptake, yield and GPC was performed. In addition, correlations were performed between N_{min} and the N uptake, yield, and GPC. The dynamics of soil N_{min} throughout the cropping season were variable, and thus, the crop N necessities were variable. The soil N_{min} values in the early wheat growth stages were well correlated with the yield, and in the late stages, they were well correlated with GPC. N_0 was correlated with the late N uptake and GPC. However, the chemical methods that avoid the long periods required for N_0 determinations were not correlated with the N uptake in the late wheat growth stages or GPC. Conversely, clay was positively correlated with the late N_{min} values and GPC. Chemical methods were unable to estimate the available soil N in the later stages of the growing cycle. Consequently, as incubation methods are too laborious for their widespread use, further research must be conducted.

Keywords: soil N supply; soil N mineralization; N fertilization; potentially mineralizable N; humid Mediterranean climate

---

1. Introduction

Few agroecosystems supply enough nitrogen (N) to sustain satisfactory crop production without fertilizers. Throughout agricultural history, agriculturists have attempted to maintain fertility levels in the soil, depending on biologically fixed N, through the application of organic amendments and the decomposition of soil organic matter (SOM) to provide N to crops. In cereal cropping systems, N is one of the most important elements controlling crop development [1]. Thus, to assure that the potential yield is achieved each year, N is frequently applied in excessive amounts without determining the appropriate N fertilization rate, which usually leads to N losses. To comply with economic and ecological regulations in recent years, concerns about the need for improving nitrogen use efficiency

(NUE) in cereal production have increased. The N fertilizer demand is dependent on the plant available N supplied by soils and the potential yield, which varies from year to year. Available soil resources should be taken into account for the determination of appropriate N fertilization rates to enhance the efficiency of agricultural systems and ecosystem health.

It is necessary to provide better insight into the capacity of soils to provide N to crops (soil N supply) to understand the factors that control N mineralization in soils and therefore improve N fertilization recommendations for cereal [2]. In the field, different climatic and agronomic parameters affect wheat yield production, and among them, the contribution of soil N dynamics is very relevant. Nitrogen cycling in the soil–plant system is very complex and involves interactions between soil and plant factors. Only a small portion of N is biologically active, serving as a substrate for N mineralization [3]. To determine the rates of N fertilizer application, it is necessary to take into account the inorganic N of the soil and the organic N mineralized during crop growth [4,5].

The method used most in Western Europe for N fertilization application in cereals is the N_{min} method, which is based on the measured amount of soil mineral N in the main rooting depth before N fertilizer application at the beginning of the rapid period of crop growth. The calculation of the N fertilizer recommended rate is made using the predicted N demand for the target yield minus the measured soil N_{min} value. However, with this method, the N that has been mineralized during the remainder of the growing cycle is not taken into account. In addition, taking soil samples in a narrow period of time and analysing the mineral N in each individual field and each season is not practical. Therefore, with the aim of measuring the N supply capacity of the soil during the entire growing season, the potentially mineralizable N (N_o) or bioavailable N is measured [6]. The standard method for measuring N_o was defined by Stanford and Smith [7], who developed a method based on long-term aerobic incubations, in which soil was maintained under optimum conditions (35 °C and field capacity). However, this method is impractical for routine laboratory analysis due to the long incubation periods required (32 weeks). With the purpose of avoiding the long period required, several chemical methods have been developed to estimate N_o, such as extractions with different saline solutions. Different methodologies have been used to build a global strategy with the aim of estimating crop N availability [4,8–11]. However, no one method has yet obtained general approval [12]. In a previous article, in the area where this study was carried out (Araba, Basque Country, northern Spain), with calcareous soils and under humid Mediterranean conditions, Villar et al. [13] determined that the most appropriate laboratory technique to estimate the amount of available N that soils are able to provide to wheat was hot KCl extraction.

In Araba (Basque Country, northern Spain), traditionally, two applications of N fertilizer are supplied at following growing stages (GS): GS21 (beginning of tillering) and GS30 (stem elongation) for wheat, according to the Zadoks scale [14]. As soils differ in their composition and mineralization patterns, the N rate applied in these two main periods should be specific with respect to the available N[15]. Furthermore, as in many other wheat-producing countries, the demand for high grain protein content (GPC) has also increased. Therefore, it is essential to estimate the amount of N mineralized from SOM to adjust the rate of N fertilizer required to optimize crop yield and quality, reducing the negative impacts of excessive N on the environment.

The main objective of this study was to assess the soil N_{min} dynamics throughout the wheat growing season at crucial stages for plant yield and GPC. To this aim, we analyzed the utility of different characteristics of the soil before sowing: (i) the commonly used routine soil physicochemical analyses (SOM, Ntot, pH, $CaCO_3$ and texture), (ii) potentially mineralizable N (N_o), analyzed using aerobic incubation, and (iii) different extraction methods to estimate N_o.

2. Materials and Methods

2.1. Experimental Setup

A greenhouse experiment was established in Derio (Bizkaia, Basque Country, Spain) at NEIKER-tecnalia experimental facilities using 16 field soils collected from Villanañe, Soportilla, Lantaron, Arangiz, Betolaza, Gauna, and Tuesta (Araba, Basque Country, Spain) from the 0–30 cm layers (Table 1). No organic amendments had been applied to the studied soils for several years before sample collection or N fertilization in the previous months. Moreover, no leguminous crops were grown in the fields preceding soil collection. Soils were air-dried and sieved though a 2 mm mesh and poured into pots (height of 30 cm, and diameter of 22 cm). Before sowing, 300 mL of a nutrient solution without N [10] was added to the soil to ensure that there were no nutrient limitations to the plants. Twenty seeds of soft red winter wheat (*Triticum aestivum* L., var Soissons) per pot were sown (30 May 2011), and after germination, the number of plants was reduced to 14 seedlings to simulate the common sowing dose of 240 kg ha^{-1}. There were three replicates per soil that were distributed in a completely randomized experimental design. Pots were kept at field capacity during the whole experiment. The experiment was extensively described in a previous article [13].

Table 1. Location, and physical and chemical characteristics, of the soils from Araba (Basque Country, northern Spain) used in the greenhouse experiment.

Soil	Location	Soil Texture	Sand [a] %	Silt [a] %	Clay [a] %	SOM [b] %	Ntot [c] %	pH [d] %	$CaCO_3$ [e] %
1	Villanañe	Clay-loam	30.5	37.1	32.4	1.8	0.12	8.3	29.0
2	Soportilla	Loam	35.9	39.3	24.8	1.0	0.07	8.5	29.4
3	Lantaron	Loam	31.7	43.0	25.4	1.3	0.09	8.4	56.5
4	Arangiz	Silty-loam	19.4	55.9	24.6	1.6	0.12	8.3	56.3
5	Arangiz	Clay-loam	26.1	45.6	28.3	2.0	0.16	8.3	21.7
6	Arangiz	Silty-clay-loam	16.2	56.9	27.0	2.0	0.13	8.3	53.0
7	Betolaza	Silty-clay-loam	12.1	58.6	29.3	3.1	0.24	8.3	29.3
8	Gauna	Sandy-clay-loam	47.4	24.9	27.6	2.0	0.15	8.1	8.0
9	Tuesta	Silty-loam	18.8	54.8	26.4	1.4	0.11	8.3	55.8
10	Arangiz	Silty-clay-loam	17.5	55.3	27.2	2.0	0.12	8.3	56.8
11	Arangiz	Silty-clay-loam	18.0	52.8	28.5	1.8	0.11	8.4	40.9
12	Gauna	Clay-loam	38.5	32.6	29.0	2.0	0.12	8.1	9.2
13	Arangiz	Silty-loam	19.8	54.8	25.4	1.5	0.10	8.4	53.8
14	Tuesta	Clay-loam	36.8	28.4	34.8	1.9	0.15	8.2	16.4
15	Gauna	Sandy-clay-loam	47.6	24.5	27.9	2.2	0.17	8.0	12.1
16	Gauna	Clay loam	44	25.9	30.1	2.1	0.16	8.0	7.2

[a] Texture using a pipette method [16]; [b] Soil organic matter [17]; [c] Soil total N (dry combustion using a LECO TruSpec CHN); [d] pH (1:2.5 soil:water); [e] $CaCO_3$ (NH_4AcO; [18]).

2.2. Plant Sampling

Wheat aboveground biomass was sampled at GS30 (stem elongation), GS37 (leaf flag emergence), and harvest (19 December 2011). Fresh biomass samples were weighed and oven dried, and the dried biomass samples were again weighed for dry matter content determination. Biomass was estimated, and the N concentration was determined using dry combustion with LECO equipment (TrueSpec® CHN-S, LECO Corporation, Michigan, USA). At harvest, grain and straw were separated and dried at 70 °C for two days to obtain the dry matter content. Grain yield was measured, and grain and straw N concentration were determined using dry combustion with LECO equipment (TrueSpec® CHN-S). Nitrogen uptake was calculated, and GPC was determined by multiplying the total grain N concentration by 5.7 [19].

2.3. Soil Samples

Soil was sampled with a soil sampling rod (full depth from each pot) at sowing, GS30 (stem elongation), GS37 (leaf flag emergence), GS60 (beginning of flowering), harvest, post-harvest, and pre-sowing to determine the ammonium and nitrate levels. NH_4-N and NO_3-N were spectrophotometrically determined [20,21]. N_{min} was calculated as the sum of NH_4-N plus NO_3-N.

2.4. Aerobic Incubation

Aerobic incubation was performed following the method described by Stanford and Smith [7] and modified by Campbell et al. [22]. Fifteen grams of each soil sample was air-dried and sieved (2 mm mesh) and then mixed with an equal amount of quartz sand. Soils were incubated aerobically at field capacity for 32 weeks in a culture chamber at 35 °C. Samples were leached every 2 weeks during the first 8 weeks and every 4 weeks thereafter with a 0.01 M $CaCl_2$ solution. Mineral N was determined in each sample spectrophotometrically, and N_o was estimated by fitting the accumulated N_{min} against time to a first-order kinetic exponential model. N mineralized in the first two weeks (N2wk) and accumulated after 30 weeks of incubation (N30wk), and potentially mineralizable N (N_o) were calculated. The procedure of the experiment was extensively described in a previous article [13].

2.5. Chemical Extractions

2.5.1. The 0.01 M $CaCl_2$ Extraction

Calcium chloride extraction was performed using the method described by Houba et al. [23] and modified by Velthof and Oenema [10]. Soil samples were divided into three parts: the first part was air-dried (30 °C), the second part was dried at 40 °C, and the last part was dried at 105 °C. After drying, 6 g of soil was extracted with 60 mL of 0.01 M $CaCl_2$ via shaking for 2 h. After extraction, the samples were centrifuged for 5 min at 2328× *g*. Nitrate and ammonium were determined spectrophotometrically. Two mineralization indices were calculated: MI-$CaCl_2$ I and MI-$CaCl_2$ II. MI-$CaCl_2$ I was calculated as the difference between the ammonium extracted at 105 °C and the ammonium extracted from the air-dried samples. MI-$CaCl_2$ II was calculated as the difference between the ammonium extracted at 105 °C and the ammonium extracted at 40 °C.

2.5.2. The KCl and HotKCl Extraction

Ten grams of soil were extracted with 20 mL of 2 M KCl at room temperature [24]. After extraction, the samples were centrifuged at 2328× *g* for 5 min and analyzed spectrophotometrically to determine the concentrations of NO_3^--N and NH_4^+-N.

The value of N_{min} obtained using hotKCl was determined by heating 1.5 g of soil with 10 mL of 2 M KCl solution on a digestion block set at 100 °C for 4 h. After extraction, the NO_3^--N and NH_4^+-N concentrations were determined as described for the room temperature procedure.

The N mineralization index (MI-hotKCl) was calculated as the difference between the amount of NH_4^+-N extracted using hotKCl and that extracted at room temperature.

2.5.3. The 0.01 M $NaHCO_3$ Extraction

$NaHCO_3$ extraction was carried out using the method described by MacLean [25] and modified by Serna and Pomares [8]. Two grams of soil was mixed with 40 mL of 0.01 M $NaHCO_3$; the samples were then centrifuged at 2328× *g* for 5 min and filtered through Whatman No. 42 filters (Whatman International Ltd., Maidstone, England) and the absorbance of the filtrate was then measured at 205 and 260 nm (205ABS and 260ABS, respectively). The procedure of the experiment was extensively described in a previous article [13].

2.6. Statistical Analysis

For analyzing N_{min} differences among soils, a one-way ANOVA was performed for each growing stage: sowing, GS30, GS37, GS60, harvest, post-harvest, and pre-sowing [26]. For analyzing differences among the soils' final N uptake, yield, and GPC, one-way ANOVA was performed [26]. To separate the means, Duncan's test was used ($p \leq 0.05$) using the R package *agricolae V. 1.2-4* [27].

Pearson's correlation analysis of the soil properties, aerobic incubation and chemical extractions with 1) N_{min} values at sowing, GS30, GS37, GS60, harvest, post-harvest, and pre-sowing and with the 2) nitrogen uptake by the plant between sowing and GS30, GS30–GS37, GS37-harvest, and yield and GPC was performed. In addition, Pearson's correlation analysis was performed between N_{min} and N uptake by the plant, sowing and GS30, GS30–GS37, GS37-harvest, and yield and GPC.

3. Results

3.1. *Range of Soil Physical and Chemical Characteristics*

The soils varied widely in their physical and chemical characteristics. Sixteen soils were classified into five texture classes: clay-loam (S-1, S-5, S-12, S-14, and S-16), loam (S-2 and S-3), silty-loam (S-4, S-9, and S-13), silty-clay-loam (S-6, S-7, S-10, and S-11), and sandy-clay-loam (S-8 and S-15). The pH values were high (8.0–8.5). Twelve soils were calcareous (>15% $CaCO_3$), and the remaining four soil values varied between 7.0 and 12.1%. Soil organic matter values were low (SOM $\leq$ 1.9%; S-1, S-2, S-3, S-4, S-9, S-11, S-13, and S-14), moderate (SOM 1.9–2.2%; S-5, S-6, S-8, S-10, S-12, S-15, S-16, and S-17), or high (SOM 3.1%; S-7).

3.2. *Availability of Mineral N in Soil*

Soil mineral N availability values were different among soils and phenological stages (Table 2). Remarkably, N_{min} did not follow the same dynamic pattern throughout the crop cycle in different soils. Soil N_{min} values decreased from sowing to GS37 in the vast majority of the soils, except for S-9, S-3, and S-5, which increased their N_{min} values from sowing to GS30. N_{min} values increased from GS37 to GS60 in each soil and then remained similar. Differences were especially evident at GS30, where S-9 had the highest value and S-2, S-8, S-10, S-12, S-14, and S-16 had the lowest values. Soils with the highest N availability in GS37 were S-1, S-6, S-7, S-8, S-12, and S-14, whereas at GS60, S-7, S-8, and S-14 had the highest N availability.

Table 2. Soil mineral nitrogen (N_{min}; mg kg^{-1}) evolution in 16 soils from Araba throughout the wheat growing season in a greenhouse experiment. Different letters represent significant differences ($p \leq 0.05$) among soils for each growing stage: sowing, stem elongation (GS30), leaf flag emergence (GS37), flowering (GS60), harvest, post-harvest, and pre-sowing.

Soil	Sowing		GS30		GS37		GS60		Harvest		Post-Harvest		Pre-Sowing	
	Mean	sd	Mean	sd	Mean	sd	Mean	sd	Mean	sd	Mean	sd	Mean	sd
1	16.6 AB	4.0	10.2 ABC	3.2	2.6 A	0.5	7.5 ABC	0.7	4.7 BC	0.9	7.2 AB	1.2	3.3 ABC	1.0
2	8.5 EFG	2.1	3.7 C	0.9	0.4 B	0.1	5.2 C	1.0	3.8 CD	0.6	3.0 C	0.7	2.5 C	0.9
3	10.3 ABC	1.8	12.3 AB	1.7	1.7 AB	0.5	6.2 BC	2.2	3.7 CD	0.8	5.6 BC	0.3	3.2 ABC	0.6
4	12.6 ABC	6.1	6.1 BC	1.9	2.2 AB	0.7	7.1 ABC	2.0	4.9 B	0.9	8.0 AB	1.2	3.2 ABC	0.6
5	11.3 ABC	2.3	9.0 ABC	1.9	1.7 AB	0.9	8.2 ABC	2.2	4.6 BC	0.8	6.5 BC	0.9	3.3 ABC	0.6
6	11.4 ABC	5.9	5.3 BC	1.7	2.9 A	0.3	8.0 ABC	0.8	5.1 B	0.5	6.8 BC	0.8	3.6 ABC	0.5
7	6.5 FG	0.4	6.6 BC	0.9	3.0 A	0.7	8.1 A	0.9	6.6 A	0.9	8.9 A	0.6	3.8 ABC	1.0
8	7.3 EFG	2.6	3.3 C	0.9	2.9 A	0.6	7.5 A	1.8	3.1 D	0.7	4.3 BC	1.0	2.7 C	0.2
9	9.7 CDE	3.5	16.4 A	1.2	0.9 AB	0.1	5.9 AB	1.0	4.4 BC	0.6	4.4 BC	0.8	3.2 ABC	0.2
10	12.7 ABC	2.9	4.2 C	0.7	2.2 AB	0.4	8.3 ABC	0.7	4.9 BC	0.9	5.5 ABC	1.0	3.9 A	0.5
11	11.6 ABC	3.2	4.9 BC	1.1	1.6 AB	0.7	7.4 ABC	1.4	5.3 BC	1.7	7.0 AB	1.3	3.6 ABC	0.7
12	5.7 G	2.0	4.4 C	0.7	2.7 A	0.7	8.2 ABC	1.9	5.1 BC	1.2	5.2 ABC	0.7	3.3 ABC	0.8
13	9.4 CDE	1.8	5.0 BC	0.9	1.3 BC	0.6	9.6 AB	1.4	4.7 BC	1.2	5.7 ABC	0.8	3.5 ABC	1.7
14	10.3 CDE	6.0	3.7 C	0.8	2.8 A	0.3	10.2 A	1.2	5.2 BC	1.3	7.1 AB	0.7	4.0 AB	0.9
15	13.7 A	3.5	5.0 BC	1.3	1.5 AB	0.4	8.1 ABC	2.0	4.3 BC	1.0	5.8 ABC	1.7	3.5 ABC	0.2
16	7.9 EFG	1.3	2.8 C	0.8	1.2 AB	0.5	4.9 C	2.3	4.5 BC	1.4	5.4 ABC	0.7	3.0 ABC	0.8

3.3. *Wheat N Uptake, Yield, and GPC*

There were significant differences in wheat N uptake at harvest, yield, and GPC depending on the soil (Figure 1). The soil with the highest N uptake (mg pot^{-1}) was S-1. The lowest N uptake was in S-2 and S-8. In the case of wheat yield (g pot^{-1}), the highest values were achieved in S-3, S-9, and S-15 and the lowest in S-2, S-8, and S-16. Regarding GPC (%), the highest values were achieved in S-14, and the lowest were in S-2.

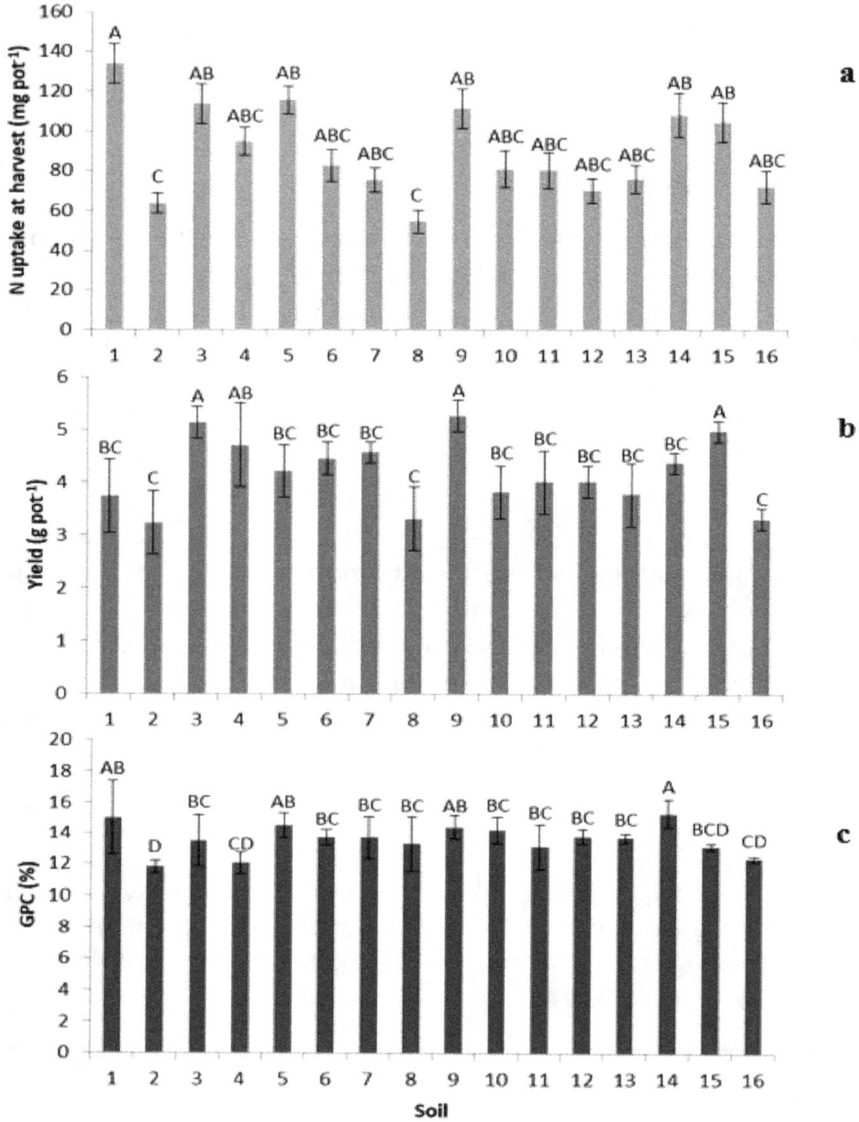

Figure 1. (**a**) Wheat N uptake at harvest (mg pot^{-1}), (**b**) yield (g pot^{-1}), and (**c**) GPC (%) in 16 soils from Araba. Different letters represent significant differences ($p \leq 0.05$) among soils. Values are the mean of three replicates ± SD.

3.4. Relationships between Initial Soil Characteristics with N_{min} throughout the Growing Cycle

Regarding the soil physicochemical properties (Table 3), sand had a negative correlation with N_{min} values at harvest, silt had a positive correlation with N_{min} values at harvest, and clay had a positive correlation with N_{min} values at GS37 and GS60. Ntot was positively correlated with N_{min} at GS37, harvest, and post-harvest. SOM had a positive correlation with N_{min} values at GS37, GS60, harvest, and post-harvest. In the case of aerobic incubations, N2wk had a positive correlation with the sowing values, and N30wk had a positive and significant correlation with the GS60, harvest, post-harvest, and pre-sowing N_{min}. N_o was positively correlated with the GS60 and pre-sowing N_{min} values. Regarding the chemical extractants used to estimate N_o, MI $CaCl_2$ I was correlated with the N_{min} values at GS60. MI-hotKCl was positively correlated with N_{min} values from GS37 to pre-sowing. For $NaHCO_3$, only 205 ABS was correlated with the sowing and pre-sowing N_{min} values. Remarkably, SOM, N30wk, and hotKCl were correlated with the N_{min} values after harvest.

Table 3. Pearson correlation coefficients (r) between soil properties, N mineralization indices calculated from the aerobic incubations and chemical extractions with Nmin values at sowing, stem elongation (GS30), leaf flag emergence (GS37), flowering (GS60), harvest, post-harvest, and pre-sowing.

Soil Characteristics		Sowing		GS30		GS37		GS60		Harvest		Post-Harvest		Pre-Sowing	
		r	*p*	r	*p*	r	*p*	r	*p*	r	*p*	r	*p*	r	*p*
Soil properties	Sand	−0.11	*ns*	−0.25	*ns*	0.00	*ns*	−0.01	*ns*	−0.63	**	−0.43	*ns*	−0.47	*ns*
	Silt	0.09	*ns*	0.27	*ns*	−0.12	*ns*	−0.11	*ns*	0.50	*	0.36	*ns*	0.33	*ns*
	Clay	0.05	*ns*	−0.11	*ns*	0.55	*	0.56	*	0.38	*ns*	0.40	*ns*	0.47	*ns*
	SOM	−0.16	*ns*	−0.24	*ns*	0.70	**	0.50	*	0.60	**	0.58	*	0.49	*ns*
	Ntot	−0.16	*ns*	−0.12	*ns*	0.57	*	0.43	*ns*	0.54	*	0.59	**	0.44	*ns*
	pH	0.04	*ns*	0.22	*ns*	−0.47	*ns*	−0.38	*ns*	0.06	*ns*	−0.03	*ns*	−0.07	*ns*
	CaCO3	0.33	*ns*	0.40	*ns*	−0.29	*ns*	−0.25	*ns*	0.09	*ns*	0.10	*ns*	0.22	*ns*
Aerobic incubations	N2wk	0.54	*	0.02	*ns*	0.25	*ns*	0.30	*ns*	0.05	*ns*	0.14	*ns*	0.46	*ns*
	N30wk	0.37	*ns*	0.15	*ns*	0.43	*ns*	0.57	**	0.51	*	0.61	**	0.83	***
	No	0.29	*ns*	0.16	*ns*	0.43	*ns*	0.65	**	0.38	*ns*	0.49	*ns*	0.73	**
Chemical extractions	MI $CaCl_2$ I	0.19	*ns*	−0.44	*ns*	0.35	*ns*	0.55	*	0.16	*ns*	0.37	*ns*	0.43	*ns*
	MI $CaCl_2$II	0.34	*ns*	−0.41	*ns*	0.31	*ns*	0.52	*	−0.03	*ns*	0.29	*ns*	0.38	*ns*
	MI-HotKCl	0.36	*ns*	−0.29	*ns*	0.62	**	0.70	*	0.50	*	0.69	**	0.64	**
	205ABS	0.59	**	0.07	*ns*	−0.16	*ns*	0.23	*ns*	0.18	*ns*	0.30	*ns*	0.59	**
	260ABS	0.31	*ns*	0.08	*ns*	−0.15	*ns*	0.15	*ns*	0.26	*ns*	0.32	*ns*	0.33	*ns*

ns, not significant ($p > 0.05$). *, **, *** Significant at the 0.05, 0.01, and 0.001 probability levels, respectively.

3.5. Relationships between Soil Characteristics and Plant N Uptake, Yield, and GPC

Regarding the soil initial properties (Table 4), clay had a positive correlation with GPC. Ntot had a positive correlation with the N uptake between sowing and GS30 and with the N uptake between GS30 and GS37. In the aerobic incubations, N30wk and N_0 showed a positive and significant correlation with the N uptake between GS37 and harvest and GPC. Concerning the chemical extractants, only 205ABS had a positive correlation with yield.

Table 4. Pearson correlation coefficients (r) among soil properties, N mineralization indices calculated from the aerobic incubations and chemical extractions with N uptake values between sowing and stem elongation (GS30), GS30 and leaf flag emergence (GS37), GS37 and flowering (GS60), and yield and GPC (grain protein content).

Soil Characteristics		N Uptake						Yield		GPC	
		Sowing-GS30		GS30-GS37		GS37-Harvest					
		r	*p*	r	*p*	r	*p*	r	*p*	r	*p*
Soil properties	Sand	−0.06	*ns*	−0.04	*ns*	0.01	*ns*	−0.22	*ns*	−0.06	*ns*
	Silt	−0.02	*ns*	0.09	*ns*	−0.10	*ns*	0.22	*ns*	−0.10	*ns*
	Clay	0.39	*ns*	−0.26	*ns*	0.45	*ns*	−0.07	*ns*	0.73	***
	SOM	0.46	*ns*	−0.47	*ns*	−0.10	*ns*	0.09	*ns*	0.31	*ns*
	Ntot	0.52	*	0.49	*	0.06	*ns*	0.24	*ns*	0.26	*ns*
	pH	−0.43	*ns*	0.33	*ns*	−0.06	*ns*	−0.16	*ns*	−0.22	*ns*
	$CaCO_3$	−0.24	*ns*	0.33	*ns*	0.08	*ns*	0.32	*ns*	−0.13	*ns*
Aerobic incubations	N2wk	0.30	*ns*	−0.24	*ns*	0.39	*ns*	0.26	*ns*	0.40	*ns*
	N30wk	0.39	*ns*	−0.25	*ns*	0.64	**	0.34	*ns*	0.75	**
	No	0.47	*ns*	−0.38	*ns*	0.64	**	0.43	*ns*	0.74	**
Chemical Extractions	MI $CaCl_2$ I	−0.01	*ns*	−0.21	*ns*	−0.11	*ns*	−0.07	*ns*	0.04	*ns*
	MI $CaCl_2$ II	−0.06	*ns*	−0.14	*ns*	−0.05	*ns*	−0.07	*ns*	0.03	*ns*
	MI-HotKCl	0.43	*ns*	−0.35	*ns*	0.21	*ns*	−0.07	*ns*	0.05	*ns*
	205ABS	0.29	*ns*	−0.24	*ns*	0.24	*ns*	0.57	*	0.00	*ns*
	260ABS	−0.11	*ns*	0.10	*ns*	−0.04	*ns*	0.10	ns	−0.08	*ns*

ns, not significant ($p > 0.05$). *, **, *** Significant at the 0.05, 0.01 and 0.001 probability levels, respectively.

The soil N_{min} values (Table 5) at sowing and GS30 were correlated positively with N uptake between GS37 and harvest. The N_{min} values at GS30 and GS60 had positive correlations with yield and GPC, respectively.

Table 5. Pearson correlation coefficients (r) among the soil N_{min} values, N uptake values between sowing and stem elongation (GS30), GS30 and leaf flag emergence (GS37), GS37 and flowering (GS60), and yield and GPC (grain protein content).

Soil N_{min}	N Uptake						Yield		GPC	
	Sowing-GS30		GS30-GS37		GS37-Harvest					
	r	*p*	r	*p*	r	*p*	r	*p*	r	*p*
Sowing	0.01	*ns*	0.18	*ns*	0.53	*	0.22	*ns*	0.09	*ns*
GS30	−0.02	*ns*	0.31	*ns*	0.65	**	0.59	**	0.31	*ns*
GS37	0.27	*ns*	−0.25	*ns*	0.00	*ns*	−0.04	*ns*	0.38	*ns*
GS60	0.22	*ns*	−0.31	*ns*	0.09	*ns*	−0.07	*ns*	0.53	**
Harvest	0.38	*ns*	−0.25	*ns*	0.00	*ns*	0.22	*ns*	0.24	*ns*

ns, not significant ($p > 0.05$). *, ** Significant at the 0.05 and 0.01 probability levels, respectively.

4. Discussion

The tested soils differed in their initial N_{min} values and their physical and chemical properties [13], which significantly influenced the N mineralization patterns. Depending on their characteristics, soils mineralize in different ways, making available different N_{min} values (Table 2). The first step in mineralization is ammonification, which is the conversion of organic N into ammonium by soil microbes. This process is carried out exclusively by heterotrophic microorganisms that utilize C as an energy source. Nitrate production is mediated via two groups of autotrophic bacteria (*Nitrosomonas* and *Nitrobacter*) that convert ammonium into nitrate by the process called nitrification. Nitrogen availability relies on both the initial availability of N_{min} and the rate of mineralization or immobilization [28], as well as the previous N uptake by the crop, influencing the yield and GPC.

There were no correlations between the soil properties and N_{min} values at sowing or GS30. It should be mentioned that the soil preparation (drying, sieving, and rewetting) prior to the experiment could have affected soil structure and functioning. This could explain the lack of correlation in the early stages. It should be mentioned that soil rewetting often causes abundant mineralization because microorganisms recover their activity [29]. However, Mikha et al. [30] suggested that N immobilization occurred in response to the easily accessible C due to the rapid increase in microbial activity. Later, in the growing cycle, from GS37 onwards, SOM, Ntot, sand, silt, and clay were relatively effective predictors of soil N_{min} dynamics.

Soil organic matter (SOM) is a heterogeneous mixture of organic compounds that vary in their nutrient composition, molecular characteristics, age, and biological stability. Increasing SOM by adding carbon is a beneficial agronomic practice that stimulates microbial communities and enhances soil N and C pools [31]. The youngest compounds are the most biologically active compounds, and the materials with intermediate ages contribute to soil physical characteristics [28]. We found positive correlations from GS37 to post-harvest with the N_{min} values (Table 3). Ros et al. [6] found that SOM explained 78% of the variation in mineralizable N, whereas other soil properties only explained 8%. In some studies, SOM fractions have been preferred to total SOM for predicting N_o due to the easy release of labile compounds during the extractions [32]. However, other studies suggested that none of those SOM fractions is an a priori preferable indicator of N_o [6,33].

Debosz and Kristensen [34] found that Ntot content had a positive relationship with N mineralization. Similarly, Dessureault-Rompré et al. [9] showed that Ntot was one of the best predictors of soil mineralizable N pools. However, in our case, Ntot only had positive correlations with the N_{min} values at harvest and post-harvest (Table 3). It is remarkable that only approximately 1–4% of the Ntot is mineralized as plant-available N (NH_4-N and NO_3-N) each year [34]. Many authors have found that soil Ntot and SOM were the best predictors of N_o [9,11].

The mineralization of N is often affected by the clay content, likely due to SOM binding to mineral particles. Clay was correlated with N_{min} at the end of the growth cycle and with the GPC. In this experiment, soils with the highest GPC were S-1, S-5, S-9, and S-14 (15–16%), where the clay values were 32.4, 28.3, 26.4, and 34.8%, respectively. Clay has indirect affects through the formation of aggregates that protect SOM, and therefore microbial biomass, and direct effects with the stabilization of organic N [35]. Hassink [36] found that the mineralization of organic N was negatively affected by a high clay content due to SOM binding to mineral particles. Similarly, Ros et al. [6] found that clay had a negative influence on mineralizable N because mineralization in clayey soils was lower than that in sandy soils. In contrast, in this experiment, clay was positively correlated with N_{min} at GS37 and GS60, and sand was negatively correlated with N_{min} at GS60 (Table 3). Some clayey soils are able to fix and release ammonium, but the regulation of N availability is not fully understood [37]. Chantigny et al. [38] showed that in clay soils, the fixation of the recently added ammonium was higher than that in sandy soil (34% and 11%, respectively). The recently fixed ammonium that can be derived from added fertilizer or from soil organic matter [39] is quickly fixed by clay minerals and later released slowly during the crop growth season due to the increased crop demand. In a greenhouse experiment, Dou and Steffens [40] found that 90–95% of the recently fixed ammonium was released during a 14-week period. Under field conditions, 66% of the recently fixed ammonium was released 86 days after fixation [41]. Provision of root exudates by plants improved the activity of heterotrophic microorganisms, which foster the release of fixed ammonium [37], retarding nitrification. The silt fraction has also been reported to bind NH_4^+ in a non-exchangeable form [37]; in our study, it was correlated with N_{min} at harvest. In S-9, where high values of GPC were achieved (15%), the silt content was 54.8.

The pH and $CaCO_3$ did not present any effect on N mineralization. Dessureault-Rompré et al. [11] found that the effect of pH on soil mineralization was very low. In other studies, soil pH, moisture and temperature were often non-linearly related to the dynamics of N [42,43]. The pH range among our soils was very low.

With respect to wheat N uptake (Table 4), among the soil properties, only soil Ntot was positively correlated with the wheat N uptake at two times: from sowing to GS30 and from GS30 to GS37. In the case of aerobic incubation, N30wk and N_0 were correlated with the N uptake from GS37 to harvest and with GPC. However, only soil N_{min} at sowing and GS30 was correlated with the N uptake from GS37 to harvest (Table 5). Historically, soil N availability has been seen as an inaccurate indicator of plant N availability because plant roots are considered poor competitors for inorganic N against soil microorganisms [44]. This idea could explain the lack of correlation between the soil N availability and the N uptake from the crop. Conversely, it has been determined that a cereal crop was able to accumulate a greater amount of added inorganic N than microorganisms [45]. The results showed that different soils followed different mineralization patterns affecting yield and GPC. The soil N_{min} at GS30 and at GS60 was positively correlated with the yield and GPC, respectively (Table 5). This suggests that the N status at those times is essential for determining the yield [15] or GPC [46,47]. Soils with the lowest N availability at GS30 and GS60 as S-2 and S-16 presented the lowest yields and GPC, respectively. However, when the N availability was high in these growth stages, the yields and GPC values were high. Remarkably, none of the soil properties was correlated with yield, but clay was positively correlated with GPC (Table 4). However, one of the methods of chemical extraction (205 ABS) was correlated with the yield.

The key to optimizing high yields, wheat quality (GPC), and environmental protection is to achieve synchronicity between the N supply and crop demand, while accounting for spatial and temporal variability in soil N. As previously observed, many factors affect soil N mineralization, and therefore wheat N uptake. In Western Europe, the soil N_{min} at the end of winter is used to correct the values of N fertilizer rates calculated from the potential yield. Nevertheless, it implies laborious and expensive sampling and analysis. In Araba (Basque Country, northern Spain), the usual last and greater N dressing application occurred at GS30, but there were no correlations between soil initial properties and N_{min} values at GS30 (Table 3). This is remarkable because a high N availability at GS30 is key for achieving high yields [15]. As the last N dressing is at GS30, it is common to have low N in wheat plants at the end of the growing cycle (GS60–harvest) in Araba [46]. Moreover, the climatic conditions in this area, humid Mediterranean, can lead to very high yields and therefore low protein concentrations in grain. Fuertes-Mendizabal et al. [47] showed that late N availability (GS37 onwards) in wheat under humid Mediterranean conditions increased GPC, especially when no N was applied in the late stages, as in the area of study. As stated above, clay apparently allowed higher N availability at GS37 and GS60. This could be explained by the positive correlation between N_{min} at GS60 and GPC or the correlation between clay and GPC. However, soils presented a narrow clay range (24.6–34.8%) to confirm that finding. Inside the aerobic incubation, N30wk and N_o were able to estimate the N available for the wheat crop at the end of the growing cycle and thus the GPC. Identifying soils in Araba where it would be possible to have late N availability with the aim of improving GPC would be interesting. In the humid Mediterranean climatic conditions of Araba, a third application at GS37 is possible since rain water usually allows the utilization by wheat of this N applied late [46,47]. However, the chemical methods that avoid the long periods required for aerobic incubation did not correlate properly with the N uptake values in the late growth stages.

In order to make N recommendations that guarantee adequate levels of GPC, methods for the diagnosis of soil available N must be improved, especially in the later stages of the growing cycle. In this sense, it is necessary to explore quick and simple methods because the most effective ones require periods of incubations that are too long. This is even more important in certain circumstances, such as organic farming, where it is difficult to make late applications of N with authorized fertilizers (organic fertilizers).

5. Conclusions

Even in a relatively small cropping area where the variability of soil properties is narrow, the dynamics of soil nitrate and ammonium throughout the cropping season were variable, and therefore, so was the crop N uptake. The soil N_{min} values at early wheat growth stages were well-correlated with yield, and at late stages, they were well-correlated with GPC.

N_0 was correlated with late N uptake and GPC. However, the chemical methods that avoid the long periods required for N_0 determinations were not correlated properly with the N uptake in the late wheat growth stages or GPC. Conversely, clay was positively correlated with the late N_{min} values and GPC, although the clay range was not very wide. Chemical methods were unable to estimate the available soil N in the later stages of the growing cycle. Consequently, as incubation methods are too laborious for their widespread use, further research must be conducted.

References

1. Samborski, S.M.; Tremblay, N.; Fallon, E. Strategies to make use of plant sensors-based diagnostic information for nitrogen recommendations. *Agron. J.* **2009**, *101*, 800–816. [CrossRef]

2. Zebarth, B.J.; Drury, C.F.; Tremblay, N.; Cambouris, A.N. Opportunities for improved fertilizer nitrogen management in production of arable crops in eastern Canada: A review. *Can. J. Soil Sci.* **2009**, *89*, 113–132. [CrossRef]

3. Parton, W.J.; Schimel, D.S.; Cole, C.V.; Ojima, D.S. Analysis of factors controlling soil organic matter levels in Great Plains grassland. *Soil Sci. Soc. Am. J.* **1987**, *51*, 1173–1179. [CrossRef]

4. Martínez, J.M.; Galantini, J.A.; Duval, M.E. Contribution of nitrogen mineralization indices, labile organic matter and soil properties in predicting nitrogen mineralization. *J. Soil Sci. Plant Nutr.* **2018**, *18*, 73–89. [CrossRef]

5. Van Groenigen, J.W.; Huygens, D.; Boeckx, P.; Kuyper, T.W.; Lubbers, I.M.; Rütting, T.; Groffman, P.M. The soil n cycle: New insights and key challenges. *Soil* **2015**, *1*, 235–256. [CrossRef]
6. Ros, G.H.; Hanegraaf, M.C.; Hoffland, E.; van Riemsdijk, W.H. Predicting soil N mineralization: Relevance of organic matter fractions and soil properties. *Soil Biol. Biochem.* **2011**, *43*, 1714–1722. [CrossRef]
7. Standford, G.; Smith, S.J. Nitrogen mineralization potentials of soils. *Soil Sci. Soc. Am. J.* **1972**, *36*, 465–472. [CrossRef]
8. Serna, M.D.; Pomares, F. Nitrogen mineralization of sludge-amended soil. *Bioresour. Technol.* **1992**, *39*, 285–290. [CrossRef]
9. Dessureault-Rompré, J.; Zebarth, B.J.; Burton, D.L.; Sharifi, M.; Cooper, J.; Grant, C.A.; Drury, C.F. Relationships among Mineralizable Soil Nitrogen, Soil Properties, and Climatic Indices. *Soil Sci. Soc. Am. J.* **2010**, *74*, 1218–1227. [CrossRef]
10. Velthof, G.; Oenema, O. Estimation of plant-available nitrogen in soils using rapid chemical and biological methods. *Commun. Soil Sci. Plant Anal.* **2010**, *41*, 52–71. [CrossRef]
11. Dessureault-Rompré, J.; Zebarth, B.J.; Burton, D.L.; Georgallas, A. Predicting soil nitrogen supply from soil properties. *Can. J. Soil Sci.* **2015**, *95*, 63–75. [CrossRef]
12. Luce, M.; Whalen, J.K.; Ziadi, N.; Zebarth, B.J. Nitrogen dynamics and indices to predict soil nitrogen supply in humid temperate soils. *Adv. Agron.* **2011**, *112*, 55–102.
13. Villar, N.; Aizpurua, A.; Castellón, A.; Ortuzar, M.A.; González-Moro, M.B.; Besga, G. Laboratory Methods for the Estimation of Soil Apparent N Mineralization and Wheat N Uptake in Calcareous Soils. *Soil Sci.* **2014**, *179*, 84–94. [CrossRef]
14. Zadoks, J.C.; Chang, T.T.; Konzak, C.F. A decimal code for growth stages of cereals. *Weed Res.* **1974**, *4*, 415–421. [CrossRef]
15. Aranguren, M.; Castellón, A.; Aizpurua, A. Topdressing nitrogen recommendation in wheat after applying organic manures: The use of field diagnostic tools. *Nutr. Cycl. Agroecosyst.* **2018**, *110*, 89–103. [CrossRef]
16. Gee, G.W.; Bauder, J.W. Particle size analysis. In *Methods of Soil Analysis: Part I. Physical and Mineralogical Methods*, 2nd ed.; Klute, A., Ed.; Soil Science Society of America: Madison, WI, USA, 1886; pp. 383–411.
17. Walkey, A.; Black, I.A. An examination of the Degtjareff method for determining soil organic matter and a proposed modification of the chromic acid titration method. *Soil Sci.* **1934**, *37*, 29–38. [CrossRef]
18. MAPA. *Métodos Oficiales de Análisis. Tomo III*; Ministerio de Agricultura, Pesca y Alimentación: Madrid, Spain, 1994.
19. Teller, G.L. Non-protein nitrogen compounds in cereals and their relation to the nitrogen factor for protein in cereals and bread. *Cereal Chem.* **1932**, *9*, 261–267.
20. Cawse, P.A. The determination of nitrate in soil solutions by ultraviolet spectrophotometry. *Analyst* **1967**, *92*, 311–315. [CrossRef]
21. Nelson, D.W. Determination of ammonium in KCl extracts of soils by the salicylate method. *Commun. Soil Sci. Plant Anal.* **1983**, *14*, 1051–1062. [CrossRef]
22. Campbell, C.A.; Ellert, B.H.; Jame, Y.W. Nitrogen mineralization potential in soils. In *Soil Sampling and Methods of Analysis*; Carter, M.R., Ed.; Canadian Society of Soil Science/Lewis Publishers: Boca Raton, FL, USA, 1993; pp. 341–349.
23. Houba, V.J.G.; Novozamsky, I.; Huybregts, A.W.M.; van der Lee, J.J. Comparison of soil extractions by 0.01 M $CaCl_2$ by EUF and by some conventional extraction procedures. *Plant Soil* **1986**, *96*, 433–437. [CrossRef]
24. Gianello, C.; Bremmer, J.M. Comparisons of chemical methods of assessing potentially available organic nitrogen in soil. *Commum. Soil Sci. Plant Anal.* **1986**, *31*, 1299–1396. [CrossRef]
25. MacLean, A.A. Measurement of nitrogen supplying power of soils by extraction with sodium bicarbonate. *Nature* **1964**, *203*, 1307–1308. [CrossRef]
26. R Core Team. *R: A Language and Environment for Statistical Computing*; R Foundation for Statistical Computing: Vienna, Austria, 2013.
27. De Mendiburu, F. Una Herramienta de Análisis Estadístico para la Investigación Agrícola. Ph.D. Thesis, Facultad de Economía y Planificación Departamento Académico de Estadística e Informática, Universidad Nacional de Ingeniería (UNI-PERU), Universidad Nacional Agraria La Molina, Lima, Peru, 2009.
28. Mohanty, M.; Sinha, N.K.; Reddy, K.S.; Chaudhary, R.S.; Rao, A.S.; Dalal, R.C.; Menzies, N.W. How important is the quality of organic amendments in relation to mineral N availability in soils? *Agric. Res.* **2013**, *2*, 99–110. [CrossRef]

29. Griffin, T.S. Nitrogen availability. In *Nitrogen in Agricultural Systems. Agronomy Monograph 49*; Schepers, J.S., Raun, W.R., Eds.; ASA, CSSA, and SSSA: Madison, WI, USA, 2008; pp. 613–646.
30. Mikha, M.M.; Rice, C.; Milliken, G. Carbon and nitrogen mineralization as affected by drying and wetting cycles. *Soil Biol. Biochem.* **2005**, *37*, 339–347. [CrossRef]
31. Urra, J.; Mijangos, I.; Lanzén, A.; Lloveras, J.; Garbisu, C. Effects of corn stover management on soil quality. *Eur. J. Soil Biol.* **2018**, *88*, 57–64. [CrossRef]
32. Wander, M. Soil organic matter fractions and their relevance to soil function. In *Soil Organic Matter in Sustainable Agriculture*; Magdoff, F., Weil, R.R., Eds.; CRC Press: Boca Raton, FL, USA, 2004; pp. 67–102.
33. Haynes, R.J. Labile organic matter fractions as central components of the quality of agricultural soils: An overview. *Adv. Agron.* **2005**, *85*, 221–268.
34. Debosz, K.; Kristensen, K. Spatial covariability of N mineralisation and textural fractions in two agricultural fields. *Semin. Site Specif. Farming* **1995**, *26*, 174–180.
35. Tisdale, S.L.; Nelson, W.L.; Beaton, J.D. *Soil Fertility and Fertilizers*, 4th ed.; Macmillon Publishing Company: New York, NY, USA, 1985.
36. Hassink, J. The capacity of soils to preserve organic C and N by their association with clay and silt particles. *Plant Soil* **1997**, *191*, 77–87. [CrossRef]
37. Nieder, R.; Benbi, D.K.; Scherer, W. Fixation and defixation of ammonium in soils: A review. *Biol. Fertil. Soils* **2011**, *47*, 1–14. [CrossRef]
38. Chantigny, M.H.; Angers, D.A.; Morvan, T.; Pomar, C. Dynamics of pig slurry nitrogen in soil and plant as determined with 15 N. *Soil Sci. Soc. Am. J.* **2004**, *68*, 637–643. [CrossRef]
39. Nieder, R.; Benbi, D.K. Carbon and nitrogen transformations in soils. In *Carbon and Nitrogen in the Terrestrial Environment*; Nieder, R., Benbi, D.K., Eds.; Springer: Heidelberg, Germany, 2008; pp. 137–159.
40. Dou, H.; Steffens, D. Recovery of 15N labelled urea as affected by fixation of ammonium by clay minerals. *Pflanzenernahr. Bodenkd.* **1995**, *158*, 351–354. [CrossRef]
41. Kowalenko, C.G. Nitrogen transformations and transport over 17 months in field fallow microplots using 15 N. *Can. J. Soil Sci.* **1978**, *58*, 69–76. [CrossRef]
42. Rodrigo, A.; Recous, S.; Neel, C.; Mary, B. Modelling temperature and moisture effects on C-N transformations in soils: Comparison of nine models. *Ecol. Model.* **1997**, *102*, 325–339. [CrossRef]
43. Paul, E.A. The nature and dynamics of soil organic matter: Plant inputs, microbial transformations, and organic matter stabilization. *Soil Biol. Biochem.* **2016**, *98*, 109–126. [CrossRef]
44. Schimel, J.P.; Bennett, J. Nitrogen Mineralization: Challenges of a Changing Paradigm. *Ecology* **2004**, *85*, 591–602. [CrossRef]
45. Inselsbacher, E.; Umana, N.H.N.; Stange, F.C.; Gorfer, M.; Schüller, E.; Ripka, K.; Zechmeister-Boltenstern, S.; Hood-Novotny, R.; Strauss, J.; Wanek, W. Short-term competition between crop plants and soil microbes for inorganic N fertilizer. *Soil Biol. Biochem.* **2010**, *42*, 360–372. [CrossRef]
46. Ortuzar-Iragorri, M.A.; Aizpurua, A.; Castellón, A.; Alonso, A.; José, M.; Estavillo, J.M.; Besga, G. Use of an N-tester chlorophyll meter to tune a late third nitrogen application to wheat under humid Mediterranean conditions. *J. Plant Nutr.* **2017**, *41*, 627–635. [CrossRef]
47. Fuertes-Mendizabal, T.; Aizpurua, A.; González-Moro, M.B.; Estavillo, J.M. Improving wheat breadmaking quality by splitting the N fertilizer rate. *Eur. J. Agron.* **2010**, *33*, 52–61. [CrossRef]

8

Are Nitrogen Fertilizers Deleterious to Soil Health?

Bijay-Singh

Department of Soil Science, Punjab Agricultural University, Ludhiana 141 004, India; bijaysingh20@hotmail.com;

Abstract: Soil is one of the most important natural resources and medium for plant growth. Anthropogenic interventions such as tillage, irrigation, and fertilizer application can affect the health of the soil. Use of fertilizer nitrogen (N) for crop production influences soil health primarily through changes in organic matter content, microbial life, and acidity in the soil. Soil organic matter (SOM) constitutes the storehouse of soil N. Studies with ^{15}N-labelled fertilizers show that in a cropping season, plants take more N from the soil than from the fertilizer. A large number of long-term field experiments prove that optimum fertilizer N application to crops neither resulted in loss of organic matter nor adversely affected microbial activity in the soil. Fertilizer N, when applied at or below the level at which maximum yields are achieved, resulted in the build-up of SOM and microbial biomass by promoting plant growth and increasing the amount of litter and root biomass added to soil. Only when fertilizer N was applied at rates more than the optimum, increased residual inorganic N accelerated the loss of SOM through its mineralization. Soil microbial life was also adversely affected at very high fertilizers rates. Optimum fertilizer use on agricultural crops reduces soil erosion but repeated application of high fertilizer N doses may lead to soil acidity, a negative soil health trait. Site-specific management strategies based on principles of synchronization of N demand by crops with N supply from all sources including soil and fertilizer could ensure high yields, along with maintenance of soil health. Balanced application of different nutrients and integrated nutrient management based on organic manures and mineral fertilizers also contributed to soil health maintenance and improvement. Thus, fertilizer N, when applied as per the need of the field crops in a balanced proportion with other nutrients and along with organic manures, if available with the farmer, maintains or improves soil health rather than being deleterious.

Keywords: soil organic matter; soil biota; soil acidity; soil erosion; fertilizer management; site-specific nutrient management; balanced use of fertilizers; integrated nutrient management

---

1. Introduction

Soil is fundamental to crop production and constitutes a natural resource that provides humans with most of their food and nutrients. However, it is finite and fragile, and requires special care and conservation so that it can be used indefinitely by future generations. Doran and Parkin [1] defined soil quality or soil health as its capacity to function within ecosystem and land-use boundaries, sustain biological productivity, maintain environmental quality, and promote plant and animal health. Soil as a medium for plant growth constitutes a living system and a habitat for many organisms and is characterized mainly by its biological functions, which operate through complex interactions with the abiotic, physical, and chemical environment. Soil health often reflects the condition of the soil in terms of management-sensitive properties and provides an idea of its overall fitness for carrying out ecosystem functions and responding to environmental stresses [2]. According to Kibblewhite et al. [3], a healthy agricultural soil is one that is capable of supporting the production of food and fiber to a level, and with regard to quality, it is sufficient to meet human requirements and can continue to sustain those functions that are essential to maintaining the quality of life for humans and the conservation of biodiversity. This definition implies that soil health is an integrative property that reflects the capacity of the soil to respond to agricultural interventions and circumvent processes that degrade it.

The main driver for anthropogenic interventions in the functioning of soils over the past century has been the quadrupling of the world's population, which has demanded a fundamental change in soil and crop management in order to produce more food from land already in cultivation [4]. Cultivation of soil to prepare the seed bed possibly constituted the first human intervention. In regions receiving little rainfall, irrigation represented another major external influence on the soil. Additionally, during the last 70 years or so, the application of mineral fertilizers has constituted an important human intervention that has influenced the functioning of agricultural soils, although the widespread use of mineral fertilizers has been one of the major factors in ensuring global food security. Every human intervention invariably represents major and sometimes irrevocable change in the nature and properties of the original soil. The key issue is to minimize the negative effects of such changes. Otherwise, the history of agriculture is replete with examples in which civilizations waned or disappeared because of failure to minimize the impact of human interventions on the soil resource.

Mineral fertilizers are applied to the soil to supplement or substitute for biological functions that are considered inadequate or inefficient for achieving the required levels of production. As per FAO's revised projection regarding world agriculture, global agricultural production in 2050 should be 60% higher than in 2005/2007 [5]. To close this gap through agricultural production increases alone, total crop production would need to increase even more from 2006 to 2050 than it did in the same number of years from 1962 to 2006—an 11% larger increase [6]. The bulk of the projected increases in crop production will come from high yields, which normally demand high fertilizer application rates, and will lead to an increase in fertilizer use [5]. According to Erisman et al. [7], over 48% of the more than 7 billion people alive today are living because of increased crop production made possible by applying fertilizer nitrogen (N). However, fertilizers being chemicals can potentially disturb the natural functioning of the soil and may also affect the output of other ecosystem services.

The challenge ahead is to manage fertilizers and soil in such a way that not only food demands are continuously met, but soil also remains healthy to support adequate food production with minimal environmental impact. The objective of this paper is to examine how fertilizer N use affects important and crucial soil health parameters such as soil organic matter (SOM), carbon (C), N, soil microorganisms, and soil acidity. As mineral fertilizers can potentially affect normal functioning of the soil, important management aspects of fertilizer N have also been discussed in terms of supplying adequate amounts of nutrients to crop plants, as well as maintenance of soil health.

2. Fertilizer Use—Soil Health Linkages

The major impact of fertilizers on the soil health and ecosystem functions is regulated through their effect on primary productivity. There are hardly any direct toxic effects even when fertilizers are applied in somewhat excessive quantities; the effects are on rates of different processes in the soil. Prior to the development of Haber-Bosch process in the early 1900s and introduction of N fertilizers around middle of the last century, organic manures (mainly animal manures) containing large amount of organic materials and legume crops used to be the major source of N for crops. An important indirect consequence of the increasing use of N fertilizers was a reduction in the use of organic manures; decoupling of animal farming from arable farming and availability of sewage sludges were also factors in the reduced use of organic manures. Subsequently, after a couple of decades, there was a revival of interest in organic manures due to their increasing supplies and their perceived role in soil health and nutrient recycling. Nevertheless, in several developing countries, particularly in Asia, crop production still relies more on fertilizers because of limited availability of animal manures and crop residues. For example, in South Asia, which accounted for more than 18% of the global fertilizer consumption in 2015 [8], a significant proportion of animal excreta are used as household fuel rather than for making organic manure for crops.

Soil organic matter is a relatively small component of the soil in terms of volume, but it constitutes the single most important soil property in relation to soil health. It exerts profound influence on the chemical, physical, and biological properties of the soil. Rate of decomposition of 'low quality' or high C:N ratio organic inputs and SOM increases when fertilizers, particularly N, are applied to the soil [9]. Fertilizer application increases microbial decomposer activity, which has been limited due to low nutrient concentrations in the organic materials. Thus, application of fertilizer N may lead to accelerated decomposition of organic matter in the soil and adversely affect the soil health.

Soil microbial life and associated microbial transformations constitute another important soil health parameter that may be affected by application of fertilizers. While net primary production in agricultural ecosystems is generally N limited, activity of soil microorganisms may be C and/or N limited [10]. The response of soil microbes to fertilizer N application may, therefore, differ from the response of the plants. That the soil biota are adversely affected due to application of N fertilizers is one of the notions that has been put forth many times to support the argument against fertilizers. However, N fertilizers may lead to increased acidity and adversely affect many soil functions. On the other hand, fertilizer use may reduce soil erosion and may have a positive impact on soil health.

3. Fertilizer Use Effects on Soil Organic Matter

Soil organic matter is a key indicator of soil health because of its vital functions that affect soil fertility, productivity, and the environment. In low-fertility ecosystems, application of nutrients through fertilizers regulates net primary productivity and SOM cycling [11,12]. Build-up of SOM definitely leads to improvement in soil health. However, over time, if the SOM level declines by soil microbial mineralization and/or other losses such as leaching and soil erosion, the soil health deteriorates not only in terms of many benefits including improvement in soil structure, increased soil C storage, and water holding capacity but also N nutrition of crop plants. Because of the fundamental coupling of microbial C and N cycling and the close correlation between soil C and N mineralization, the management practices that lead to loss of soil organic C (SOC) also have serious implications for the storage of N in soil. Thus loss of SOM can be inherently detrimental to crop productivity.

Dourado-Neto et al. [13] conducted a ^{15}N-recovery experiment in 13 diverse tropical agro-ecosystems and estimated the total recovery of one single ^{15}N application of inorganic N during three to six growing seasons. Between 7 and 58% (average of 21%) of crop N uptake (mean 147 $\pm$ 6 kg N ha^{-1}) during the first growing season was derived from fertilizer. On average, 79% of crop N was derived from the soil (Table 1). Average recoveries of ^{15}N-labeled fertilizer and residue in crops after the first growing season were 33 and 7%, respectively. Corresponding recoveries in the soil were 38 and 71%. After five growing seasons, more residue N (40%) than fertilizer N (18%) was recovered in the soil, better sustaining the N content in SOM. Making a worldwide evaluation of fertilizer N use efficiency in cereals, Ladha et al. [14] used data from 93 published studies and concluded that average ^{15}N fertilizer recovery in the grain and straw in maize, rice, and wheat in the first growing season was 40, 44, and 45%, respectively. Overall recovery based on ^{15}N dilution method among regions and crops was 44% (572 data points). The International Atomic Energy Agency [15] reported that the average percentage of single applications of ^{15}N fertilizer recovered in above-ground portion of the crop plants in the subsequent five growing seasons (excluding the crop to which ^{15}N fertilizer was applied) across all locations was 5.7 to 7.1%. Thus, with an average ^{15}N fertilizer recovery of 44% in the first crop of a cropping system [14], the total recovery of ^{15}N fertilizer in the first and the five subsequent crops is approximately 50%. Assuming that amount of ^{15}N in the roots becomes negligible in the sixth growing season, large portion of remaining 50% of the ^{15}N fertilizer will become part of the large soil N pool and some portion may get lost from the cropping system [16]. Thus, N bound to C in the SOM is not only the largest source of N for the crop plants but also the largest sink of N fertilizer inputs in modern cereal cropping systems, so that SOC impacts both crop yield and N losses to the environment.

Table 1. Total above-ground N accumulation and contribution of fertilizer N and soil N as estimated by applying ^{15}N labelled fertilizers for crops grown under diverse soil and climatic conditions.

Country	Soil Order	Crop	Fertilizer N Applied (kg N ha^{-1})	Total Crop N (kg N ha^{-1})	Derived from Fertilizer N (%)	Derived from Soil N (%)
Bangladesh	Haplaquepts	Wheat	60	60 ± 3	43 ± 1	57 ± 1
Brazil	Ultisol	Sugarcane	63	251 ± 7	16 ± 1	84 ± 1
Chile	Andisol	Maize	300	178 ± 7	31 ± 2	69 ± 2
Chile	Andisol	Wheat	160	124 ± 4	16 ± 2	84 ± 2
China	Inceptisol	Rice	60	292 ± 7	7 ± <1	93 ± <1
Egypt	Entisol	Wheat	60	80 ± 6	20 ± 1	80 ± 1
Malaysia	Ultisol	Maize	60	53 ± 2	23 ± 1	77 ± 1
Morocco	Aridisol	Wheat	42	161 ± 7	18 ± 1	82 ± 1
Morocco	Inceptisol	Sunflower	35	129 ± 7	7 ± <1	93 ± <1
Morocco	Inceptisol	Bean	85	225 ± 6	7 ± <1	93 ± <1
Sri Lanka	Ultisol	Maize	60	139 ± 6	11 ± <1	89 ± <1
Sri Lanka	Ultisol	Maize	60	139 ± 6	18 ± 1	82 ± 1
Vietnam	Ultisol	Maize	120	92 ± 3	58 ± 1	42 ± 1
		Mean		147 ± 6	21 ± 1	79 ± 1

Modified from Dourado-Neto et al. [13].

Plant uptake of native soil N is boosted either through increase in mineralization of soil N or by plant-mediated processes such as increased root growth and rhizosphere N priming [17,18]. Native soil N priming dynamics are influenced by soil type, fertilizer type, and environmental factors [19–21]. Using a meta-analysis based on 43 ^{15}N studies from all over the globe, Liu et al. [22] revealed fertilizer N effects on mineralization and plant uptake of native soil N were not influenced by study type (laboratory or field), location and duration, soil texture, C and N content, and pH. Although fertilizer tended to increase N priming through variable effects on native soil N mineralization, plant uptake of native soil N increased consistently. This inconsistency suggested that there exists a complex interaction between fertilizer N addition and microbial immobilization-mineralization of N and C, but not that fertilizer N application results in loss of SOM.

Potentially, fertilizer N application can affect SOM in two ways: (i) it may increase SOM by promoting plant growth and increasing the amount of litter and root biomass added to soil compared with the soil not receiving fertilizer N; and (ii) it may accelerate SOM loss through decay or microbial transformation of litter (leaves, straw, manures) and indigenous forms of organic C already present in the soil [9]. The first mechanism is widely accepted, but the second mechanism has not been demonstrated indisputably. Normally, SOM decreases with cultivation [3,23,24] when no N fertilizer is applied. Application of fertilizer N often increases SOM level and C sequestration in soils of intensively managed multiple cropping systems [25–30]. Ghimire et al. [26] have cited a number of long-term fertility experiments from India and Nepal in which SOC in control plots after 20 years ranged from 1.9 to 7.3 g kg^{-1}, but in all the experiments application of optimum N, P and K fertilizers registered an increase in SOC over control ranging from 0.2 to 3.5 g kg^{-1}. Also, fertilizer use could promote aggregate formation [31] and stabilization [32], and enhance the spatial inaccessibility for decomposing organisms [33].

Poffenbarger et al. [34] evaluated changes in surface SOC over 14 to 16 years by applying fertilizer N rates empirically determined to be insufficient, optimum, or excessive for maximum maize yield. It was observed that SOC balances were negative when no N was applied. For continuous maize, the rate of SOC storage increased with increasing N rate, reaching a maximum at the optimum N rate but decreasing above the optimum N rate. When fertilizer N application rate was below the optimum, applied N stimulated crop growth, leading to increasing crop residue inputs to the soil and, in turn, increasing the rate of soil organic storage. However, when the N application rate was above the optimum, added N did not increase crop residue production beyond that observed at the optimum level but increased residual inorganic N, which enhanced SOC mineralization leading to loss of SOC. Green et al. [35] also observed that annual additions of more N than needed to maximize yields of maize could cause losses of SOM and suggested that reducing unnecessary fertilization could help conserve

SOM. Conceptual understanding of the SOC response to N fertilization is illustrated in Figure 1 [34]. Residual soil inorganic N produced due to application of fertilizer N beyond the optimum level may enhance mineralization of SOC by eliminating N limitation on microbial growth [35,36] or by adversely affecting soil aggregation [37,38], which makes previously protected SOM more susceptible to decay. Excessive N fertilization may also decrease the C:N ratio of crop residues [39] and enhance their decomposition rate. There may be multiple processes controlling the SOC response to N fertilization, but the extent of increased C inputs vis-à-vis SOC mineralization depends on the N sufficiency level.

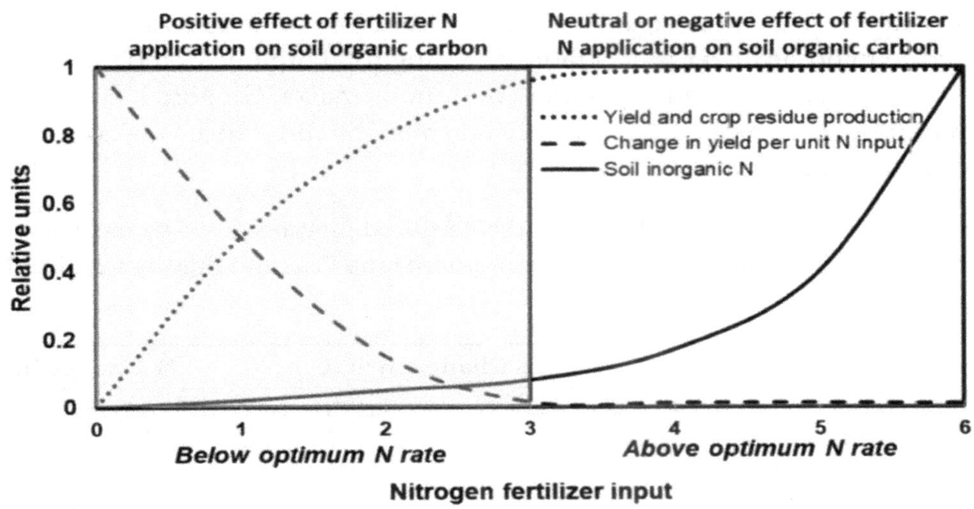

Figure 1. Conceptual diagram showing possible effects of fertilizer application to crops on SOC as defined by relationships between increasing fertilizer N application levels and (i) yield and crop residue production, (ii) change in yield per unit N input, and (iii) residual soil inorganic N. Maximum yield of the crop is obtained at the optimum N rate. Expected SOC responses to fertilizer N application below and above optimum N rate are shown above the grey and white areas of the plots, respectively (Modified from Poffenbarger et al. [34]).

Glendining and Powlson [40] found that in 84% comparisons in 45 long-term experiments in temperate regions, applications of fertilizer N on long-term basis increased total soil organic N (SON) as compared to in the treatments receiving no fertilizer. However, Khan et al. [41] and Mulvaney et al. [36] reported that in long-term experiments located in both temperate and tropical regions, continuous application of fertilizer N induced a net loss of SOC in 73% sites and reduction in soil N at 92% of the sites examined. Powlson et al. [42] argued that data sets used by these authors were not comprehensive enough, and long-term changes in soil N and C in the zero-N control plots were not taken into consideration. Ladha et al. [43] resolved this controversy using data from 135 studies of 114 long-term experiments located at 100 sites located all over the world. The data pertaining to SOC and SON were analyzed following time-response ratio and time by fertilizer N response ratio. The time-response ratio is a percentage change in total SOC or N compared with the initial amount, and it was calculated separately for both zero-N and N-fertilized treatments. Khan et al. [41] and Mulvaney [36] used this approach, and like them Ladha [43] also observed an average decline in SOC to the tune of 16% and 10% in zero-N and fertilizer N amended plots; corresponding decline in SON was 11% and 4% (Table 2). These decreases were confounded with decrease in SOM content occurring independently of the use of fertilizer N. Ladha et al. [43] separated the two processes by following the change over time in SOM content with or without fertilizer, and this was done by analyzing the data using time by fertilizer N response ratio. While the time-response ratio addressed the impact of the whole system (tillage, residue management, erosion, fertilizer amendment) on changes in SOC or SON, the time by fertilizer N response ratio specifically assessed the impact of fertilizer N amendment, and it is defined as the percentage difference between the change in SOC or N in the N-fertilized treatments compared with the changes in zero-N treatment. Using the time by fertilizer ratio, which is based on changes in the paired comparisons at the initiation of the long-term experiments and final sampling period, Ladha et al. [43] observed overall averages of 8% higher SOC and 10% higher SON with fertilizer N than with zero-N (Table 2). Furthermore, the positive effect of fertilizer N in tropical, humid subtropical, and temperate soils ranged from 3 to 16% for SOC and 8 to 15% for SON, with the highest increases observed in the tropical environment (Table 2). Due to inherently lower status of SOC and N than in temperate soils, the relatively higher positive effect of fertilizer N application is expected in tropical soils. Recently, Geiseller and Scow [44] and Körschens et al. [45] also observed that in long-term experiments from all over the world, application of mineral fertilizers leads to increase in SOM as compared to in no-fertilizer plots (Table 3). Using total organic C and natural ^{13}C abundance measurements in a long-term experiment under continuous maize, Gregorich et al. [46] observed that fertilized soils had more organic C than unfertilized soils; the difference was accounted for by more C4-derived C in the fertilized soils.

Table 2. Changes in SOC and SON in zero-N and N fertilized plots observed by meta-analysis of data from 114 long-term experiments following time-response ratio (TR) and time by fertilizer N response ratio (TNR).

	% Change in SOC		% Change in SON	
	Zero-N	Fertilizer N	Zero-N	Fertilizer N
TR: overall changes	−16	−10	−11	−4
TNR: overall changes	-	8	-	10
TNR: changes in tropical soils	-	16	-	15
TNR: changes in humid tropical soils	-	11	-	11
TNR: changes in temperate soils	-	3	-	8

Data source: Ladha et al. [43].

Table 3. Increase in SOC due to fertilizer application as compared to in the unfertilized controls in meta-analysis conducted on long-term experiments from all over the world.

Crops	Region	Duration of Long-Term Experiments (years)	Increase in SOC (%)	Reference
Non-lowland rice crops	World	5–130	12.8	Geiseller and Scow [44]
Cereal crops	World	6–158	8	Ladha et al. [43]
Wheat, barley, oats, sugar beets, potato, maize, sorghum, rye	Europe	16–108	10	Körschens et al. [45]

The North Indian state of Punjab is the most intensively cultivated region in India, with a cropping intensity of 190%, predominantly of a rice–wheat cropping system. A study based on 0.319 million soil samples of the 0–20 cm plough layer analyzed during 25 year period between 1981/82 to 2005/06 revealed that as a weighted average for the whole state, SOC increased from 2.9 g kg^{-1} in 1981/82 to 4.0 g kg^{-1} in 2005/06, an increase of 38% [47]. A close relationship (R^2 = 0.79) between SOC stocks in the plough layer and total rice and wheat grain yield during the 25-year period was observed. Increased productivity of rice and wheat resulted in enhanced C accumulation in the plough layer by 0.8 t C ha^{-1} t^{-1} of increased grain production. The increased productivity of both rice and wheat in the Punjab was achieved through increasing fertilizer (N, P, and K) use from 0.762 Mt in 1980/81 to 1.687 Mt in 2005/06 or from 112.5 kg ha^{-1} in 1980/81 to 214.0 kg ha^{-1}. Soil pH declined by 0.8 pH units from 8.5 in 1981/82 to 7.7 in 2005/06. This pH decline has positive implications for availability of P and micronutrients such as Zn, Fe, and Mn. Tian et al. [48] conducted a meta-analysis of paired-treatment data from 95 long-term field experiments published from 1980 to 2012 to characterize the changes in SOC in paddy soils in China. While significant increase in the SOC was observed in the optimum fertilizer N, P, and K fertilizer treatment as compared to in the no-fertilizer treatment; the mean difference in SOC change rates between the two treatments was measured to be 0.140 ± 0.023 g kg^{-1} $year^{-1}$. Using a meta-analysis based on 257 published studies, Lu et al. [49] revealed that despite increased soil respiration, there was a significant 3.5% increase in C storage in agricultural ecosystems due to application of N. The N-induced change in soil C storage was related to changes in below-ground production rather than above-ground growth. Russel et al. [39] also observed that quantity of below-ground organic C inputs was the best predictor of long-term soil C storage. Shang et al. [50] conducted a meta-analysis based on published data on crop yields and soil parameters from long-term experiments in maize-wheat, rice-rice, and rice-wheat cropping systems in China. Although conservation of SOC in upland maize-wheat system was conspicuously less than in the rice based cropping systems, application of optimum rate of N, P, and K fertilizers resulted in build-up of SOC over no-fertilizer control in all the three cropping systems (Table 4). Decrease in SOC content in the no-fertilizer control from the initial values in the completely aerobic maize-wheat cropping system should be due to cultivation of the soil.

Table 4. Average SOC content at the start (initial) of long-term experiments on maize-wheat, rice-wheat, and rice-rice cropping systems and in no-fertilizer (N, P, and K) control and optimum N, P, and K fertilizer level treatments at the end of the experiments in different locations in China.

Cropping System	Number of Experiments	Duration (years)	SOC (g kg^{-1})		
			Initial	No-Fertilizer Control	Optimum N, P and K Fertilizer Levels
Maize-wheat	12	6–25	6.4	5.8	6.8
Rice-wheat	10	9–27	14.3	14.9	16.3
Rice-rice	23	6–26	16.7	18.1	19.6

Data Source: Shang et al. [50].

Cultivation invariably reduces SOM levels to an extent that depends on management and inputs. In well managed cultivated soils, SOC fluctuated between a low steady state value of SOM in the heavily cultivated soil and the highest value observed in the uncultivated soil [51]. Cultivation of the soil leads to lower equilibrium soil C levels, but the addition of fertilizers reduces the extent of SOM decline observed with cultivation. Katyal et al. [52] critically analyzed data from several long-term fertility experiments in India and documented such changes. Twenty years after initiation of a long-term experiment in a virgin soil, SOM content in the no-fertilizer control reached 34% of the initial value and seemed to have stabilized at a lower equilibrium level as defined by Buyanovsky and Wagner [51]. Loss in SOM was obviously due to cultivation of the virgin soil. Buyanovsky and Wagner [51] reported a decline in native organic matter between 20 and 40% within 5 years after opening of virgin land. However, when optimum level of fertilizers was applied, SOM remained stable over the first decade, but in the next 3 years fell to about 40% of the initial value. In contrast to a virgin soil, already cultivated soil implies that the soil had already shifted to a new dynamic equilibrium but had probably not yet reached the steady state low value of SOM in the heavily cultivated soil. In long-term experiments initiated in soil already under cultivation, SOM declined without any fertilizer application. However, SOM levels were either maintained or increased when adequate amount of N, P, and K fertilizers was applied [52]. This conclusion was valid, irrespective of the location or the cropping system. That soil health in terms of SOC and SON declines when soil is tilled year after year is now an established fact [3,23,24]. Therefore, interaction between tillage and fertilizer use should be taken into account when interpreting changes with time in the SOM in long-term experiments.

4. Effect of Fertilizer Use on Microbial Life Ion the Soil

Several ecosystem services or the beneficial functions provided by soil are driven by many interrelated and complex biological processes. The concept of soil health takes into account not only the soil biota and the myriad of biotic interactions that occur, but also considers that the soil provides a living space for the biota. Microorganisms and various by-products of their metabolism play an important role in the formation of soil aggregates and in soil structure maintenance. Since soil constitutes an open system, its integrity or health is affected by external environmental and anthropogenic pressures. Recently, Hermans et al. [53] observed that soil bacterial communities and their relative abundances varied more in response to changing soil environments than in response to changes in climate or increasing geographic distance. As microorganisms play an important role in maintaining fertile and productive soils, the effect of fertilizers on microbial communities has potentially important implications for sustainable agriculture. Applied nutrients constitute a controlling input to the soil system and the processes within it, but adequate knowledge is lacking about the impacts of nutrient additions on the condition of different assemblages of soil organisms. According to O'Donnell et al. [54], fertilizers do affect microbial community structure, but the relationship between diversity, community structure, and function remains complex and difficult to interpret using currently available chemical and molecular fingerprinting techniques. Mineral fertilizers interact with microbial communities in the soil in a number of ways and affect the population, composition, and function of soil microorganisms [55]. These may promote growth of microbes directly by providing nutrients and indirectly by stimulating plant growth and enhancing root C flow [56]. However, fertilizers, particularly N, when applied to soil may result in soil acidification limiting microbial growth and activity in soils [57]. Several studies conducted during last 2–3 decades have revealed that fertilizer application usually favours the accumulation of bacterial residues [58] and increases soil microbial biomass [59–63]. In some studies, fertilizer application increased biomass C and N [64–66]. Significant improvement in soil quality in terms of increased SOC and soil microbial biomass due to long-term application of fertilizers in maize–wheat cropping systems has been reported by Li et al. [67] and Liu et al. [68].

Mbuthia et al. [69] observed that fertilizer N application to cotton continuously for 31 years significantly increased soil microbial biomass N, mycorrhizae fungi biomarkers, b-glucosaminidase (N-cycling) activity, and basal microbial respiration rates. In a study in which inorganic fertilizers were continuously applied for 13 years to flooded double rice crop, Zhong and Cai [70] found that stimulation of microbial biomass and community functional diversity by fertilizer N could be achieved only after improvement of the P supply. However, most microbial parameters were correlated with SOC content, indicating that the application of nutrients through fertilizers affected microbial parameters in the soil indirectly by increasing the accumulation of SOM. It is generally considered that the primary limiting factor for microbial activity in soils is the availability of C substrate. However, soil microbes may frequently be limited by the supply of N in the soil [71]. When demand for N exceeds its supply, the functional capacity of the soil system is strongly influenced by N availability. Under such situations in agro-ecosystems, soil health declines without additional inputs of N via fertilizers or organic manures, and particularly without due consideration of the associated C requirements of the biomass [37].

Effect of fertilizer application on the soil biota can be positive or negative and vary in duration, depending upon the type and amount of fertilizer used and mode of application. For example, potential damage to soil microorganisms from high concentration of ammonia fertilizer applied in bands is usually short-term, and only in the zone of application. Angus et al. [72] reported that injection of urea and ammonia in bands generally exhibited a short-term effect on microbial activity in the soil. Total microbial activity was reduced in narrow bands of application for a period of 5 weeks, after which levels returned to normal. However, an 80% reduction in the number of protozoa did not return to normal after 5 weeks. On the other hand, there was a large increase in the number of nitrifying bacteria in the soil 5 weeks after application of urea/ammonia in bands. Geiseller and Scow [44] carried out a meta-analysis based on 107 data sets from 64 long-term experiments from around the world and revealed that application of mineral fertilizers resulted in a significant increase (15.1%) in the microbial biomass above levels in the no-fertilizer control treatments. Where soil pH was 7 or higher, the fertilizer induced increase in microbial biomass averaged 48%, but fertilizer application tended to reduce microbial biomass in soils with a pH below 5 (Table 5). Furthermore, the increase in microbial biomass was the highest in experiments that were in place for at least 20 years. Biederbeck et al. [73] also reported little impact on soil microbial populations when urea and anhydrous ammonia were applied continuously for 10 years. The arbuscular mycorrhizal fungi biomass was increased by application of N and P fertilizer in the N- and P-deficient sites, respectively [74].

Table 5. Unweighted averages of soil microbial biomass C (mg kg^{-1}) in fertilizer N (+N) and no-N treatments in 64 non-lowland rice long-term experiments from all over the world.

	Number of Data Sets	Soil Microbial Biomass C (mg kg^{-1})	
		no-N	+N
All data sets	107	238	268
pH in +N treatment: <5	17	240	213
pH in +N treatment: 5–7	39	234	253
pH in +N treatment: 7 or higher	17	139	205
Duration of long-term experiment: 5–10 years	18	300	239
Duration of long-term experiment: 10–20 years	34	227	270
Duration of long-term experiment: 20 years or longer	55	224	276

Modified from Geiseller and Scow [44].

That tilling of soil leads to decline of its health is also revealed by changes in microbial community structure assessed using phospholipid fatty acid analysis and automated ribosomal intergenic spacer analysis [75,76]. In a study conducted by Doran [77], microbial biomass and potentially mineralizable N levels of no-tillage soils averaged 54% and 37% higher, respectively, than those in the ploughed soils. In a meta-analysis based on 139 observations from 62 studies, Zuber and Villamil [78] inferred that microbial biomass and enzyme activities were greater under no-till as compared to in the tilled soils. Therefore, in conventionally tilled fertilized soils the reduced microbial activity is due to cultivation of soils rather than the effect of fertilizer application.

Over-use of mineral fertilizers and excessive tillage can affect biological communities in the soil by damaging their habitats and disrupting their functions [37]. Over-use of fertilizer, particularly N, is like enrichment of ecosystems with reactive N. Using a meta-analysis based on 82 published field studies, Treseder [79] reported that microbial biomass declined 15% on average under heavy N fertilization, but fungi and bacteria were not significantly altered in studies that examined each group separately. Declines in abundance of microbes and fungi were more evident in studies of longer durations and with higher total amounts of N added.

5. Potential Contribution of Nitrogen Fertilizers to Soil Acidity

Nitrogen fertilizers can exert indirect negative effects on soil health arising through lowering of soil pH due to natural transformations of N in the soil. Soil pH is one of the most influential factors affecting the microbial community in soil. As shown in Table 5, while fertilizer-induced increase in microbial population in long-term experiments was observed at soil pH 7 or higher, a reduction in microbial biomass was observed in soils with a pH below 5. In a silt loam soil on which barley has been continuously grown for more than 100 years, Rousk et al. [80] observed a fivefold decrease in bacterial growth and a fivefold increase in fungal growth due to lowering of pH from 8.3 to 4.0.

Form of fertilizer N applied (NO_3^-, NH_4^+, urea), fertilizer product type (for example, ammonium nitrate, calcium ammonium nitrate), the net balance between proton-producing and consuming processes, and the buffering capacity of the soil dictate the extent of soil acidification due to application of fertilizer N. Buffering capacity of the soil as determined by the presence of solid-phase calcium carbonate resists change in soil pH due to N transformations [81]. In arid and semi-arid areas of the world, soils are generally calcareous and thus highly buffered. In temperate regions, soils are generally neutral or slightly acidic in reaction, whereas tropical soils are usually highly weathered and generally acidic with little or no buffering capacity. During the acidification process, base cations such as calcium and magnesium are released from the soil. With continued addition of fertilizer N, the base cations get depleted and aluminum (Al^{3+}) is released from soil minerals, often reaching toxic levels that induce nutrient disorders in plants. Guo et al. [82] reported severe soil acidification in large crop production areas in China following application of high fertilizer N rates between the 1980s and 2000s. Based on strictly paired data available from 154 agricultural fields, top soils were significantly acidified with an average pH decline of 0.50. Fertilizer N application released 20 to 221 kg hydrogen ion (H^+) ha^{-1} year^{-1}, and base cations uptake contributed a further 15 to 20 kg H^+ ha^{-1} year^{-1} to soil acidification. In Southern China, Lu et al. [83] observed that after application of ammonium nitrate for 6 years, the site was showing high acidification [pH(H_2O) < 4.0], negative water-extracted acid neutralizing capacity, and low base saturation (<8%) throughout soil profiles.

6. Rational Use of Fertilizers Enhances Soil Health by Reducing Soil Erosion

Role of anthropogenic activities in causing soil erosion is very well documented [84], but the connection between erodibility of the soil (defined as the susceptibility of a soil to become detached and transported by wind, water, or ice) and crop production practices, especially the use of fertilizers, is not well documented. Soil erosion is a problem when there is insufficient ground cover to protect the soil and reduce the impact of rainfall and wind on the soil surface and when aggregate stability is reduced due to limited SOC. Adequately fertilized crops will have extensive root system and top growth. A well-developed canopy reduces the pounding effect of water drops from rain so that runoff is reduced and erosion is minimized. Also, extensive root system developed in the well fertilized soil helps hold soil in place and decreases the potential for soil loss in runoff water. Bhattacharyya et al. [85] reported reduced loss of soil due to erosion by applying fertilizers to crops as compared to when no fertilizer was applied. At 2% slope, soil loss by erosion was reduced by 7.2% and 11.7% by applying fertilizer to sorghum (*Sorghum bicolor*) and chickpea (*Cicer arietinum*), respectively. According to Portch and Jin [86], balanced fertilization of crops in China could reduce soil erosion. They further reported that work conducted by International Board for Soil Research and Management (IBSRAM) in late 1980s in several Asian countries showed that fertilizer use alone could reduce soil erosion from 50 to 15 t ha^{-1} year^{-1}. Biological N fixation and manure recycling are the only local nutrient sources that are not always optimally exploited. The inability to match crop harvests with sufficient nutrient inputs leads to depletion of nutrients and SOM, declining soil health, and increased risk of land degradation through erosion.

7. Optimizing Fertilizer Management to Maintain Soil Health

A sustainable agricultural production system with good soil health having the capacity to produce high yields with fewer external nutrient inputs can be developed using the correct combination of ecosystem processes and appropriate use of fertilizers. Soils in agro-ecosystems should be able to supply a certain minimum level of plant-available N and other essential nutrients at different growth stages of crop plants. In principle, the concept of optimum fertilization aims at a dynamic balance between nutrient requirement to obtain high yields and nutrient uptake by crops. This is achieved by maintaining synchrony between nutrient demand of the crop and the supply of nutrients from all sources including fertilizer and soil throughout the growing season of the crop.

Application of optimum doses of all nutrients is important, but due to fundamental coupling of C and N cycles, optimization of fertilizer N management is more closely linked to build-up of SOC and soil health. Concepts emerging from the work of Poffenbarger et al. [34] and depicted in Figure 1 suggest that when N inputs are below the optimum rate at which maximum yield is obtained, applied N stimulates crop growth, increasing crop residue inputs to the soil and thereby increasing SOC. Additionally, when fertilizer N inputs are above the optimum level, added N imparts no change in crop residue production but increases residual inorganic N, which alleviates microbial N limitation and thereby enhances mineralization of SOC [35]. However, crop response to N fertilization is site-specific because there exists large spatial and temporal variability in soil N supply, which is in part due to historical differences in management. Regional blanket fertilization recommendations cannot account for this variability. Thus, site-specific nutrient management strategies based on principles of synchronization of crop N demand with N supply from all sources including soil and fertilizer N can ensure high yields along with maintenance of soil health. These can not only account for site-to-site variability in optimum fertilizer rate but also resolve uncertainty regarding response of SOC build-up to fertilizer application.

In the last two decades, site-specific real-time methods of N management that utilize crop simulation models, remote sensing, or on-the-go crop sensing/variable-rate N spreaders to determine the spatially variable needs for N at critical growth stages are increasingly being used to apply optimum doses of fertilizer N to crops following synchrony principles. Whether implemented for crops in small fields with little or no mechanization in developing countries or practiced as precision agriculture for variable rate adjustment using on-the-go canopy reflectance spectra in large fields of developed countries [87], the principles and objectives of site-specific N management are the same.

The first report of the Status of the World's Soil Resources prepared by the Intergovernmental Technical Panel on Soils lists nutrient imbalances (both nutrient deficiency and nutrient excess) as one of the specific threats to soil functions [88]. In a long-term field trial with spring barley, Johnston et al. [89] demonstrated that the grain yield increased by more than 50% with the same amount of fertilizer N only when the plants were grown on a soil well supplied with K. Similarly, barley cultivated on a P-deficient soil yielded only half of the crop, which was grown on a soil with adequate P, although receiving the same amount of fertilizer N. Haerdter and Fairhurst [90] showed that the recovery of N from fertilizers increased from 16% at traditional N and P fertilization levels to 76% at balanced application of N, P, and K fertilizers. Kumar and Yadav [91] reported higher SOM content in plots in which N, P, and K were applied in a balanced proportion on a long-term basis than in treatments receiving only N or inadequate amounts of P (Figure 2). Similarly, Belay et al. [92] observed more SOC and soil microbial biomass in the N, P, and K fertilizer treatment rather than in N, P, or K alone fertilizer treatments in a long-term field experiment on maize-field pea rotation initiated in 1939 in South Africa.

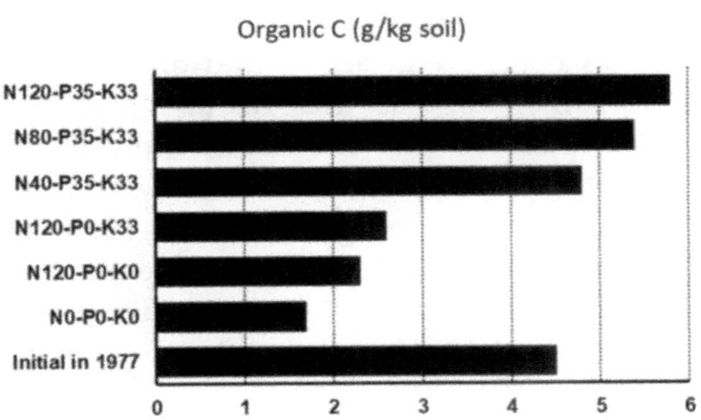

Figure 2. Effect of application of different combinations of N, P, and K fertilizers to rice–wheat cropping system for 20 years on organic C content in the soil in a long-term experiment at Faizabad, India. The numbers after N, P, and K indicate kg ha^{-1}. Data source: Kumar and Yadav [91].

In a 16-year long-term field experiment, Chu et al. [93] observed that balanced application of N, P, and K fertilizers had a higher microbial biomass and activity than in the P- and N-deficient treatments. Balanced fertilization resulted in higher dehydrogenase activity than under nutrient-deficiency fertilization. In a 33-year long-term experiment in a brown soil in China, long-term N and P, as well as N, P, and K, fertilizer application treatments exhibited greatly increased soil microbial biomass C and dehydrogenase activity compared to in the only N treatment [94]. Similarly, in a 21-year long-term experiment, Zhong et al. [95] observed that balanced fertilization with N, P, and K promoted the soil microbial biomass, activity, and diversity and thus enhanced soil health, crop growth, and production. In a wheat-maize cropping system in a fluvo-aquic soil in the North China Plain, Gong et al. [27] reported that balanced application of N, P, and K fertilizers for 18 years showed higher C and N contents of the light and heavy fractions, as well as more culturable microbial counts, than in unbalanced N and P, P and K, or N and K fertilizer treatments.

8. Integrated Management of Fertilizers and Organic Manures for Improvement of Soil Health

With increasing awareness about soil health and sustainability in agriculture, organic manures have regained importance, because these can supply precious organic matter, along with many different nutrients, including micronutrients to the soil. Organic manures also influence the availability of plant nutrients in the soil for plants by changing both the physical and biological characteristics of the soil. The concept of integrated management of mineral fertilizers and organic manures became the mainstay of soil fertility management practices at the turn of the 20th century, because it strives to maintain/improve the fertility and health of the soil for sustained crop productivity on a long-term basis [96]. Nutrients supplied through fertilizers are used to supplement those supplied by the different organic sources available to farmers. In Sub-Saharan Africa, where the traditional farming systems depend primarily on mining soil nutrients, the concept of integrated soil fertility management based on the use of mineral fertilizers, organic inputs, and improved germplasm, combined with the knowledge of adapting these practices to local conditions, has been introduced to intensify agriculture. Fertilizers constitute an entry point for practicing integrated soil fertility management, which is a field-specific strategy for increasing productivity, improving soil health, and a sustainable cropping system [97].

In several long-term experiments initiated in 1970s with different cropping systems in various agro-climatic zones in India, along with several other treatments, the two consisted of application of optimum level of N, P, and K fertilizers with and without farmyard manure. Soil organic C in different treatments estimated at the initiation of the experiments and 20 years later is shown in Table 6. The data convincingly proves that integrated management of mineral fertilizers and farmyard manure resulted in build-up of SOC more than in the fertilizer only treatment. Nevertheless, as already discussed, application of optimum levels of N, P, and K fertilizers resulted in accumulation of SOC more than in the control treatment to which neither fertilizer nor manure was applied. In recent years, several other workers [27,29,32,98–102] have reported that the application of organic manures along with mineral fertilizers increases SOM and different fractions of SOC more effectively than the application of mineral fertilizers alone. Integrated management of organic manures and mineral fertilizers rather than application of fertilizers alone not only has a positive impact on build-up of SOC but also on soil health related microbial indicators like soil microbial biomass, soil bacterial community diversities, and soil enzyme activities [67,103,104].

Table 6. Changes in SOC due to application of optimum N, P, and K fertilizer levels with and without farmyard manure for 20 years to different cropping systems in long-term experiments established in different soil types in India.

Cropping System	Location	Soil	SOC at Initiation (%)	SOC after 20 Years (%)		
				Control	N, P and K Fertilizers	N, P and K Fertilizers + Farmyard Manure
Rice-rice	Bhubaneshwar	Inceptisol	0.27	0.41	0.59	0.76
Rice-wheat	Pantnagar	Mollisol	1.48	0.50	0.95	1.51
Rice-wheat	Faizabad	Inceptisol	0.37	0.19	0.40	0.50
Rice-wheat-jute	Barrackpore	Inceptisol	0.71	0.42	0.45	0.52
Rice-wheat-cowpea	Pantnagar	Mollisol	1.48	0.60	0.90	1.44
Maize-wheat	Palampur	Alfisol	0.79	0.62	0.83	1.20
Rice-wheat	Karnal	Alfisol	0.23	0.30	0.32	0.35
Cassava	Trivandrum	Ultisol	0.70	0.26	0.60	0.98

Data source: Nambiar [105], Swarup et al. [106].

In Sub-Saharan Africa, two types of soils have been recognized in terms of responsiveness to mineral fertilizers. One type of soils are termed as responsive soils, because, due to nutrient mining, crops grown in these soils respond to fertilizer application in a normal way. The other type of soils are referred to as poor, less-responsive soils because these are highly degraded in terms of both extensive nutrient mining and loss of SOM, and crops grown in these respond to fertilizer use minimally or do not respond [107]. The degradation of soil to non-responsive state occurs due to discontinuous, insufficient, or no fertilizer application over a certain period of time. When a certain threshold of soil degradation is exceeded, this condition may not be reversible and soils may not respond immediately to fertilizer or organic manure application so that crop productivity may not return to the level attained before fertilizer use was discontinued. In a study conducted by Zingore et al. [108], response to fertilizer application on less-responsive soils was observed only after application of 17 t ha^{-1} $year^{-1}$ of farmyard manure during three consecutive years. Once the soil became responsive to fertilizers, improvement in agronomic efficiency and soil health could be achieved through integrated nutrient management of fertilizers and farmyard manure. This unique interaction of organic manures and fertilizers seems to be very valuable in dealing with soils degraded due to long history of nutrient depletion.

9. Conclusions and Policy Implications

Nitrogen fertilizers, when applied at rates less than the optimum at which maximum yields are obtained, stimulate crop growth, leading to increasing crop residue inputs to the soil and, in turn, increasing the rate of soil organic storage. Until and unless fertilizer N acidifies the soil to pH < 5, the application of fertilizer at optimal rate generally has a positive effect on soil biota. The balanced application of N, P, and K fertilizers results in further significant improvement in the soil health in terms of increased SOC and soil microbial biomass. The uptake of N by crop plants is generally greater from native soil N than from N applied as fertilizers. As a decline in SOM following the application of fertilizer N is not a general phenomenon, a spiral of decline in soil functioning and crop productivity due to fertilizer N use is not expected. Application of fertilizers more than the optimum level can not only adversely influence biological communities in the soil but may also result in increased residual inorganic N, which can enhance SOC mineralization and loss of SOC. Because there exists large spatial and temporal variability in soil N supply, crop response to N fertilization is site-specific. Thus, site-specific nutrient management strategies based on principles of synchronization of crop N demand with N supply from all sources including soil and fertilizer N hold great potential for ensuring high yields of crops along with maintenance or improvement in soil health.

Soil and agronomic research reviewed and analysed in this paper shows that sustainable agricultural intensification through application of fertilizer N and healthy soils are compatible goals. The extent to which fertilizer N can contribute to economic and efficient crop production, and concomitantly benefit the soil in terms of quality or health, is dictated by the adoption of management practices that ensure that fertilizer N is not applied indiscriminately to agricultural crops. Fertilizer N should never be applied in amounts greater than what is required to obtain optimum yields. Ideally, fertilizer N should be managed on a site-specific basis, whether based on the nutrient status of soil or plants in a given field, so that N is applied in the right amount and at a right time according to the needs of the soil-plant system. The application of fertilizer N in a balanced proportion with other nutrients and integrated nutrient management based on organic manures and fertilizers can lead to further improvements in soil health.

The effect of temperature and moisture on SOM decomposition is very well documented in the literature. However, hardly any studies are available in which the interaction effects of fertilizer N and temperature and moisture on SOM decomposition are reported. This information is needed to evaluate the effect of fertilizer use on soil health under different temperature and moisture regimes. While studies related to soil health and fertilizer N are being reported from different climatic regions of the world, models can be usefully employed to define the specific effects of rainfall or soil moisture and soil temperature on fertilizer N-related soil health issues. The response of different microbial groups to repeated applications of fertilizer N varies and depends on environmental and crop management-related factors. As enough data are not available to understand the interactions among environmental factors, fertilizer N rates and types, and specific groups of soil microorganisms, there is a need to conduct studies to understand these complex interactions. Also, there is a need for adequate documentation of the effect of fertilizer N on the stability of SOM and the fate of organic residues in the long-term in different cropping systems. Long-term agronomic experiments involving the application of fertilizers in different agro-ecological zones across the world can be used to generate information on these lines. Increased soil salinity due to application of mineral fertilizers can deteriorate soil health, but N fertilizers based on sodium salts are no longer applied to field crops. In the quest to reduce the cost of cultivation and possibly maintain and/or improve soil health, in many parts of the world conservation agriculture systems are being adopted. In these systems, soil is tilled to a minimum extent and crop residues are retained in the soil so as to help build up of SOM. There is a need to establish appropriate fertilizer management strategies in such systems so that soil health is maintained or improved.

References

1. Doran, J.W.; Parkin, T.B. Defining and assessing soil quality. In *Defining Soil Quality for a Sustainable Environment*; SSSA Special Publication 35; Doran, J.W., Coleman, D.C., Bezdicek, D.F., Stewart, B.A., Eds.; Soil Science Society of America: Madison, WI, USA, 1994; pp. 1–21. [CrossRef]
2. Lewandowski, A.; Zumwinkle, M.; Fish, A. *Assessing the Soil System: A Review of Soil Quality Literature*; Minnesota Department of Agriculture, Energy and Sustainable Agriculture Program: St. Paul, MN, USA, 1999.
3. Kibblewhite, M.G.; Ritz, K.; Swift, M.J. Soil health in agricultural systems. *Philos. Trans. R. Soc. B* **2008**, *363*, 685–701. [CrossRef] [PubMed]
4. Lal, R.; Stewart, B.A. *Food Security and Soil Quality*; CRC Press: Boca Raton, FL, USA, 2010.
5. Alexandratos, N.; Bruinsma, J. *World Agriculture towards 2030/2050: The 2012 Revision*; ESA Working Paper No. 12-03; Food and Agriculture Organization of the United Nations: Rome, Italy, 2012.
6. Searchinger, T.; Hanson, C.; Ranganathan, J.; Lipinski, B.; Waite, R.; Winterbottom, R.; Dinshaw, A.; Heimlich, R. *Creating a Sustainable Food Future. A Menu of Solutions to Sustainably Feed More Than 9 Billion People by 2050*; World Resources Report 2013–14: Interim Findings; World Resources Institute: Washington, DC, USA, 2014.
7. Erisman, J.W.; Sutton, M.A.; Galloway, J.N.; Klimont, Z.; Winiwarter, W. How a century of ammonia synthesis changed the world? *Nat. Geosci.* **2008**, *1*, 636–639. [CrossRef]
8. IFADATA. Available online: http://ifadata.fertilizer.org/ucSearch.aspx (accessed on 20 February 2018).
9. Recous, S.; Robin, D.; Darwis, D.; Mary, B. Soil inorganic nitrogen availability: Effect on maize residue decomposition. *Soil Biol. Biochem.* **1995**, *27*, 1529–1538. [CrossRef]
10. Wardle, D.A. A comparative assessment of factors which influence microbial biomass carbon and nitrogen levels in soil. *Biol. Rev.* **1992**, *67*, 321–358. [CrossRef]
11. Kirkby, C.A.; Richardson, A.E.; Wade, L.J.; Batten, G.D.; Blanchard, C.; Kirkegaard, J.A. Carbon-nutrient stoichiometry to increase soil carbon sequestration. *Soil Biol. Biochem.* **2013**, *60*, 77–86. [CrossRef]
12. Zhang, H.; Ding, W.; Yu, H.; He, X. Linking organic carbon accumulation to microbial community dynamics in a sandy loam soil: Result of 20 years compost and inorganic fertilizers repeated application experiment. *Biol. Fertil. Soils* **2015**, *51*, 137–150. [CrossRef]
13. Dourado-Neto, D.; Powlson, D.; Abu Bakar, R.; Bacchi, O.O.S.; Basanta, M.V.; thi Cong, P.; Keerthisinghe, G.; Ismaili, M.; Rahman, S.M.; Reichardt, K.; et al. Multiseason recoveries of organic and inorganic nitrogen-15 in tropical cropping systems. *Soil Sci. Soc. Am. J.* **2010**, *74*, 139–152. [CrossRef]
14. Ladha, J.K.; Pathak, H.; Krupnik, T.J.; Six, J.; van Kessel, C. Efficiency of fertilizer nitrogen in cereal production: Retrospects and prospects. *Adv. Agron.* **2005**, *87*, 85–156. [CrossRef]
15. IAEA (International Atomic, Energy Agency). *Management of Crop Residues for Sustainable Crop Production*; IAEA TECHDOC-1354; International Atomic, Energy Agency: Vienna, Austria, 2003.

16. Jansson, S.L.; Persson, J. Mineralization and immobilization of soil nitrogen. In *Nitrogen in Agricultural Soils*; Agronomy Monograph 22; Stevenson, F.J., Ed.; ASA, CSSA, and SSSA: Madison, WI, USA, 1982; pp. 229–252.
17. Jenkinson, D.S.; Fox, R.H.; Rayner, J.H. Interactions between fertilizer nitrogen and soil nitrogen—The so-called priming effect. *Eur. J. Soil Sci.* **1985**, *36*, 425–444. [CrossRef]
18. Schimel, J.P.; Bennett, J. Nitrogen mineralization: Challenges of a changing paradigm. *Ecology* **2004**, *85*, 591–602. [CrossRef]
19. Glendining, M.J.; Poulton, P.R.; Powlson, D.S.; Jenkinson, D.S. Fate of ^{15}N-labelled fertilizer applied to spring barley grown on soils of contrasting nutrient status. *Plant Soil* **1997**, *195*, 83–98. [CrossRef]
20. Liu, X.-J.A.; Sun, J.; Mau, R.L.; Finley, B.K.; Compson, Z.G.; van Gestel, N.; Brown, J.R.; Schwartz, E.; Dijkstra, P.; Hungate, B.A. Labile carbon input determines the direction and magnitude of the priming effect. *Appl. Soil Ecol.* **2017**, *109*, 7–13. [CrossRef]
21. Recous, S.; Fresneau, C.; Faurie, G.; Mary, B. The fate of labelled ^{15}N urea and ammonium nitrate applied to a winter wheat crop I. Nitrogen transformations in the soil. *Plant Soil* **1988**, *112*, 205–214. [CrossRef]
22. Liu, X.-J.A.; van Groenigen, K.J.; Dijkstra, P.; Hungate, B.A. Increased plant uptake of native soil nitrogen following fertilizer addition—Not a priming effect? *Appl. Soil Ecol.* **2017**, *114*, 105–110. [CrossRef]
23. Haddaway, N.R.; Hedlund, K.; Jackson, L.E.; Kätterer, T.; Lugato, E.; Thomsen, I.K.; Jørgensen, H.B.; Isberg, P.E. How does tillage intensity affect soil organic carbon? A systematic review. *Environ. Evid.* **2017**, *6*, 30. [CrossRef]
24. Liu, X.; Herbert, S.J.; Hashemi, A.M.; Zhang, X.; Ding, G. Effects of agricultural management on soil organic matter and carbon transformation—A review. *Plant Soil Environ.* **2006**, *52*, 531–543. [CrossRef]
25. Cong, R.H.; Xu, M.G.; Wang, X.J.; Zhang, W.J.; Yang, X.Y.; Huang, S.M.; Wang, B.R. An analysis of soil carbon dynamics in long-term soil fertility trials in China. *Nutr. Cycl. Agroecosyst.* **2012**, *93*, 201–213. [CrossRef]
26. Ghimire, R.; Lamichhane, S.; Acharya, B.S.; Bista, P.; Sainju, U.M. Tillage, crop residue, and nutrient management effects on soil organic carbon in rice-based cropping systems: A review. *J. Integr. Agric.* **2017**, *16*, 1–15. [CrossRef]
27. Gong, W.; Yan, X.; Wang, J.; Hu, T.; Gong, Y. Long-term manure and fertilizer effects on soil organic matter fractions and microbes under a wheat–maize cropping system in northern China. *Geoderma* **2009**, *149*, 318–324. [CrossRef]
28. Manna, M.C.; Swarup, A.; Wanjari, R.H.; Mishra, B.; Shahi, D.K. Long-term fertilization, manure and liming effects on soil organic matter and crop yields. *Soil Tillage Res.* **2007**, *94*, 397–409. [CrossRef]
29. Purakayastha, T.J.; Rudrappa, L.; Singh, D.; Swarup, A.; Bhadraray, S. Long-term impact of fertilizers on soil organic carbon pools and sequestration rates in maize–wheat–cowpea cropping system. *Geoderma* **2008**, *144*, 370–378. [CrossRef]
30. Tian, J.; Lou, Y.; Gao, Y.; Fang, H.; Liu, S.; Xu, M.; Blagodatskaya, E.; Kuzyakov, Y. Response of soil organic matter fractions and composition of microbial community to long-term organic and mineral fertilization. *Biol. Fertil. Soils* **2017**, *53*, 523–532. [CrossRef]
31. Sleutel, S.; Neve, S.D.; Németh, T.; Tóth, T.; Hofmana, G. Effect of manure and fertilizer application on the distribution of organic carbon in different soil fractions in long-term field experiments. *Eur. J. Agron.* **2006**, *25*, 280–288. [CrossRef]
32. Blair, N.; Faulkner, R.D.; Till, A.R.; Poulton, P.R. Long-term management impacts on soil C, N and physical fertility Part I: Broadbalk experiment. *Soil Tillage Res.* **2006**, *91*, 30–38. [CrossRef]
33. Kögel-Knabner, I.; Ekschmitt, K.; Flessa, H.; Guggenberger, G.; Matzner, E.; Marschner, B.; Lützow, M. An integrative approach of organic matter stabilization in temperate soils: Linking chemistry, physics, and biology. *J. Plant Nutr. Soil Sci.* **2008**, *171*, 5–13. [CrossRef]
34. Poffenbarger, H.J.; Barker, D.W.; Helmers, M.J.; Miguez, F.E.; Olk, D.C.; Sawyer, J.E.; Six, J.; Castellano, M.J. Maximum soil organic carbon storage in Midwest US cropping systems when crops are optimally nitrogen-fertilized. *PLoS ONE* **2017**, *12*, e0172293. [CrossRef] [PubMed]
35. Green, C.J.; Blackmer, A.M.; Horton, R. Nitrogen effects on conservation of carbon during corn residue decomposition in soil. *Soil Sci. Soc. Am. J.* **1995**, *59*, 453–459. [CrossRef]
36. Mulvaney, R.L.; Khan, S.A.; Ellsworth, T.R. Synthetic nitrogen fertilizers deplete soil nitrogen: A global dilemma for sustainable cereal production. *J. Environ. Qual.* **2009**, *38*, 2295–2314. [CrossRef] [PubMed]
37. Chivenge, P.; Vanlauwe, B.; Gentile, R.; Six, J. Comparison of organic versus mineral resource effects on short-term aggregate carbon and nitrogen dynamics in a sandy soil versus a fine textured soil. *Agric. Ecosyst. Environ.* **2011**, *140*, 361–371. [CrossRef]

38. Fonte, S.J.; Quansah, G.W.; Six, J. Fertilizer and residue quality effects on organic matter stabilization in soil aggregates. *Soil Sci. Soc. Am. J.* **2009**, *73*, 961–966. [CrossRef]
39. Russell, A.E.; Cambardella, C.A.; Laird, D.A.; Jaynes, D.B.; Meek, D.W. Nitrogen fertilizer effects on soil carbon balances in Midwestern U.S. agricultural systems. *Ecol. Appl.* **2009**, *19*, 1102–1113. [CrossRef] [PubMed]
40. Glendining, M.J.; Powlson, D.S. The effects of long-continued applications of inorganic nitrogen fertilizer on soil organic nitrogen—A review. In *Soil Management: Experimental Basis for Sustainability and Environmental Quality*; Advances in Soil Science Series; Lal, R., Stewart, B.A., Eds.; Lewis: Boca Raton, FL, USA, 1995; pp. 385–446.
41. Khan, S.A.; Mulvaney, R.L.; Ellsworth, T.R.; Boast, C.W. The myth of nitrogen fertilization for soil carbon sequestration. *J. Environ. Qual.* **2007**, *36*, 1821–1832. [CrossRef] [PubMed]
42. Powlson, D.S.; Jenkinson, D.S.; Johnston, A.E.; Poulton, P.R.; Glendining, M.J.; Goulding, K.W.T. Comments on "Synthetic nitrogen fertilizers deplete soil nitrogen: A global dilemma for sustainable cereal production," by R.L. Mulvaney, S.A. Khan, and T.R. Ellsworth in the Journal of Environmental Quality (2009, 38:2295–2314). *J. Environ. Qual.* **2010**, *39*, 749–752. [CrossRef] [PubMed]
43. Ladha, J.K.; Kesava Reddy, C.; Padre, A.T.; van Kessel, C. Role of nitrogen fertilization in sustaining organic matter in cultivated soils. *J. Environ. Qual.* **2011**, *40*, 1756–1766. [CrossRef] [PubMed]
44. Geiseller, D.; Scow, K.M. Long-term effects of mineral fertilizers on soil microorganisms—A review. *Soil Biol. Biochem.* **2014**, *75*, 54–63. [CrossRef]
45. Körschens, M.; Albert, E.; Armbruster, M.; Barkusky, D.; Baumecker, M.; Behle-Schalk, L.; Bischoff, R.; Čergan, Z.; Ellmer, F.; Herbst, F.; et al. Effect of mineral and organic fertilization on crop yield, nitrogen uptake, carbon and nitrogen balances, as well as soil organic carbon content and dynamics: Results from 20 European long-term field experiments of the twenty-first century. *Arch. Agron. Soil Sci.* **2013**, *59*, 1017–1040. [CrossRef]
46. Gregorich, E.G.; Liang, B.C.; Ellert, B.H.; Drury, C.F. Fertilization effects on soil organic matter turnover and corn residue C storage. *Soil Sci. Soc. Am. J.* **1996**, *60*, 472–476. [CrossRef]
47. Benbi, D.K.; Brar, J.S. A 25-year record of carbon sequestration and soil properties in intensive agriculture. *Agron. Sustain. Dev.* **2009**, *29*, 257–265. [CrossRef]
48. Tian, K.; Zhao, Y.; Xu, X.; Hai, N.; Huang, B.; Deng, W. Effects of long-term fertilization and residue management on soil organic carbon changes in paddy soils of China: A meta-analysis. *Agric. Ecosyst. Environ.* **2015**, *204*, 40–50. [CrossRef]
49. Lu, M.; Zhou, X.; Luo, Y.; Yang, Y.; Fang, C.; Chen, J.; Li, B. Minor stimulation of soil carbon storage by nitrogen addition: A meta-analysis. *Agric. Ecosyst. Environ.* **2011**, *140*, 234–244. [CrossRef]
50. Shang, Q.; Ling, N.; Feng, X.; Yang, X.; Wu, P.; Zou, J.; Shen, Q.; Guo, S. Soil fertility and its significance to crop productivity and sustainability in typical agroecosystem: A summary of long-term fertilizer experiments in China. *Plant Soil* **2014**, *381*, 13–23. [CrossRef]
51. Buyanovsky, G.A.; Wagner, G.H. Changing role of cultivated land in the global carbon cycle. *Biol. Fertil. Soils* **1998**, *27*, 242–245. [CrossRef]
52. Katyal, J.C.; Rao, N.H.; Reddy, M.N. Critical aspects of organic matter management in the Tropics: The example of India. *Nutr. Cycl. Agroecosyst.* **2001**, *61*, 77–88. [CrossRef]
53. Hermans, S.M.; Buckley, H.L.; Case, B.S.; Curran-Cournane, F.; Taylor, M.; Lear, G. Bacteria as emerging indicators of soil condition. *Appl. Environ. Microbiol.* **2017**, *83*, e02826-16. [CrossRef] [PubMed]
54. O'Donnell, A.G.; Seasman, M.; Macrae, A.; Waite, I.; Davies, J.T. Plants and fertilisers as drivers of change in microbial community structure and function in soils. *Plant Soil* **2001**, *232*, 135–145. [CrossRef]
55. Marschner, P.; Kandeler, E.; Marschner, B. Structure and function of the soil microbial community in a long-term fertilizer experiment. *Soil Biol. Biochem.* **2003**, *35*, 453–461. [CrossRef]
56. Buyanovsky, G.A.; Wagner, G.H. Carbon transfer in a winter wheat (*Triticum aestivum*) ecosystem. *Biol. Fertil. Soils* **1987**, *5*, 76–82. [CrossRef]
57. Khonje, D.J.; Varsa, E.C.; Klubek, B. The acidulation effects of nitrogenous fertilisers on selected chemical and microbiological properties of soil. *Commun. Soil Sci. Plant Anal.* **1989**, *20*, 1377–1395. [CrossRef]
58. Murugan, R.; Kumar, S. Influence of long-term fertilisation and crop rotation on changes in fungal and bacterial residues in a tropical rice field soil. *Biol. Fertil. Soils* **2013**, *49*, 847–856. [CrossRef]
59. Ge, Y.; Zhang, J.B.; Zhang, L.M.; Yang, M.; He, J.Z. Long-term fertilization regimes affect bacterial community structure and diversity of an agricultural soil in northern China. *J. Soils Sediments* **2008**, *8*, 43–50. [CrossRef]

60. Girvan, M.S.; Bullimore, J.; Ball, A.S.; Pretty, J.N.; Osborn, A.M. Responses of active bacterial and fungal communities in soils under winter wheat to different fertilizer and pesticide regimens. *Appl. Environ. Microbiol.* **2004**, *70*, 2692–2701. [CrossRef] [PubMed]
61. Kumar, U.; Shahid, M.; Tripathi, R.; Mohanty, S.; Kumar, A.; Bhattacharyya, P.; Lal, B.; Gautam, P.; Raja, R.; Panda, B.B.; et al. Variation of functional diversity of soil microbial community in sub-humid tropical rice-rice cropping system under long-term organic and inorganic fertilization. *Ecol. Indic.* **2017**, *73*, 536–543. [CrossRef]
62. Mandal, A.; Patra, A.K.; Singh, D.; Swarup, A.; Masto, R.E. Effect of long-term application of manure and fertilizer on biological and biochemical activities in soil during crop development stages. *Bioresour. Technol.* **2007**, *98*, 3585–3592. [CrossRef] [PubMed]
63. Zhao, J.; Ni, T.; Li, Y.; Xiong, W.; Ran, W.; Shen, B.; Shen, Q.; Zhang, R. Responses of bacterial communities in arable soils in a rice-wheat cropping system to different fertilizer regimes and sampling times. *PLoS ONE* **2014**, *9*, e85301. [CrossRef] [PubMed]
64. Goyal, S.; Mishra, M.M.; Hooda, I.S.; Singh, R. Organic matter-microbial biomass relationships in field experiments under tropical conditions: Effects of inorganic fertilization and organic amendments. *Soil Biol. Biochem.* **1992**, *24*, 1081–1084. [CrossRef]
65. Kanazawa, S.; Asakawa, S.; Takai, Y. Effect on fertilizer and manure application on microbial numbers, biomass, and enzyme activities in volcanic ash soils. I. Microbial numbers and biomass carbon. *Soil Sci. Plant Nutr.* **1988**, *34*, 429–439. [CrossRef]
66. Lynch, J.M.; Panting, L.M. Effects of season, cultivation and nitrogen fertiliser on the size of the soil microbial biomass. *J. Sci. Food Agric.* **1982**, *33*, 249–252. [CrossRef]
67. Li, J.; Cooper, J.M.; Lin, Z.A.; Li, Y.; Yang, X.; Zhao, B. Soil microbial community structure and function are significantly affected by long-term organic and mineral fertilization regimes in the North China Plain. *Appl. Soil Ecol.* **2015**, *96*, 75–87. [CrossRef]
68. Liu, E.; Yan, C.; Mei, X.; He, W.; Bing, S.H.; Ding, L.; Liu, Q.; Liu, S.; Fan, T. Long-term effect of chemical fertilizer, straw, and manure on soil chemical and biological properties in northwest China. *Geoderma* **2010**, *158*, 173–180. [CrossRef]
69. Mbuthia, L.W.; Acosta-Martínez, V.; DeBruyn, J.; Schaeffer, S.; Tyler, D.; Odoi, E.; Mpheshea, M.; Walker, F.; Eash, N. Long term tillage, cover crop, and fertilization effects on microbial community structure, activity: Implications for soil quality. *Soil Biol. Biochem.* **2015**, *89*, 24–34. [CrossRef]
70. Zhong, W.H.; Cai, Z.C. Long-term effects of inorganic fertilizers on microbial biomass and community functional diversity in a paddy soil derived from quaternary red clay. *Appl. Soil Ecol.* **2007**, *36*, 84–91. [CrossRef]
71. Schimel, J.P.; Bennett, J.; Fierer, N. Microbial community composition and soil nitrogen cycling: Is there really a connection? In *Biological Diversity and Function in Soils*; Bardgett, R.D., Usher, M.B., Hopkins, D.W., Eds.; Cambridge University Press: Cambridge, UK, 2005; pp. 172–188.
72. Angus, J.J.; Gupta, V.V.S.R.; Good, A.J.; Pitson, G.D. *Wheat Yield and Protein Responses to Anhydrous Ammonia (coldflo) and Urea, and their Effects on Soil*; Final Report of Project CSP 169 for the Grain Research and Development Corporation; CSIRO: Canberra, Australia, 1999; p. 17.
73. Biederbeck, V.O.; Campbell, C.A.; Ukrainetz, H.; Curtin, D.; Bouman, O.T. Soil microbial and biochemical properties after 10 years of fertilization with urea and anhydrous ammonia. *Can. J. Soil Sci.* **1996**, *76*, 7–14. [CrossRef]
74. Treseder, K.K.; Allen, M.F. Direct nitrogen and phosphorus limitation of arbuscular mycorrhizal fungi: A model and field test. *New Phytol.* **2002**, *155*, 507–515. [CrossRef]
75. Jackson, L.E.; Calderon, F.J.; Steenwerth, K.L.; Scow, K.M.; Rolston, D.E. Responses of soil microbial processes and community structure to tillage events and implications for soil quality. *Geoderma* **2003**, *114*, 305–317. [CrossRef]
76. Mathew, R.P.; Feng, Y.; Githinji, L.; Ankumah, R.; Balkcom, K.S. Impact of no-tillage and conventional tillage systems on soil microbial communities. *Appl. Environ. Soil Sci.* **2012**. [CrossRef]
77. Doran, J.W. Microbial biomass and mineralizable nitrogen distributions in no-tillage and plowed soils. *Biol. Fertil. Soils* **1987**, *5*, 68–75. [CrossRef]
78. Zuber, S.M.; Villamil, M.B. Meta-analysis approach to assess effect of tillage on microbial biomass and enzyme activities. *Soil Biol. Biochem.* **2016**, *97*, 176–187. [CrossRef]

79. Treseder, K.K. Nitrogen additions and microbial biomass: A meta-analysis of ecosystem studies. *Ecol. Lett.* **2008**, *11*, 1111–1120. [CrossRef] [PubMed]
80. Rousk, J.; Brookes, P.C.; Bååth, E. Contrasting soil pH effects on fungal and bacterial growth suggest functional redundancy in carbon mineralization. *Appl. Environ. Microbiol.* **2009**, *75*, 1589–1596. [CrossRef] [PubMed]
81. Bolan, N.S.; Adriano, D.C.; Curtin, D. Soil acidification and liming interactions with nutrient and heavy metal transformation and bioavailability. *Adv. Agron.* **2003**, *78*, 215–272. [CrossRef]
82. Guo, J.H.; Liu, X.J.; Zhang, Y.; Shen, J.L.; Han, W.X.; Zhang, W.F.; Christie, P.; Goulding, K.W.T.; Vitousek, P.M.; Zhang, F.S. Significant acidification in major Chinese croplands. *Science* **2010**, *327*, 1008–1010. [CrossRef] [PubMed]
83. Lu, X.; Mao, Q.; Gilliam, F.S.; Luo, Y.; Mo, J. Nitrogen deposition contributes to soil acidification in tropical ecosystems. *Glob. Chang. Biol.* **2014**, *20*, 3790–3801. [CrossRef] [PubMed]
84. Lal, R. Anthropogenic influences in world soil and implications for global food security. *Adv. Agron.* **2007**, *93*, 69–93. [CrossRef]
85. Bhattacharyya, R.; Ghosh, B.N.; Mishra, P.K.; Mandal, B.; Rao, C.S.; Sarkar, D.; Das, K.; Anil, K.S.; Lalitha, M.; Hati, K.M.; et al. Soil degradation in India: Challenges and potential solutions. *Sustainability* **2015**, *7*, 3528–3570. [CrossRef]
86. Portch, S.; Jin, J.Y. Fertilizer use in China: Types and amounts. In *Encyclopaedia of Life Support Systems, Agricultural Sciences*; Lal, R., Ed.; EOLSS Publishers Co. Ltd.: Oxford, UK, 2009; Volume II, pp. 247–256.
87. Buresh, R.J.; Witt, C. Site-specific nutrient management. In *Fertilizer Best Management Practices*; International Fertilizer Industry Association (IFA): Paris, France, 2007; pp. 47–55.
88. Montanarella, L.; Pennock, D.J.; McKenzie, N.; Badraoui, M.; Chude, V.; Baptista, I.; Mamo, T.; Yemefack, M.; Aulakh, M.S.; Yagi, K.; et al. World's soils are under threat. *Soil* **2016**, *2*, 79–82. [CrossRef]
89. Johnston, A.E.; Poulton, P.R.; Syers, J.K. *Phosphorus, Potassium and Sulphur Cycles in Agricultural Soils*; Proceedings No. 465; The International Fertiliser Society: York, UK, 2001.
90. Haerdter, R.; Fairhurst, T. Nutrient use efficiency in upland cropping systems of Asia. In Proceedings of the IFA Regional Conference, Cheju Island, Korea, 6–8 October 2003.
91. Kumar, A.; Yadav, D.S. Long-term effects of fertilizers on the soil fertility and productivity of a rice-wheat system. *J. Agron. Crop Sci.* **2001**, *186*, 47–54. [CrossRef]
92. Belay, A.; Claassens, A.; Wehner, F.C. Effect of direct nitrogen and potassium and residual phosphorus fertilizers on soil chemical properties, microbial components and maize yield under long-term crop rotation. *Biol. Fertil. Soils* **2002**, *35*, 420–427. [CrossRef]
93. Chu, H.; Lin, X.; Fujii, T.; Morimoto, S.; Yagi, K.; Hu, J.; Zhang, J. Soil microbial biomass, dehydrogenase activity, bacterial community structure in response to long-term fertilizer management. *Soil Biol. Biochem.* **2007**, *39*, 2971–2976. [CrossRef]
94. Luo, P.; Han, X.; Wang, Y.; Han, M.; Shi, H.; Liu, N.; Bai, H. Influence of long-term fertilization on soil microbial biomass, dehydrogenase activity, and bacterial and fungal community structure in a brown soil of northeast China. *Ann. Microbiol.* **2015**, *65*, 533–542. [CrossRef] [PubMed]
95. Zhong, W.; Gu, T.; Wang, W.; Zhang, B.; Lin, X.; Huang, Q.; Shen, W. The effects of mineral fertilizer and organic manure on soil microbial community and diversity. *Plant Soil* **2010**, *326*, 511–522. [CrossRef]
96. Palm, C.A.; Myers, R.J.K.; Nandwa, S.M. Combined use of organic and inorganic nutrient sources for soil fertility maintenance and replenishment. In *Replenishing Soil Fertility in Africa*; Buresh, R.J., Sanchez, P.A., Calhoun, F., Eds.; American Society of Agronomy and Soil Science Society of America: Madison, WI, USA, 1997; pp. 193–217. [CrossRef]
97. Vanlauwe, B.; Bationo, A.; Chianu, J.; Giller, K.E.; Merckx, R.; Mokwunye, U.; Ohiokpehai, O.; Pypers, P.; Tabo, R.; Shepherd, K.; et al. Integrated soil fertility management: Operational definition and consequences for implementation and dissemination. *Outlook Agric.* **2010**, *39*, 17–24. [CrossRef]
98. Wu, T.Y.; Schoenau, J.J.; Li, F.M.; Qian, P.Y.; Malhi, S.S.; Shi, Y.C.; Xue, F.L. Influence of cultivation and fertilization on total organic carbon and carbon fractions in soils from the Loess Plateau of China. *Soil Tillage Res.* **2004**, *77*, 59–68. [CrossRef]
99. Rudrappa, L.; Purakayastha, T.J.; Singh, D.; Bhadraray, S. Long-term manuring and fertilization effects on soil organic carbon pools in a Typic Haplustept of semi-arid sub-tropical India. *Soil Tillage Res.* **2006**, *88*, 180–192. [CrossRef]

100. Brar, B.S.; Kamalbir-Singh; Dheri, G.S.; Balwinder-Kumar. Carbon sequestration and soil carbon pools in a rice–wheat cropping system: Effect of long-term use of inorganic fertilizers and organic manure. *Soil Tillage Res.* **2013**, *128*, 30–36. [CrossRef]
101. Manna, M.C.; Swarup, A.; Wanjari, R.H.; Singh, Y.V.; Ghosh, P.K.; Singh, K.N.; Tripathi, A.K.; Saha, M.N. Soil organic matter in a West Bengal Inceptisol after 30 years of multiple cropping and fertilization. *Soil Sci. Soc. Am. J.* **2006**, *70*, 121–129. [CrossRef]
102. Yadvinder-Singh; Bijay-Singh; Ladha, J.K.; Khind, C.S.; Gupta, R.K.; Meelu, O.P.; Pasuquin, E. Long-term effects of organic inputs on yield and soil fertility in the rice–wheat rotation. *Soil Sci. Soc. Am. J.* **2004**, *68*, 845–853. [CrossRef]
103. Gu, Y.; Zhang, X.; Tu, S.; Lindström, K. Soil microbial biomass, crop yields, and bacterial community structure as affected by long-term fertilizer treatments under wheat-rice cropping. *Eur. J. Soil Biol.* **2009**, *45*, 239–246. [CrossRef]
104. Hao, X.H.; Liu, S.L.; Wu, J.S.; Hu, R.G.; Tong, C.L.; Su, Y.Y. Effect of long-term application of inorganic fertilizer and organic amendments on soil organic matter and microbial biomass in three subtropical paddy soils. *Nutr. Cycl. Agroecosyst.* **2008**, *81*, 17–24. [CrossRef]
105. Nambiar, K.K.M. Major cropping systems in India. In *Agricultural Sustainability-Economic Environment and Statistical Considerations*; Barnett, V., Pyne, R., Steiner, R., Eds.; John Wiley and Sons: New York, NY, USA, 1995; pp. 133–168.
106. Swamp, A.; Reddy, D.D.; Prasad, R.N. *Proceedings of a National Workshop on Long Term Soil Fertility Management through Integrated Plant Nutrient Supply*; Indian Institute of Soil Science: Bhopal, India, 1998.
107. Vanlauwe, B.; Kihara, J.; Chivenge, P.; Pypers, P.; Coe, R.; Six, J. Agronomic use efficiency of N fertilizer in maize-based systems in sub-Saharan Africa within the context of integrated soil fertility management. *Plant Soil* **2011**, *339*, 35–50. [CrossRef]
108. Zingore, S.; Murwira, H.K.; Delve, R.J.; Giller, K.E. Soil type, management history and current resource allocation: Three dimensions regulating variability in crop productivity on African smallholder farms. *Field Crops Res.* **2007**, *101*, 296–305. [CrossRef]

9

Green Manuring Effect on Changes of Soil Nitrogen Fractions, Maize Growth and Nutrient Uptake

Lu Yang [1,2], Jinshun Bai [2], Jia Liu [3], Naohua Zeng [2] and Weidong Cao [2,*]

1 Graduate School, Chinese Academy of Agricultural Sciences, Beijing 100081, China; luckyyl_520@163.com
2 Key Laboratory of Plant Nutrition and Fertilizer, Ministry of Agriculture and Rural Affairs/Institute of Agricultural Resources and Regional Planning, Chinese Academy of Agricultural Sciences, Beijing 100081, China; baijinshun@caas.cn (J.B.); zengnaohua@caas.cn (N.Z.)
3 Institute of Soil & Fertilizer and Resources & Environment, Jiangxi Academy of Agricultural Sciences, Nanchang 330200, China; liujia422@126.com
* Correspondence: caoweidong@caas.cn

Abstract: Green manure is a promising, at least partial, substitution for chemical fertilizer in agriculture, especially for nitrogen (N), which in soil can be radically changed by exogenous input. However, it is not well understood how, after green manure incorporation, soil N changes coordinate with crop N uptake and consequently contribute to fertilizer reduction in a maize–green manure rotation. A four-year field study was performed consisting of (1) control, no fertilization; (2) F_{100}, recommended inorganic fertilization alone; (3) G, green manure incorporation alone; (4) F_{70} + G (70% of F_{100} plus G); (5) F_{85} + G; and (6) F_{100} + G. The results show that treatments with 15–30% reduction of inorganic fertilizer (i.e., F_{70} + G and F_{85} + G) had similar grain yield, dry matter (DM) accumulation, and N uptake as F_{100} treatment. F_{100} + G maize had 17% greater DM and 15% more N uptake at maturity relative to F_{100}. Of the five soil N fractions examined, dissolved organic N (DON) and mineral N (N_{min}) explained over 70% of the variation of maize DM and N accumulation. Partial least squares path modeling further revealed that soil N fractions had positive indirect effects on DM production through N uptake, which might be coordinated with improved DON and N_{min} status at both early and mid-late stages of maize growth. Overall, the results highlight enhanced maize production with reduced fertilizer inputs based on green manure incorporation in temperate regions.

Keywords: green manure; nitrogen uptake; *Orychophragmus violaceus* L.; soil nitrogen pools; grain yield; *Zea mays* L.

---

1. Introduction

Improving crop yield and nutrient use efficiency simultaneously is challenging due to the increasing demand for food and intensifying environmental issues [1,2]. To pursue higher crop yields, farmers in some intensive agriculture areas of China apply excessively high rates of nitrogen (N) and phosphorous (P) fertilizer. The annual N fertilizer input in a maize–wheat cropping system in North China is up to 588 kg ha^{-1} $year^{-1}$, which far exceeds levels in the United States and Northern European countries [3]. However, the substantial inputs do not reliably maintain the expected yields [3,4]. The ratio of an average farmer's yield-to-yield potential is 0.41 [5], which illuminates a large yield gap for maize in this region. Much of the applied N fertilizer is lost to the environment [6], through nitrates leaching to groundwater [7] and greenhouse gas emissions [8]. Thus, it is of great importance to produce more grain yield with less environmental impact, e.g., lower nutrient losses [3,4].

Substitution of inorganic fertilizer by green manure has been adopted to reduce chemical inputs in agriculture [9,10]. Winter green manure *Orychophragmus violaceus* (OV)-maize rotation is an innovative eco-agricultural practice in the North China Plain, where OV is cultivated during the winter–spring fallow season (September to April) and well matched with the sowing of maize [11,12]. It is an alternative to maize-fallow or maize-wheat rotation that consumes plenty of groundwater and causes severe nutrient losses during the winter wheat season [13]. Integrated incorporation of OV residues with chemical fertilizers has been practiced to reduce nitrogen losses and improve grain yield and N economy in maize [11,12]. OV belongs to the Brassicaceae family, without the ability to fix atmospheric N_2, but it can trap residual nitrate and reduce N leaching losses. Other members of the Brassicaceae family have been found to have extensive rooting depth, which has been shown to correlate with soil N depletion [10,14]. Therefore, in green manure-based systems, both inputs and losses of inorganic N fertilizer could be decreased [15–17]. In addition, previous studies have shown that, although OV crop is less capable than legume green manure *Vicia villosa* of enhancing soil N availability, the total ecological service value is slightly greater for OV crops (by 3.5%) in terms of succeeding crop production, greenhouse gas and air pollutant reduction, water and soil conservation, and soil nutrient improvement [18]. The green manure-maize rotation may give farmers less gross return relative to wheat-maize rotation, while it is ecologically beneficial for sustainable development, at least outperforming maize-fallow rotation [18].

The effects of green manuring on crop growth and nutrient utilization are associated with an improvement in soil physiochemical properties, such as bulk density, water conductivity, and carbon and N levels [10,19]. Increasing N use efficiency in crop production highly depends on the synchrony between crop N demand and supply from various sources through the growing season [4,20]. A model-driven integrated soil-crop management system has been well established for maize production in the North China Plain based on the timing of in-season soil N monitoring and application [4], while many farmers are still applying N as basal fertilizer for maize in practice, leading to N losses and a mismatch between N supply and crop demand [6]. A previous study using ^{15}N isotope technique showed that with the combination of green manure and fertilizer, N supply is greatly consistent with maize N requirements at different stages [11]. This is associated with the concept that green manure N could potentially be retained longer in soil when combined with inorganic N [11,21], implying coordinated effects of green manure incorporation on N uptake in maize. However, it is still not well understood how the coordination is processed between maize N uptake and soil N changes after OV incorporation.

The changes of soil N pools after green manure incorporation involve transformations between soil N fractions, such as microbial biomass N, dissolved organic N, and mineral N. The positive effect of green manure incorporation on succeeding crop growth is proposed to relate to the stimulated microbial processes. Along with nitrate and ammonium, dissolved organic N is another soluble N source for microbes or plant uptake [22]. These three components are recognized as active fractions in soil N cycling and closely associated with plant N uptake [23]. Given the crucial role of soil N fractions in influencing crop N utilization, the objective of this study was to investigate the temporal changes of soil N fractions after OV incorporation and their effects on subsequent maize growth and N uptake in an OV-maize rotation system.

2. Materials and Methods

2.1. Field Experiments

The field experiments were carried out from 14 September 2008, to 13 August 2012, at the Wanzhuang Experimental Station (116°35′ E, 39°34′ N), Chinese Academy of Agricultural Sciences, China, on calcareous sandy soil. Before the start of the field experiment, maize was cultivated followed by a fallow season each year. Soil background properties at 0–20 cm depth were: organic matter 4.16 g kg^{-1}, pH (H_2O) 8.39, soil total N 0.34 g kg^{-1}, Olsen-P 4.20 mg kg^{-1}, and NH_4OAc-K 73.0 mg kg^{-1}. Green manure *Orychophragmus violaceus* L. (OV) was planted in September and incorporated into the soil the following April at its full blooming stage, followed by maize (hybrid Zhengdan 958) cultivation from May to August each year. The total yearly precipitation was 514 mm, with 70% distributed in June,

July, and August based on a 30-year average from 1981 to 2010 (http://data.cma.cn/). Starting from October 2010 (the third cycle of OV-maize rotation), in-season meteorological data were recorded until the end of the experiment (Figure 1).

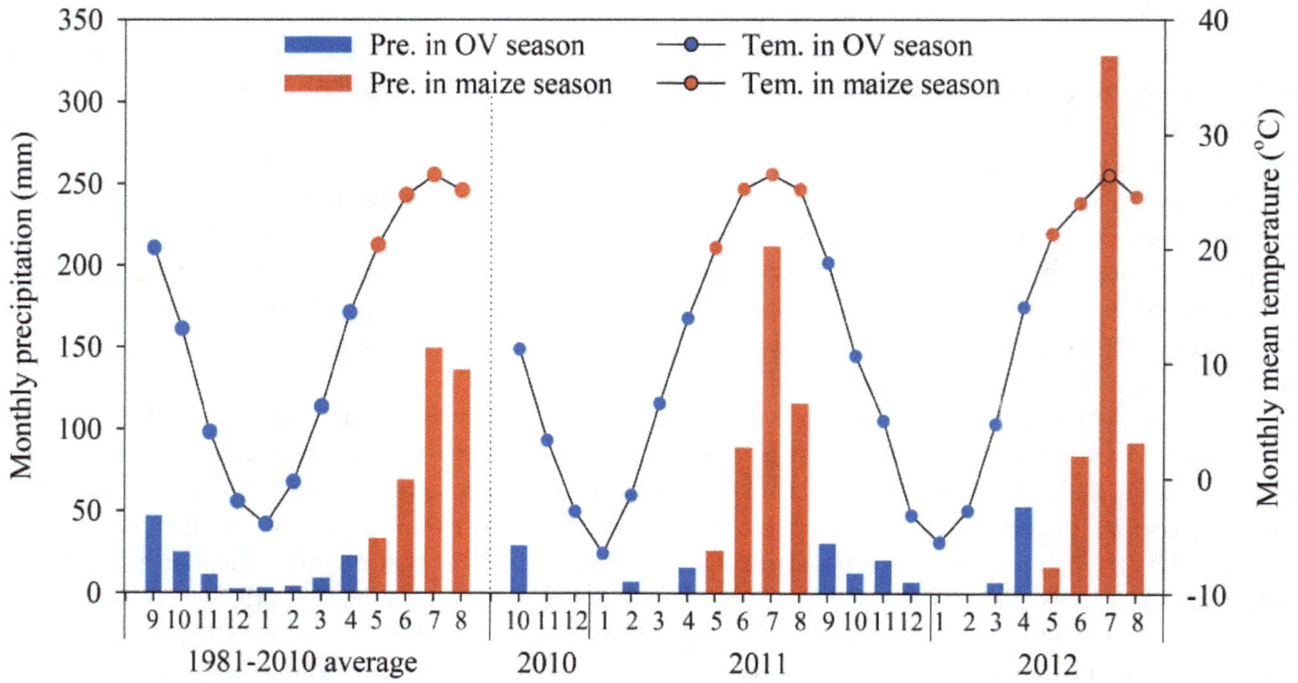

Figure 1. Monthly total precipitation (pre.) and average temperature (tem.) during the green manure (*Orychophragmus violaceus* (OV)) season (blue) and maize growing season (red) based on the 30-year average from 1981–2010 and in-season recording from October 2010 to August 2012.

The following regimes of inorganic fertilization or green manure incorporation were followed, as given in Table 1: (1) control, no fertilizer application and no OV incorporation; (2) F_{100}, the recommended fertilizer rate, alone; (3) G, incorporation of green manure OV alone; (4) F_{70} + G (70% of F_{100} plus G); (5) F_{85} + G (85% of F_{100} plus G); and (6) F_{100} + G. The 6 treatments were imposed on a randomized complete block design with 4 replications. Each treatment was in the same plot in all study years. Each plot was 4.8 m long and 2.4 m wide. The fertilizer rate of F_{100} was recommended by the National Soil Testing and Fertilization Recommendation Project on a yield basis for farmers in this region of China, which was also consistent with other reports [4,24]. Amounts of 225 kg N ha^{-1}, 49 kg P ha^{-1}, and 94 kg K ha^{-1} were applied as urea, super phosphate, and potassium sulfate, respectively. A half rate of N and full rates of P and K were applied as base fertilizer before sowing, and the remaining half rate of N was top-dressed at maize tasseling. Maize seedlings were hand-thinned to a stand of 75,000 plants ha^{-1}.

No fertilizer was applied during the OV growth. Green manure OV was grown *in situ* and incorporated from 2008 to 2011, while part of the OV residues was imported to the field with an OV incorporation amount of 22,500 kg/ha in 2012 (Table 2). That year, the biomass of green manure in some of the plots, such as in treatment G, was extremely low. To investigate the effects of green manure incorporation, certain amounts of OV residues were introduced from an adjacent field. The water content of fresh OV residues for incorporation ranged from 83.4% to 87.5%, and the carbon (C), N, P, and K concentrations were, respectively, 38.2%, 2.38–3.36%, 0.33–0.51%, and 3.32–3.53% on a dry weight basis in different years. Fresh OV residues were incorporated into 0–20 cm soil 8 days before maize sowing each year. The total amounts of OV incorporation and nutrients introduced each year are listed in Table 2. The C/N ratio of decomposing OV residues in maize season (from OV incorporation to maize maturity) ranged from 9.7 to 15.3, with an average of 13 [25].

Table 1. Cropping system and chemical fertilizer input for each treatment.

Treatment	Cropping System	Input of Chemical Fertilizer Each Year (kg/ha) [1]		
		N	P	K
Control	Maize-winter fallow	0	0	0
F_{100}	Maize-winter fallow	225	49	94
G	Maize-green manure	0	0	0
F_{70} + G	Maize-green manure	158	34	66
F_{85} + G	Maize-green manure	191	42	80
F_{100} + G	Maize-green manure	225	49	94

[1] Chemical fertilizer was applied only for maize growth, and not for green manure. F_{100}, the recommended fertilizer rate, alone; G, incorporation of green manure OV alone; F_{70} + G, 70% of F_{100} plus G; F_{85} + G, 85% of F_{100} plus G.

Table 2. Amounts of incorporated green manure and introduced nutrients for each treatment.

Treatment	Incorporated Amount of Green Manure (Fresh Weight, t ha^{-1}) [1]				Amount of Introduced Nutrient by Green Manure (N–P–K) (kg ha^{-1})			
	2009	2010	2011	2012	2009	2010	2011	2012
Control	0	0	0	0	0	0	0	0
F_{100}	0	0	0	0	0	0	0	0
G	15.6	n.d. [2]	4.6	22.5 (18.7) [3]	52.6–9.6–77.4	n.d.	21.5–2.2–21.2	94.0–9.8–99.2 (78.5–8.2–82.7) [3]
F_{70} + G	15.6	n.d.	10.2	22.5 (10.9)	52.6–9.6–77.4	n.d.	48.0–4.8–47.0	94.0–9.8–99.2 (45.8–4.8–48.2)
F_{85} + G	15.6	n.d.	19.0	22.5 (6.3)	52.6–9.6–77.4	n.d.	89.8–9.0–87.9	94.0–9.8–99.2 (26.5–2.8–27.9)
F_{100} + G	15.6	n.d.	15.2	22.5 (7.0)	52.6–9.6–77.4	n.d.	71.5–7.2–70.0	94.0–9.8–99.2 (29.4–3.1–31.0)

[1] Water content of fresh green manure residue ranged from 83.4% to 87.5%. [2] n.d., not determined. [3] Numbers in brackets indicate amounts of exogenous import of OV green manure and introduced nutrients other than that grown in situ.

2.2. Sampling and Measurements

Maize grain yield was measured based on the entire plot every year, while the dynamics of soil N fractions and their effects on maize N uptake were investigated in the final year, 2012. Plants and 0–20 cm soils were sampled at the V3 (third leaf fully expanded), V8 (eighth leaf fully expanded), VT (tasseling), R3 (milk), and R6 (physiological maturity) stages of maize growth. Soil samples at the maize tasseling stage were taken before the second N application. At each harvest, 3 uniform and randomly selected maize plants from each plot were cut at the stem base. The shoot parts were dried at 70 °C to a constant weight, weighed, and ground to a fine powder. Around 0.2 g of plant material was used to determine total N concentration using a modified Kjeldahl digestion method [26]. The remaining digests were used for analysis of total P concentration (molybdivanadate method) by automated colorimetry [27] and total K concentration by a flame photometer [28].

Soil N fractions were analyzed as follows: (1) Soil total N content was measured with a Kjeldahl method [29]. (2) Soil mineral N (N_{min} = NO_3^--N + NH_4^+-N) was extracted with 2 mol L^{-1} KCl and measured with continuous flow analysis (Seal AA3, Norderstedt, Germany). (3) Soil organic N: soil samples were extracted by 2 mol L^{-1} KCl (soil-to-solution ratio 1:5, *w:v*) for 60 min. The supernatant was carefully discarded, and the soil was re-extracted by 2 mol L^{-1} KCl (soil-to-solution ratio 1:2.5, *w:v*) for 20 min and centrifuged, and the supernatant was removed. The pellet soil was washed with distilled water by shaking for 20 min, dried at 60 °C, and analyzed for N content with the Kjeldahl method [29]. (4) The fumigation extraction method was used to determine soil microbial biomass N (SMBN) [30]. Briefly, each fresh soil sample was divided into 6 subsamples (equivalent to 25 g oven-dried weight). Three unfumigated subsamples were immediately extracted with 100 mL 0.5 M K_2SO_4 on a rotary shaker at 220 rpm for 30 min and then filtered through Whatman qualitative filter paper. The remaining 3 subsamples were fumigated with alcohol-free chloroform for 24 h at 25 °C, and then extracted and filtered as described above. Soil filtrates were stored at −20 °C prior to N assay. The SMBN content was calculated by the N differences between fumigated and unfumigated samples with a conversion factor of 0.54 [30]. (5) The soil dissolved organic N (DON) content was expressed as the difference between total dissolved N content and N_{min} content, extracted with 0.5 M K_2SO_4 and 2 M KCl, respectively, using the method by Jones and Willett [31].

2.3. Statistical Analysis

Data were subjected to analysis of variance using PROC ANOVA with SAS package 9.1 (SAS Institute, Cary, NC, USA). The management of OV incorporation and/or fertilizer application was treated as a fixed effect and replications as random factors. The least significant difference (LSD) was used to determine treatment differences at a $p < 0.05$ level of probability.

The relative weight analysis and partial least squares path modeling (PLS-PM) were performed in R with "relweights" [32] and "plspm" [33] packages, respectively. Pooled data from 120 samples across treatments and stages were used for the analysis. The PLS-PM model path used 1000 bootstraps to validate the estimates of path coefficients and the coefficients of determination (r^2). The direct effects were represented by path coefficients, indicating the direction and strength of the linear relationships between variables. Indirect effects were the sum of multiplied path coefficients between a predictor and a response variable except the direct effect. Path model was evaluated using the goodness of fit statistic to measure its overall predictive power.

3. Results

3.1. Effects of OV Incorporation on Maize Growth, Yield, and Nutrient Uptake

High variability was observed between years (p = 0.0013), as well as between treatments ($p < 0.0001$) and years by treatment ($p < 0.0001$) (Figure 2). Maize grain yield in the control without fertilization declined from 6.1 to 3.2 t ha^{-1} over the years, implying an unsustainable development of maize production (Figure 2). Compared to the control, the average grain yield was 22% greater in the G treatment, and was further increased by the F plus G combinations ($p < 0.05$; Figure 2). On average, similar grain yield around 9 t ha^{-1} was observed in the F_{100}, F_{70} + G, and F_{85} + G treatments.

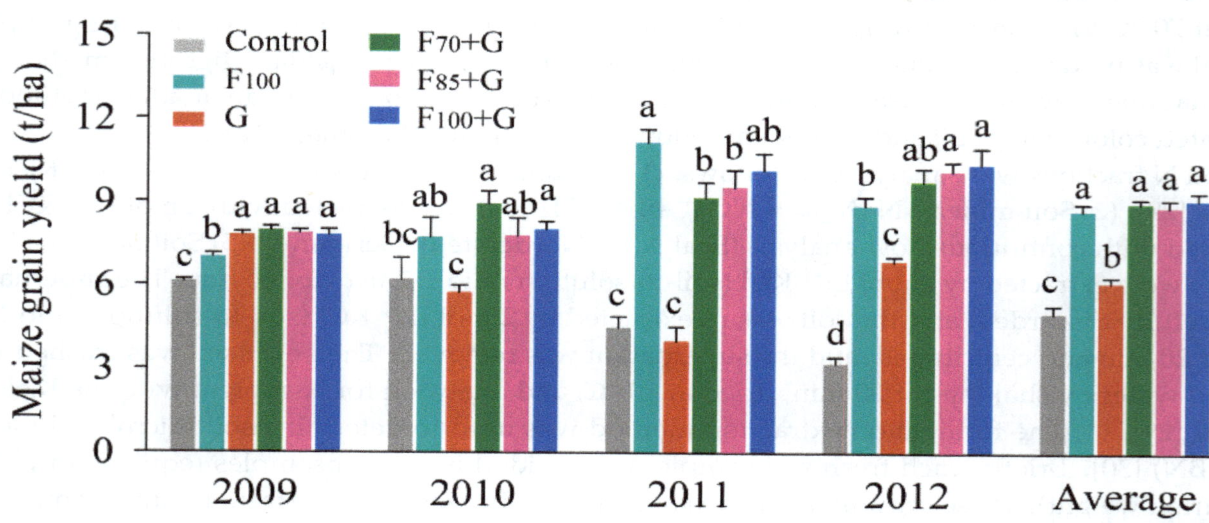

Figure 2. Maize grain yield and average yield from 2009 to 2012. Data are mean ± standard error (vertical bars, n = 4). Different letters above columns indicate statistical differences between treatments by least significant difference (LSD) test ($p < 0.05$). Control, no fertilization; F_{100}, recommended inorganic fertilization alone; G, green manure incorporation alone; F_{70} + G, F_{85} + G, and F_{100} + G are 70%, 85%, and 100% of F_{100} plus G, respectively.

Dry matter (DM) accumulation and N uptake in the G treatment were 28–114% and 83% to 146% greater, respectively, than the control across maize developmental stages, but were still far lower than other F plus G treatments (Figure 3a,b). No significant differences were detected between F_{100}, F_{70} + G, and F_{85} + G treatments. Compared to F_{100}, maize plants in the F_{100} + G treatment had 17% greater DM accumulation and 15% more N uptake at maturity, respectively ($p < 0.05$; Figure 3a,b). In addition, the P and K uptake in G-treated maize was 0.7- and 1.2-fold greater, respectively, than in the control plants. F_{70} + G, F_{85} + G, and F_{100} + G treatments had similar P uptake compared to F_{100} treatment, while K uptake was significantly increased by 55%, 48%, and 50%, respectively, at maturity (Figure 3c,d).

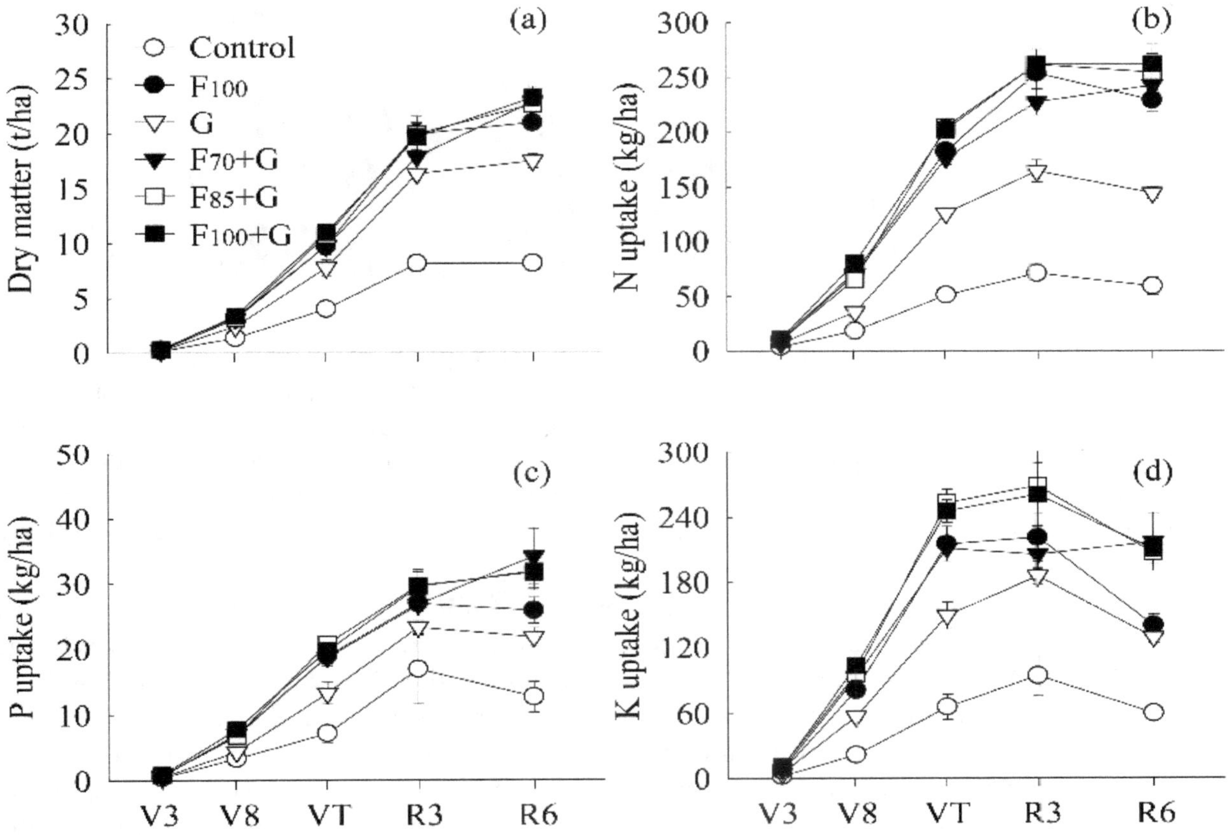

Figure 3. (**a**) Dry matter accumulation, (**b**) nitrogen (N) uptake, (**c**) phosphorus (P) uptake, and (**d**) potassium (K) uptake along maize development stages in 2012. Vertical bars indicate standard error of means ($n = 4$). V3 and V8 (the third and eighth leaf fully expanded, respectively), VT (tasseling), R3 (milk), and R6 (physiological maturity). Control, no fertilization; F_{100}, recommended inorganic fertilization alone; G, green manure incorporation alone; F_{70} + G, F_{85} + G, and F_{100} + G are 70%, 85%, and 100% of F_{100} plus G, respectively.

3.2. Temporal Changes of Soil N Fractions during Maize Growth as Influenced by Fertilizer Application and OV Incorporation

Total N and organic N contents of soil were relatively constant during maize growth, with a slight decrease after harvest (Figure 4). Compared to the initial total soil N content (0.34 g/kg) in 2008, the value in the control treatment remained fairly constant, while total soil N increased in most treatments with fertilizer and OV incorporation. F_{70} + G, F_{85} + G, and F_{100} + G treated soil had similar or greater contents of total N and organic N than F_{100} treatment through the maize growing seasons. The OV incorporation alone led to lower levels of soil total N and organic N than F plus G treatments (Figure 4).

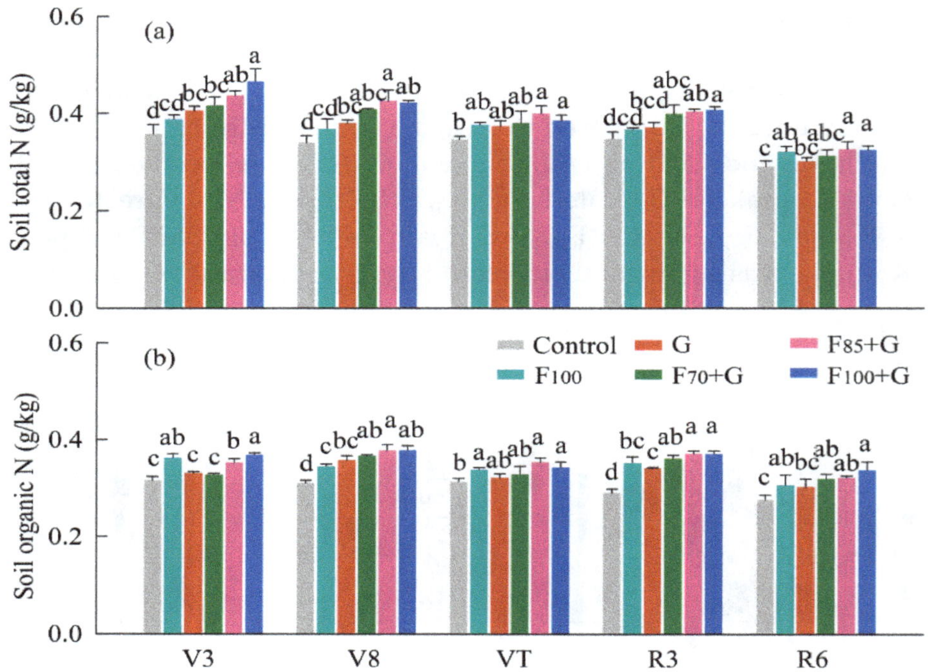

Figure 4. Changes of soil (**a**) total N and (**b**) organic N in the 0–20 cm soil layer during maize growth. Data are mean ± standard error (vertical bars, $n = 4$). Different letters above columns at each harvest indicate statistical differences between treatments by LSD test ($p < 0.05$). V3 and V8 (the third and eighth leaf fully expanded, respectively), VT (tasseling), R3 (milk), and R6 (physiological maturity). Control, no fertilization; F_{100}, recommended inorganic fertilization alone; G, green manure incorporation alone; F_{70} + G, F_{85} + G, and F_{100} + G are 70%, 85%, and 100% of F_{100} plus G, respectively.

In contrast to soil total N and organic N, the contents of soil microbial biomass N (SMBN), mineral N (N_{min}), and dissolved organic N (DON) fluctuated over maize development stages, with greater values occurring at V3 and V8 (Figure 5). At these two stages, the SMBN content was 25–44% greater in the F_{70} + G, F_{85} + G, and F_{100} + G treatments than both control and F_{100} treatment (Figure 5a). The DON content was also significantly increased in F plus G treatments over control or F_{100} during maize growth, especially at the stem elongation V8 (Figure 5b). The N_{min} content was quite sensitive to inorganic fertilizer N input and environmental conditions, where the higher fertilizer N application generally produced more N_{min} than the control or G treatment (Figure 5c). However, with 15–30% reduction of fertilizer N application, similar N_{min} content was obtained with F_{70} + G and F_{85} + G and F_{100} treatments, while it was further increased by F_{100} + G treatment. Soil N_{min} content decreased to a low level after the V8 stage, but was still greater in F and G treatments than control.

3.3. Interactions between Soil N Fractions and Maize Growth and Nutrient Uptake

A relative weights analysis was performed to quantify the relative contribution of soil N fractions to maize growth and N uptake (Figure 6). Of the five N fractions examined in the present study, DON and N_{min} together explained over 70% of the variation in DM production and N uptake, i.e., there was a greater contribution of these two N components than other variables (soil total N, organic N, and SMBN).

Further, large indirect effects (0.53, calculated as the sum of multiplied path coefficients from soil N to yield except the direct effect) of soil N status on maize grain yield were observed through plant nutrient uptake and DM production in the PLS-PM model (Figure 7). The direct effects of soil N on nutrient uptake, plant NPK uptake on dry matter production, and dry matter production on final grain yield were 0.66 ($p < 0.001$), 0.99 ($p < 0.001$), and 0.83 ($p < 0.01$), respectively, which together constituted a predominant indirect path from soil N pools to grain yield (Figure 7).

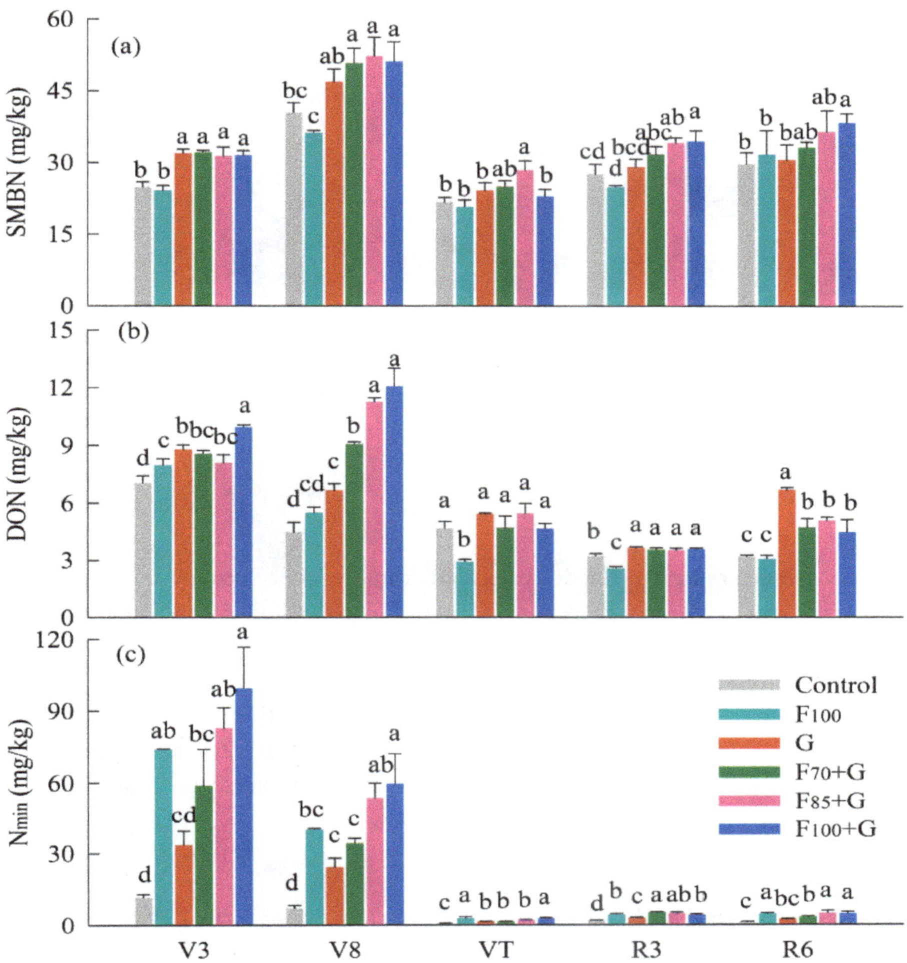

Figure 5. Changes in (**a**) soil microbial biomass N (SMBN), (**b**) dissolved organic N (DON), and (**c**) mineral N (N_{min}) in the 0–20 cm soil layer during maize growth. Data are mean ± standard error (vertical bars, $n = 4$). Different letters above columns at each harvest indicate statistical differences between treatments by LSD test ($p < 0.05$). V3 and V8 (the third and eighth leaf fully expanded, respectively), VT (tasseling), R3 (milk), and R6 (physiological maturity). Control, no fertilization; F_{100}, recommended inorganic fertilization alone; G, green manure incorporation alone; F_{70} + G, F_{85} + G, and F_{100} + G are 70%, 85%, and 100% of F_{100} plus G, respectively.

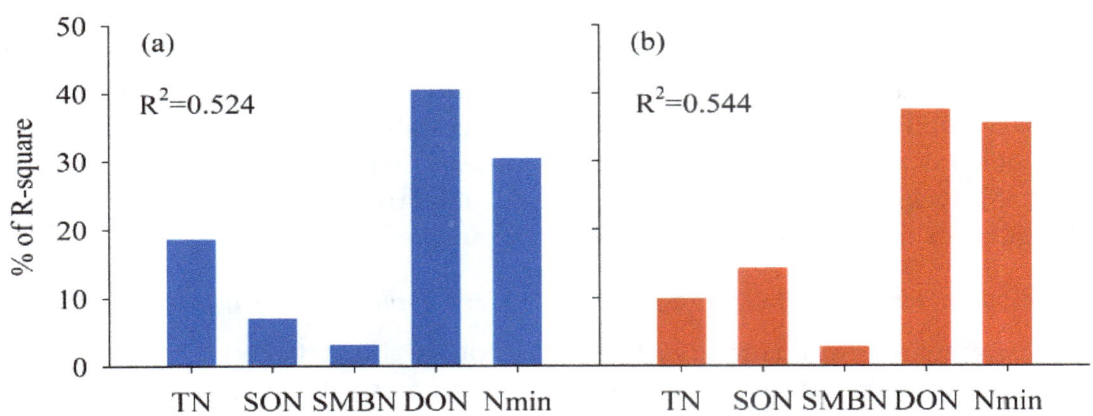

Figure 6. Relative influences of soil nitrogen (N) fractions on (**a**) dry matter production and (**b**) N uptake of maize. TN, total N; SON, soil organic N; SMBN, soil microbial biomass N; DON, dissolved organic N; N_{min}, mineral N.

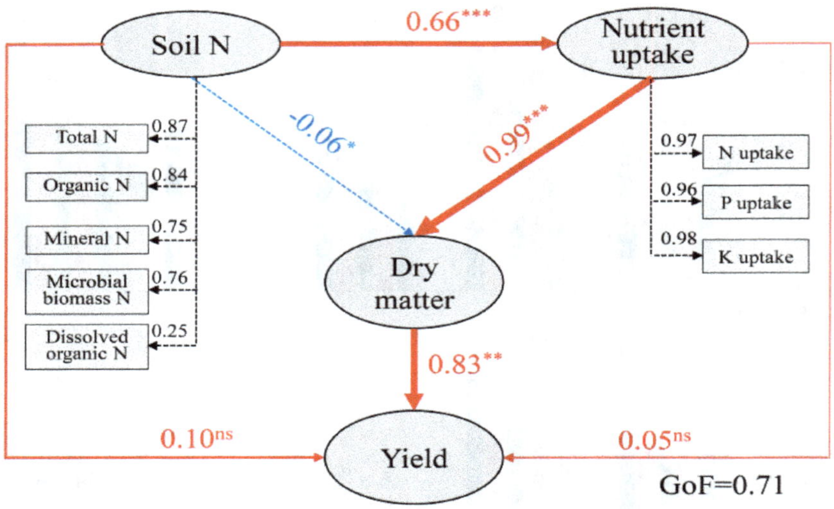

Figure 7. Partial least squares path model based on the effects of soil N fractions, nutrient uptake, and dry matter production on maize grain yield. The loading for soil N pools and plant nutrient uptake that create the latent variables are shown in rectangles. Numbers adjacent to arrows are standardized path coefficients. Continuous red and dashed blue arrows indicate positive and negative effects, respectively. Width of arrows is proportional to the strength of path coefficients. Coefficients differing significantly from zero are indicated by ** $p < 0.01$, and *** $p < 0.001$; ns, not significant. The model is assessed using the goodness of fit (GoF) statistic, a measure of overall prediction performance.

4. Discussion

4.1. At Least 15–30% of Inorganic Fertilizer Input Could Be Reduced in the OV Incorporation-Based Maize Rotation

From 1980 to 2010, the inorganic N input on cropland in China increased from 9.4 to 28.9 Tg/year [34], while the rates of cereal grain yield increased slowly [4,20]. Often twice as much fertilizer N and P is applied than is recovered in crops, and this nutrient imbalance in turn aggravates environmental issues [4,20]. Thus, it is very important to reduce the chemical fertilizer input to the agricultural system, while also maintaining or increasing crop yield [1]. One of the realistic ways is to substitute chemical N or P by green manure. Many studies have shown that either green manure alone or combined with fertilizer can stimulate the following crop growth, yield, and nutrient uptake [11,12,15]. Under the present conditions, the yield, biomass, and N uptake of F_{70} + G and F_{85} + G treatments were similar to or even greater than the recommended fertilization (F_{100}), indicating that with OV incorporation, 15–30% reduction of fertilizer input could still maintain soil N availability to sustain maize growth and yield. In addition, green manure incorporation further increased maize yield and N uptake based on the full rates of fertilizer recommendations, showing stimulated crop growth and N uptake, which was in agreement with other studies using either legume or non-legume green manures [11,12,15]. Overall, the results highlight the crucial role of OV incorporation in enhancing maize growth and nutrient uptake under reduced inorganic fertilizer input.

4.2. Integrated Management of OV Incorporation and Fertilizer Application Compensated for Soil N Pools

Leguminous green manure is mainly cultivated for its ability to fix, accumulate, and supply large amounts of N, while non-legumes are mainly used to prevent soil erosion, trap N, and reduce leaching to the water table [10,35]. Among the desired effects of green manure are improving soil fertility and increasing the stability of N supply [10]. OV incorporation with reduced chemical fertilizer application increased the soil total N content in most cases compared to control, which is in agreement with other reports that application of either inorganic fertilizer or organic substrates can increase soil total N storage [10,36]. Green manure application combined with reduced fertilization had similar or greater soil total N and organic N content than the recommended fertilization through the maize growing season, indicating that OV incorporation well compensated for soil N reserves when 15–30% of fertilizer N input was removed. This was associated with the N supplement by OV incorporation *per se*, since there was a certain amount of green manure N imported to F plus G treatments, partially contributing to total N increase. For instance, in 2012, exogenous N inputs through imported OV plants were 78.5, 45.8, 26.5, and 29.4 kg/ha in the G, F_{70} + G, F_{85} + G, and F_{100} + G treatments, respectively (Table 2). A previous study showed that 67.5% of OV residues can be decomposed during the maize growing season, and 85.3% of OV's N can be released to the soil [25]. Such substantial N release likely furnished the soil N pools, while it is hard to quantify the contribution of exogenous N input to soil total N increase in the present study.

Alternatively, the amount of OV's N input was not the only factor affecting soil total N levels, since N input in the G treatment (from OV alone) was about one-third of that in the F_{100} treatments (from fertilizer N), but it produced slightly more soil total N content during maize season, an average of 0.37 vs. 0.36 g kg^{-1} (Figure 4a). In this case, it might be attributed to enhanced mineralization of soil residual organic matter by OV incorporation, as indicated by similar or greater SMBN content in G relative to F_{100} treatment (Figure 5a). Although we claim that N, P, and K in green manure were taken up and returned *in situ* within the system from 2008 to 2011, and not considered as exogenous inputs (such as fertilizer application), the OV incorporation may affect the in-season mineralization processes of organic residues and soil nutrient availability. In addition, the pattern of maize grain yield in the first year was different from other years; for example, yield in G treatment was significantly greater than in F_{100} treatment. This further supports the point of stimulated N mineralization by OV incorporation.

4.3. Temporal Changes of Soil N Pools Highlights Dominant Contribution of Dissolved Organic N and Mineral N to Maize N Uptake and Growth

DON and N_{min} represent the soluble organic and inorganic N fractions related to quick turnover of soil N pools, which fluctuated over maize growth with higher content at the V3 or V8 stage. These results imply that even at early and fast-growing stages, OV's N could be quickly and easily released to increase soil N availability (Figure 3) due to the low-residue C/N ratio (~13) [25]. Other studies found that a lower substrate C/N ratio favors N mineralization [15,23,37]. Mineral N and DON are two major N sources for microbes and plant uptake, and play a pivotal role in crop N utilization [22,38]. Dissolved organic N mainly consists of amino acids, amino sugars, and low-molecular-mass proteins. These N-containing compounds can be taken up by plants directly through roots or mycorrhiza, especially for the amino acid groups [38,39]. There was a tight relationship between SMBN and DON after long-term fertilization, suggesting a close interaction of these soil N forms in the mineralization processes from incorporated organic residues [23,36].

Under the present conditions, F_{70} + G, F_{85} + G, and F_{100} + G treated soil had significantly greater DON content than control or F_{100} treatment at early (particularly V3 and V8) and mid-late stages. Similar N_{min} content was obtained with F_{100} and F_{70} + G, F_{85} + G and even increased in the F_{100} + G treatment (Figure 5), which was consistent with the similar or greater N uptake and yield in the F plus G treatments. The relative weight analysis showed that the variation of N uptake and dry matter production of maize was mainly contributed by DON and N_{min}, rather than other variables of soil total N, organic N, and SMBN (Figure 6). This might be because N_{min} and DON, but not SMBN, are two major inorganic and organic sources of plant-available N in soil. The changes of N_{min} and DON could directly influence plant N uptake, while the effects of SMBN have to be achieved through N mineralization (involving the turnover of organic and inorganic N fractions). Taken together with the partial least squares path modeling, soil N pools had a positive indirect effect on dry matter production through N uptake (Figure 7), and the enhanced N uptake of maize plants with integrated application of green manure and fertilizer might be coordinated with improved soil DON and N_{min} content, even at the mid-late stages.

In addition, the overall dramatic decrease of DON and N_{min} after the maize V8 stage corresponded to a period with continuous and heavy rainfall (Figure 1), implying N leaching losses by the precipitation. As with nitrate, DON leaching is also a significant pathway of N losses because of its easy movement along the soil profile [7,40]. The nitrate and DON contents in deeper soil layers were not investigated in the present study; however, since other studies have shown that N leaching losses were reduced under OV incorporation [12], it was speculated that the uptake of leached nitrate or DON below the 20 cm layer might be enhanced in the F plus G treatments, which partly contributed to the greater N uptake in maize.

5. Conclusions

In conclusion, the present study demonstrated a promising way of protecting N in fallow seasons by green manure cultivation and reducing the fertilizer inputs in maize production. Due to the enhanced dissolved organic and inorganic N levels by OV incorporation and increased nutrient uptake in soil, 15–30% of recommended fertilizer input could be reduced for maize. The results suggest that green manure–maize rotation could hold great promise in substituting a certain amount of inorganic fertilizer input by OV incorporation without sacrificing crop yield in the North China Plain and other temperate regions.

Author Contributions: Conceptualization, L.Y., J.B., J.L., N.Z., and W.C.; Funding acquisition, W.C.; Project administration, J.B. and N.Z., Supervision, W.C.; Methodology, L.Y., J.B., and J.L.; Software, L.Y.; Investigation, L.Y., J.B., and J.L.; Formal analysis, L.Y. and J.L.; Writing—original draft preparation, L.Y.; Writing—review and editing, L.Y. and W.C.

Acknowledgments: The authors would like to thank Songjuan Gao, Jing Xiong, and Danna Chang for their assistance with the field work, and Michael J.W. Maw at Abraham Baldwin Agricultural College, USA, and Charles B. Krueger at the University of Missouri, USA, for their comments and language editing.

References

1. Chen, X.; Cui, Z.; Fan, M.; Vitousek, P.; Zhao, M.; Ma, W.; Wang, Z.; Zhang, W.; Yan, X.; Yang, J.; et al. Producing more grain with lower environmental costs. *Nature* **2014**, *514*, 486. [CrossRef] [PubMed]
2. Yu, W.; Elleby, C.; Zobbe, H. Food security policies in India and China: Implications for national and global food security. *Food Secur.* **2015**, *7*, 405–414. [CrossRef]
3. Vitousek, P.M.; Naylor, R.; Crews, T.; David, M.B.; Drinkwater, L.E.; Holland, E.; Johnes, P.J.; Katzenberger, J.; Martinelli, L.A.; Matson, P.A.; et al. Nutrient imbalances in agricultural development. *Science* **2009**, *324*, 1519–1520. [CrossRef] [PubMed]
4. Chen, X.P.; Cui, Z.L.; Vitousek, P.M.; Cassman, K.G.; Matson, P.A.; Bai, J.S.; Meng, Q.F.; Hou, P.; Yue, S.C.; Römheld, V.; et al. Integrated soil-crop system management for food security. *Proc. Natl. Acad. Sci. USA* **2011**, *108*, 6399–6404. [CrossRef] [PubMed]
5. Meng, Q.; Hou, P.; Wu, L.; Chen, X.; Cui, Z.; Zhang, F. Understanding production potentials and yield gaps in intensive maize production in China. *Field Crop Res.* **2013**, *143*, 91–97. [CrossRef]
6. Ju, X.T.; Xing, G.X.; Chen, X.P.; Zhang, S.L.; Zhang, L.J.; Liu, X.J.; Cui, Z.L.; Yin, B.; Christie, P.; Zhu, Z.L.; et al. Reducing environmental risk by improving N management in intensive Chinese agricultural systems. *Proc. Natl. Acad. Sci. USA* **2009**, *106*, 3041–3046. [CrossRef] [PubMed]
7. Zhou, J.; Gu, B.; Schlesinger, W.H.; Ju, X. Significant accumulation of nitrate in Chinese semi-humid croplands. *Sci. Rep.* **2016**, *6*, 25088. [CrossRef] [PubMed]
8. Zheng, X.; Han, S.; Huang, Y.; Wang, Y.; Wang, M. Re-quantifying the emission factors based on field measurements and estimating the direct N_2O emission from Chinese croplands. *Glob. Biogeochem. Cycles* **2004**, *18*, GB2018. [CrossRef]

9. Xie, Z.; Tu, S.; Shah, F.; Xu, C.; Chen, J.; Han, D.; Liu, G.; Li, H.; Muhammad, I.; Cao, W. Substitution of fertilizer-N by green manure improves the sustainability of yield in double-rice cropping system in south China. *Field Crop Res.* **2016**, *188*, 142–149. [CrossRef]
10. Thorup-Kristensen, K.; Magid, J.; Jensen, L.S. Catch crops and green manures as biological tools in nitrogen management in temperate zones. *Adv. Agron.* **2003**, *79*, 227–302.
11. Yang, L.; Cao, W.; Thorupkristensen, K.; Bai, J.; Gao, S.; Chang, D. Effect of *Orychophragmus violaceus* incorporation on nitrogen uptake in succeeding maize. *Plant Soil Environ.* **2015**, *61*, 260–265. [CrossRef]
12. Bai, J.; Cao, W.; Xiong, J.; Zeng, N.; Gao, S.; Katsuyoshi, S. Integrated application of February Orchid (*Orychophragmus violaceus*) as green manure with chemical fertilizer for improving grain yield and reducing nitrogen losses in spring maize system in northern China. *J. Integr. Agric.* **2015**, *14*, 2490–2499. [CrossRef]
13. Li, H.; Zheng, L.; Lei, Y.; Li, C.; Liu, Z.; Zhang, S. Estimation of water consumption and crop water productivity of winter wheat in North China Plain using remote sensing technology. *Agric. Water Manag.* **2008**, *95*, 1271–1278. [CrossRef]
14. Thorup-Kristensen, K. Are differences in root growth of nitrogen catch crops important for their ability to reduce soil nitrate-N content, and how can this be measured? *Plant Soil* **2001**, *230*, 185–195. [CrossRef]
15. Zhang, D.; Yao, P.; Na, Z.; Yu, C.; Cao, W.; Gao, Y. Contribution of green manure legumes to nitrogen dynamics in traditional winter wheat cropping system in the Loess Plateau of China. *Eur. J. Agron.* **2016**, *72*, 47–55.
16. Yu, Y.; Xue, L.; Yang, L. Winter legumes in rice crop rotations reduces nitrogen loss, and improves rice yield and soil nitrogen supply. *Agron. Sustain. Dev.* **2014**, *34*, 633–640. [CrossRef]
17. Hooker, K.V.; Coxon, C.E.; Hackett, R.; Kirwan, L.E.; O'Keeffe, E.; Richards, K.G. Evaluation of cover crop and reduced cultivation for reducing nitrate leaching in Ireland. *J. Environ. Qual.* **2008**, *37*, 138–145. [CrossRef] [PubMed]
18. Zhou, Z.; Zhang, L.; Cao, W.; Huang, Y. Appraisal of agro-ecosystem services in winter green manure-spring maize. *Ecol. Environ. Sci.* **2016**, *25*, 597–604. (In Chinese with English Abstract)
19. Mandal, U.K.; Singh, G.; Victor, U.S.; Sharma, K.L. Green manuring: Its effect on soil properties and crop growth under rice-wheat cropping system. *Eur. J. Agron.* **2003**, *19*, 225–237. [CrossRef]
20. Zhang, F.; Cui, Z.; Chen, X.; Ju, X.; Shen, J.; Chen, Q.; Liu, X.; Zhang, W.; Mi, G.; Fan, M. Chapter one-Integrated nutrient management for food security and environmental quality in China. *Adv. Agron.* **2012**, *116*, 1–40.
21. Glasener, K.M.; Wagger, M.G.; Mackown, C.T.; Volk, R.J. Contributions of shoot and root nitrogen-15 labeled legume nitrogen sources to a sequence of three cereal crops. *Soil Sci. Soc. Am. J.* **2002**, *66*, 523–530. [CrossRef]
22. Kielland, K. Amino acid absorption by arctic plants: Implications for plant nutrition and nitrogen cycling. *Ecology* **1994**, *75*, 2373–2383. [CrossRef]
23. Stark, C.; Condron, L.M.; Stewart, A. Influence of organic and mineral amendments on microbial soil properties and processes. *Appl. Soil Ecol.* **2007**, *35*, 79–93. [CrossRef]
24. Wu, L.Q.; Wu, L.; Cui, Z.L.; Chen, X.P.; Zhang, F.S. Basic NPK fertilizer recommendation and fertilizer formula for maize production regions in China. *Acta Pedol. Sin.* **2015**, *52*, 802–817. (In Chinese with English Abstract)
25. Liu, J.; Chen, X.; Zhang, J.; Xu, C.; Cao, W. Study on characteristics of decomposition and nutrients release of winter green manure crop *Orychophragmus violaceus* in North China. *Chin. J. Grassl.* **2013**, *35*, 58–63. (In Chinese with English Abstract)
26. Nelson, D.W.; Sommers, L.E. Determination of total nitrogen in plant material. *Agron. J.* **1973**, *65*, 109–112. [CrossRef]
27. Soon, Y.K.; Kalra, Y.P. A comparison of plant tissue digestion methods for nitrogen and phosphorus analyses. *Can. J. Soil Sci.* **1995**, *75*, 243–245. [CrossRef]
28. Walker, J.M.; Barber, S.A. Absorption of potassium and rubidium from the soil by corn roots. *Plant Soil* **1962**, *17*, 243–259. [CrossRef]
29. Bremner, J.M. Determination of nitrogen in soil by the Kjeldahl method. *J. Agric. Sci.* **1960**, *55*, 11–33. [CrossRef]
30. Brookes, P.C.; Landman, A.; Pruden, G.; Jenkinson, D.S. Chloroform fumigation and the release of soil nitrogen: A rapid direct extraction method to measure microbial biomass nitrogen in soil. *Soil Biol. Biochem.* **1985**, *17*, 837–842. [CrossRef]

31. Jones, D.L.; Willett, V.B. Experimental evaluation of methods to quantify dissolved organic nitrogen (DON) and dissolved organic carbon (DOC) in soil. *Soil Biol. Biochem.* **2006**, *38*, 991–999. [CrossRef]
32. Kabacoff, R. *R in Action: Data Analysis and Graphics with R.*; Manning Publications Co.: Shelter Island, NY, USA, 2015.
33. Sanchez, G. Plsdepot: Partial Least Squares (PLS) Data Analysis Methods v. 0.1.17. 2012. Available online: http://cran.r-project.org/web/packages/plsdepot/index.html (accessed on 9 September 2018).
34. Gu, B.; Ju, X.; Chang, J.; Ge, Y.; Vitousek, P.M. Integrated reactive nitrogen budgets and future trends in china. *Proc. Natl. Acad. Sci. USA* **2015**, *112*, 8792. [CrossRef] [PubMed]
35. Tosti, G.; Benincasa, P.; Farneselli, M.; Pace, R.; Tei, F.; Guiducci, M.; Thorup-Kristensen, K. Green manuring effect of pure and mixed barley-hairy vetch winter cover crops on maize and processing tomato N nutrition. *Eur. J. Agron.* **2012**, *43*, 136–146. [CrossRef]
36. Liang, B.; Yang, X.; He, X.; Zhou, J. Effects of 17-year fertilization on soil microbial biomass C and N and soluble organic C and N in Loessial soil during maize growth. *Biol. Fertil. Soils* **2011**, *47*, 121–128. [CrossRef]
37. Kumar, K.; Goh, K.M. Crop Residues and Management Practices: Effects on Soil Quality, Soil Nitrogen Dynamics, Crop Yield, and Nitrogen Recovery. *Adv. Agron.* **2000**, *68*, 197–319.
38. Streeter, T.C.; Bol, R.; Bardgett, R.D. Amino acids as a nitrogen source in temperate upland grasslands: The use of dual labelled (^{13}C, ^{15}N) glycine to test for direct uptake by dominant grasses. *Rapid Commun. Mass Spectrom.* **2015**, *14*, 1351–1355. [CrossRef]
39. Jones, D.L.; Darrah, P.R. Amino-acid influx at the soil-root interface of *Zea mays* L. and its implications in the rhizosphere. *Plant Soil* **1994**, *163*, 1–12. [CrossRef]
40. Yu, Z.S.; Northup, R.R.; Dahlgren, R.A. Determination of dissolved organic nitrogen using persulfate oxidation and conductimetric quantification of nitrate-nitrogen. *Commun. Soil Sci. Plant Anal.* **1994**, *25*, 3161–3169. [CrossRef]

10

Yields, Soil Health and Farm Profits under a Rice-Wheat System: Long-Term Effect of Fertilizers and Organic Manures Applied Alone and in Combination

Vinod K. Singh [1], Brahma S. Dwivedi [1,*], Rajendra P. Mishra [2], Arvind K. Shukla [3], Jagadish Timsina [4], Pravin K. Upadhyay [1], Kapila Shekhawat [1], Kaushik Majumdar [5] and Azad S. Panwar [2]

1 ICAR-Indian Agricultural Research Institute, New Delhi-110 012, India; vkumarsingh_01@yahoo.com (V.K.S.); pravin.ndu@gmail.com (P.K.U.); drrathorekapila@gmail.com (K.S.)
2 ICAR-Indian Institute of Farming Systems Research, Modipuram-250 110, India; rp_min@yahoo.co.in (R.P.M.); director.iifsr@icar.gov.in (A.S.P.)
3 ICAR-Indian Institute of Soil Science, Bhopal-462 038, India; arvindshukla2k3@yahoo.co.in
4 Soils and Environment Research Group, Faculty of Veterinary and Agricultural Sciences, University of Melbourne, Victoria 3010, Australia; timsinaj@hotmail.com
5 International Plant Nutrition Institute, Gurgaon-122001, India; kmajumdar@ipni.net
* Correspondence: bsdwivedi@yahoo.com

Abstract: The rice-wheat system (RWS), managed over 10.5 Mha in the Indo-Gangetic Plains of India suffers from production fatigue caused by declining soil organic matter, multi-nutrient deficiencies and diminishing factor productivity. We, therefore, conducted a long-term field experiment (1998–1999 to 2017–2018) in Modipuram, India to study the effect of continuous use of farmyard manure (FYM) as an organic fertilizer (OF), mineral fertilizers applied alone (RDF) and their combination (IPNS), as well as the inclusion of forage berseem (IPNS+B) or forage cowpea (IPNS+C) on crop yield, soil health and profits. The long-term yield trends were positive ($p < 0.05$) in all treatments except the control (unfertilized) in rice, and the control and RDF in wheat. Although the yields of rice, wheat and RWS were highest under IPNS treatments (IPNS, IPNS+B, IPNS+C), the maximum annual yield increase in rice (9.2%) and wheat (13.7%) was obtained under OF. A linear regression fitted to the yield data under different IPNS options revealed a highly significant ($p < 0.001$) annual yield increase in rice (5.1 to 6.6%) and wheat (6.8 to 7.7%) crops. Continuous rice-wheat cropping with RDF brought an increase in soil bulk density (Db) over the initial Db at different soil profile depths, more so at depths of 30–45 cm, but inclusion of forage cowpea or berseem in every third year (IPNS+B or C) helped to decrease Db, not only in surface (0–15 cm) but also in sub-surface (15–30 and 30–45 cm depth) soil. Whereas soil organic carbon (SOC) increased under OF, IPNS and IPNS + legume (B or C) treatments, it remained unaffected under RDF after 20 RW cycles. The inclusion of legumes along with IPNS not only helped to trap the NO_3–N from soil layers below 45 cm but also increased its retention in the upper soil (0–15 cm depth). On the other hand, RDF had a higher NO_3–N content in the lower layers (beyond 45 cm depth), indicating downward NO_3–N leaching beyond the root zone. A build-up of Olsen-P was noticed under RDF at different time intervals. The soil exchangeable K and available S contents were maximal under OF and IPNS options, whereas a decline in DTPA extractable-Zn was recorded under OF. Overall, RWS economics revealed that OF treatment involved the maximum cost of cultivation (US$1174 ha^{-1}) with the least economic net return (US$1211 ha^{-1}). Conversely, IPNS + legume (B or C) had lowest cost of cultivation (US$707 to 765 ha^{-1}) and a significantly higher ($p < 0.05$) net return (US$2233 to 2260 ha^{-1}). The study, thus, underlines the superiority of IPNS over RDF or OF; the inclusion of legumes gives an added advantage in terms of production sustainability and soil health. Further studies involving IPNS ingredients other than FYM is needed to develop location-specific IPNS recommendations.

Keywords: rice-wheat system; organic farming; forage legume; long-term productivity; soil health; economics

1. Introduction

The rice (*Oryza sativa* L.)-wheat (*Triticum aestivum* L.) system (RWS) occupying around 13.5 million hectares area in South Asia is the spine of social and economic growth of millions of people [1,2]. The RWS remains the mainstay of cereal production by contributing 23% of food grains [3]. However, the system has begun to show signs of fatigue with yield plateau and soil health decline. Moreover, the differential ecological requirement for more than 10 Mha of agricultural area in India [4], where these crops are grown in sequence, makes the system complex, and the maintenance of sustainability remains a challenge as well as a necessity [5]. The sustainability of the system has been questioned due to yield stagnation, soil health deterioration and poor carbon and water footprints in the environment for more than two decades now [6,7]. Further, the low system diversity results in the associated problems of multi-nutrient deficiencies, higher insect-pest and disease insurgence, and infestation of some noxious weeds. The input-intensive nature of this system makes it less profitable under the ever-diminishing natural resources such as the declining water table and poor soil fertility [8]. Although there are limited options for the diversification of RWS, its productivity ought to rise with the increasing demographic pressure.

Ladha et al., (2003) [9] reviewed 33 years of long-term experiments (LTEs) and observed yield stagnation in rice and wheat in 72% and 85% of the LTEs, respectively, with the application of recommended rates of N, P and K, whereas 22% and 6% of the LTEs showed significant declining trends for rice and wheat yields, respectively. The decadal yield trend analysis also revealed that the RWS have not only suffered productivity stagnation but have also undergone a depletion of inherent nutrients and a reduction in the quantity and quality of organic matter [10]. The soil organic C (SOC) content in LTEs in the major RWS growing areas of Northwestern India decreased sharply under unbalanced fertilizer input [11].

In an intensive RWS, nutrient removal often exceeds replenishment through fertilizers [12–16]. LTEs continuing under diverse agro-ecologies in India underlined that neither the fertilizers nor the organic sources in isolation can achieve sustainable production. The superiority of the combined use of these nutrient sources over their sole application is well-documented [6,17]. Nutrient management strategies thus have an over-riding impact on RWS production through alterations of soil resilience, responsiveness and receptiveness. The nutrient management strategies for this economically and socially important cropping system should ensure optimum plant nutrient supply and desired crop productivity while sustaining or improving soil fertility. The benefits of the use of fertilizers have been documented [18,19], yet the timely availability of cheaper or subsidised fertilizers and lack of options for providing all the limiting nutrients in the right proportions and in the right amounts often prevents their use as a complete nutrient package [20]. The integrated plant nutrient supply system (IPNS), which involves applying the traditionally used organic nutrient sources in conjunction with fertilizers, provides a plant-demand synchronized and slow-release nutrient input to maintain a continuous nutrient supply, preventing losses and ensuring efficient utilization of the applied nutrients [21]. With IPNS, organic manures along with fertilizers serve as a labile source and an immediate sink of C, N, P and S in the soils which improves crop productivity, organic carbon, and soil fertility status [22]. The long-term application of organic manures alone in the form of well-rotten and good quality farmyard manure (FYM) has been reported to make nutrients available gradually, in synchrony with plant needs. Besides improving the physico-chemical properties of soil, the application of organic manures can also increase productivity while maintaining a better energy and environmental balance [23]. Nonetheless, low and variable nutrient contents, scarce availability and problems associated with handling and storage of organic manures are the major constraints in their large-scale usage [19].

However, the inclusion of legumes, especially forage legumes, as a break crop during the third or fourth cycle can benefit RWS without impairing the food security [14]. Besides providing the green fodder for cattle, there is evidence that a forage legume crop can fix 35–120 kg N ha^{-1} in the current season with a carryover effect ranging between 35–60 kg N ha^{-1} for the succeeding crop [16,24,25]. The inclusion of a legume crop in RWS adds N through biological N fixation, recycles nutrients from the deeper soil layers, minimizes soil compaction, increases soil organic matter, breaks weed and pest cycles and minimizes harmful allelopathic effects [26,27]. However, the effects of fertilizers and IPNS vis-à-vis sole organic input and the substitution of main crop with a forage legume as a break crop at 2–3 year intervals, have rarely been studied in intensively cultivated RWS. Also, the limited information on the use of sole organics in RWS makes it difficult to compare different nutrient management options. Such comparisons are, however, essentially required to aid in the understanding of the effects of different nutrient supply options on soil health and crop productivity and to formulate optimum recommendations to attain sustainability of this important cropping system.

Against this background, the present long-term experiment was conducted to study the effects of continuous application of distinct nutrient supply and crop diversification options *viz.* organics and fertilizers applied alone or in combination as well as the effect of the inclusion of forage legumes on crop yield, soil health and farm profits under RWS.

2. Materials and Methods

2.1. The Experimental Site

A long-term field experiment was established during the monsoon season of 1998–1999 and continued for twenty consecutive years, i.e., up until 2017–2018, on Typic Ustocrept soil of the Research Farm of the Indian Institute of Farming Systems Research, Modipuram, Meerut to study the changes in yield, soil health and farm profits of rice and wheat grown in sequence with different nutrient management and crop diversification options. Modipuram (29°4′ N, 77°46′ E, 237 m above sea level), located in the Western part of Uttar Pradesh, represents an irrigated, mechanised and input-intensive cropping area of the Upper Gangetic Plain (UGP) transect of the Indo-Gangetic Plain (IGP). The climate of the experimental site is semi-arid sub-tropical, with dry hot summers and cold winters. The long-term average annual rainfall is 807 mm, and nearly 80% of the total rainfall is received through northwest monsoons from July to September. The year-wise variations in temperature and rainfall are illustrated in Figure 1. Apparently, there were no remarkable changes in these weather parameters over the years of experimentation.

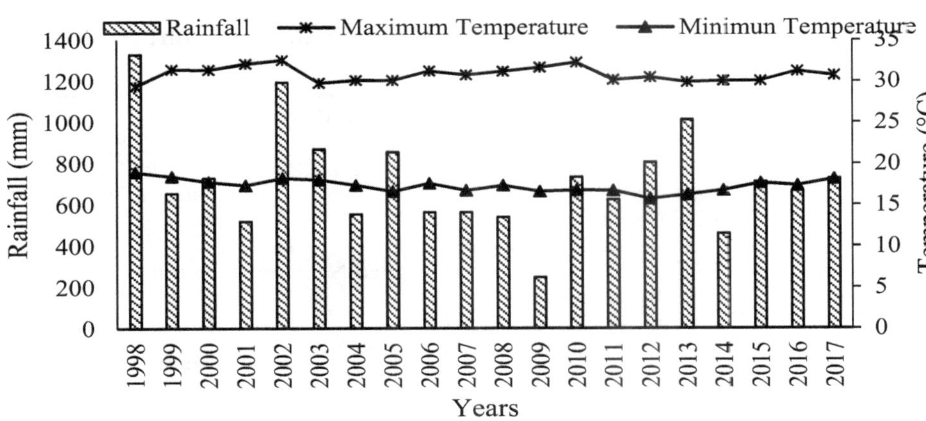

Figure 1. Average yearly temperature (minimum and maximum) and total annual rainfall during the study period.

Nonetheless, the average monthly minimum temperatures fluctuated from 4.5 to 7.8 °C in January (the coolest month) and from 23.7 to 25.4 °C in May (the hottest month). The respective maximum temperatures ranged from 15.7 to 23.4 °C in January and 37.1 to 42.9 °C in May. The soil at the long-term experimental site was sandy loam (17.8% clay, 19.3% silt and 62.9% sand) of Gangetic alluvial origin, which was very deep (>20 m), well-drained, and flat (about 1% slope), representing one of the most extensive soil series, i.e., the Sobhapur series of northwest India.

Data on the initial soil characteristics of the surface soil (0–15 cm depth) measured at the onset of the experiment in 1998–1999 revealed that the soil was mildly alkaline (pH 8.1) and non-saline (electrical conductivity (EC) 0.11 dSm^{-1}) and contained 5.1 g kg^{-1} Walkley and Black carbon (WBC), 8.3 mg kg^{-1} Olsen (0.05 M $NaHCO_3$-extractable) P, 74.1 mg kg^{-1} exchangeable (1 M ammonium acetate-extractable) K, 14.3 mg kg^{-1} available (0.15% $CaCl_2$-extractable) S and 0.54 mg kg^{-1} of DTPA-extractable Zn. The important physico-chemical characteristics of the soil at commencement of the experiment are presented in Table 1. Prior to establishment of the long-term experiment, the site was managed under a sugarcane–ratoon–wheat cropping system, with an intermittent rice (puddled–transplanted)–wheat system.

Table 1. Physico-chemical characteristics of the soil measured at commencement of the field experiment in 1998.

Soil Profile-Depth (cm)	Bulk Density (Mgm^{-3})	pH	EC (dS m^{-1})	Organic Carbon (g kg^{-1})	NO_3–N (mg kg^{-1})	NH_4–N (mg kg^{-1})	Olsen Available P (mg kg^{-1})	Available K (mg kg^{-1})	Available S (mg kg^{-1})	DTPA–Zn (mg kg^{-1})
0–15	1.49	8.01	0.11	5.1	6.9	11.3	8.3	74.1	14.3	0.54
15–30	1.52	7.89	0.12	3.6	6.8	11.6	6.1	66.3	9.1	0.46
30–45	1.58	-	-	-	6.1	11.64	-	-	-	-
45–60	1.60	-	-	-	5.42	10.8	-	-	-	-
60–75	1.61	-	-	-	5.2	10.4	-	-	-	-

2.2. Treatments and Crop Culture

The long-term experiment comprised six treatments involving the following nutrient management options: control (no-fertilizer), FYM as the sole organic fertilizer (OF), recommended doses of fertilizers (RDF) alone, combination of FYM and fertilizers (IPNS), and IPNS with cowpea (*Vigna ungiculata* L.) (IPNS+C) and berseem (*Trifolium alexandrinum* L.) (IPNS+B) as a break crop introduced every third year. These were evaluated on a permanent (undisturbed) layout with a randomised block design (Table 2). The experiment was established as a randomised block design with four replications.

Table 2. Treatment details of the long-term experiment chosen for different nutrient management options.

Treatment Code	Treatment Details	
	Monsoon (Rice)	Winter (Wheat)
Control	No chemical fertilizer or organic manure	No chemical fertilizer or organic manure
RDF	Recommended N, P and K through fertilizers	Recommended N, P and K through fertilizers
IPNS	75% of recommended N, P and K through fertilizers + 25% substitution of recommended N through FYM	Recommended N, P and K through fertilizers
IPNS+C	75% of recommended N, P and K through fertilizers + 25% substitution of recommended N through FYM + every third rice substituted with cow pea	Recommended N, P and K through fertilizers
IPNS+B	75% of recommended N, P and K through fertilizers + 25% substitution of recommended N through FYM + every third wheat substituted with berseem	Recommended N, P and K through fertilizers
OF	100% of recommended N, P and K through organic manures (FYM)	100% of recommended N, P and K through organic manures (FYM)

The plot size was 25 m × 10 m. The recommended NPK rates of 120 kg ha^{-1} N, 26 kg ha^{-1} P and 33 kg ha^{-1} K for both rice and wheat crops were applied through urea (46.4% N), diammonium phosphate (DAP, 18% N and 20.09% P) and muriate of potash (MOP, 49.6% K), respectively. Apart from the control and OF plots, 5 kg ha^{-1} Zn was applied through zinc sulfate (33% Zn) to rice in all treatments. One-third of the N and the entire quantity of P, K and Zn were applied as basal doses to each treatment, except OF at the time of transplanting or sowing, and the remaining N was top-dressed in two equal portions at the maximum tillering stage and at panicle or ear emergence. The quantity of FYM was predetermined on the basis of the N content. The average composition of FYM was C (%) 29.8 ± 6.2, N (%) 0.72 ± 0.13, P (%) 0.26 ± 0.08, K (%) 0.51 ± 0.18, S (%) 0.16 ± 0.05 and a C: N ratio of 6.2. Thus, the quantity of FYM used annually in the IPNS and OF treatments averaged 4.2 and 16.8 t ha^{-1}, respectively. FYM was incorporated into the soil 1 week before the transplantation or sowing of crops.

Each year, 25-day-old rice seedlings (cv. PD 4/PR 106/PR112) were transplanted manually at intervals of 15 cm in rows spaced at 20 cm in puddled plots during the first week of July. After rice harvesting, wheat (cv. UP 2338/ PBW 343/ HD 2967) was sown in rows spaced at 20 cm intervals during the third week of November. All crops were grown under assured irrigated conditions, and chemical weed control was used to maintain almost completely weed-free conditions in the different treatments, except for the OF plots wherein weeds were managed through manual weeding. Need-based spot-weeding was also done in other treatments at 60 days after rice transplantation/wheat sowing depending on weed intensity, despite chemical weed control. Crops were harvested manually using sickle just above the ground level, and the above ground biomass was removed from the plots. The treatments involving cowpea in place of rice (IPNS+C) and berseem in place of wheat (IPNS+B) were sown during first week of July and second week of November during every third rice-wheat cycle. Forage cowpea was harvested manually at the maximum vegetative growth stage, whereas four cuttings of berseem were taken, and green biomass was removed from the plots.

2.3. Soil Sampling and Analysis

Soil samples were collected from a profile depth of 0–75 cm at 15-cm depth intervals from four places in the experimental field using a core sampler of 8 cm diameter before commencement of the experiment in 1998–1999. The post-wheat harvest soil samples (profile depth of 0–105 cm at 15-cm depth intervals) were drawn following the same procedure after completion of every fifth crop cycle at wheat harvest (2002–2003, 2007–2008, 2012–2013 and 2017–2018) in each treatment. The sub-samples obtained were mixed and bulked, and representative soil samples for each depth were drawn for chemical analysis. The initial and post-harvest soil samples were pulverized using a wooden pestle-mortar and sieved through a 2 mm sieve. The processed samples were analysed for mineral N content [28] at all sampled depths and organic carbon [29], Olsen P (0.5 M $NaHCO_3$, pH 8.5 extractable) [30], exchangeable K (1 M NH_4OAc, pH 7.0 extract) [31], available S (0.15% $CaCl_2$ extraction) [32], and DTPA-extractable Zn (DTPA-$CaCl_2$–TEA, pH 7.3 extraction) [33] at depths of 0–15 and 15–30 cm. During 2012–2013, soils from a depth of 0–15 cm were also analysed for extractable N by the alkaline $KMNO_4$ method [34] at 15-day intervals starting from 15 days after transplantation/sowing in rice and wheat up until 120 days into the crop period. The initial samples were also analysed for pH and electrical conductivity (1:2 soil:water suspensions) and mechanical composition (international pipette method), following standard analytical procedures [35].

2.4. Soil Bulk Density

The soil bulk density (Db) at 0–15, 15–30, 30–45, 45–60, 60–75, 75–90 and 90–105 cm depths was measured using undisturbed soil cores [36] drawn just before start of the study (1998–1999) and in the terminal year (post-wheat 2017–2018). Two cores were collected at random in each plot by placing metal cores (5 cm inner diameter) in the middle of each soil layer (i.e., 6–9, 21–24, 36–39, 51–54, 66–69, cm depth in 0–15, 15–30, 30–45, 45–60 and 60–75 cm soil layers). Bulk density was obtained from the gravimetric weights of cores and from the core volume (58.9 cm^3) after oven drying for 48 h at 105 °C. The average Db of two cores collected for each layer was reported for each plot.

2.5. Economic Analysis

Partial economics based on averaged data for the final three years (2015–2016, 2016–2017 and 2017–2018) was computed by using variable costs and income from the sale of grains and straw of rice and wheat as well as berseem and cowpea fodder. The variable cost of cultivation (VCC) of rice and wheat crops included the costs involved in different operations (e.g., rice nursery raising, tillage for seed bed preparation, seeding, insect-pest and weed management, harvesting, threshing) and the inputs (seed, irrigation, fertilizers and agrochemicals and labour) used for raising the crops. The economic analysis, however, did not include the value of the land. The market price considered for different inputs was US\$0.5 kg^{-1} of rice and wheat seed, US\$1.5 kg^{-1} of berseem and US\$0.8 kg^{-1} of cowpea (1US\$ = Indian rupees, INR = 70.00). The cost of fertilizers was calculated on nutrient basis as, N = US\$0.17 kg^{-1}, P = US\$1.99 kg^{-1}, K = US\$0.46 kg^{-1}, Zn= US\$1.21 kg^{-1}. The cost of FYM, herbicide/insecticide and bio-pesticides use was US\$14.28 t^{-1}, US\$10 ha^{-1} and US\$ 8ha^{-1} $application^{-1}$, respectively. Among the field operations, the cost of nursery raising was taken as US\$0.035 m^{-2}, plowing/harrowing was US\$8.5 ha^{-1}, puddling was US\$20.00 ha^{-1}, dry planking was US\$10.00 ha^{-1}, wet planking was US\$17.00 ha^{-1}, sowing of wheat using seed drill was US\$25.00 ha^{-1} and transplantation of rice was US\$45.00 ha^{-1}. The cost of irrigation was taken as US\$7 ha^{-1}, hand weeding was US\$50 ha^{-1}, spot weeding was US\$25 ha^{-1} insecticide/ herbicide spraying was US\$8.5 ha^{-1} and labour was US\$5.0 $unit^{-1}$ day^{-1}.

Gross returns (GR) were calculated by multiplying the grain, straw yield and green fodder by their prices. The minimum support prices (MSPs) fixed by the government for rice (US\$0.18 kg^{-1}) and wheat grain (US\$0.21 kg^{-1}) were used, whereas the price of rice straw (US\$0.016 kg^{-1}) and wheat straw (US\$0.05 kg^{-1}), respectively, and green fodder of cowpea and berseem (US\$ = 0.71) were taken as the prevailing prices in local markets at the time of harvest. The net return (NR) for each crop was calculated as the GR minus the VCC. The NR of rice/cowpea was added to the NR of wheat/berseem to compute the cropping system's net returns for each treatment in respective years. Here, we report data based on the average of the three terminal years (2015–2016, 2017–2018, 2018–2019) for economic analysis.

2.6. Statistical Analysis and Computations

In Figures 2–4, the standard error (SE $\pm$) of the treatment means was computed as

$$S\,E = S\,D(\sqrt{N})^{-1} \tag{1}$$

where SD is the standard deviation of the mean, and N is the number of observations on which the mean is based.

For comparison of treatment means in the field experiments, a multivariate analysis was adapted for multiple comparison of the means, following the procedures of randomized block design [37], as shown in Tables 4–9 and Figure 5. In order to compare the treatments over the years, the yield data of the initial years (1998–1999 and 1999–2000) and final years (2016–2017 and 2017–2018) were pooled, and an analysis of variance (ANOVA) was performed (Table 4).

A least-squares linear regression of yield versus time (years) was computed for all treatments to test the hypothesis that yield trends throughout the experimentation period are not significantly different from zero. In order to understand the yield stability in rice and wheat over time, the yields of 20 years from the start of the experiment were analysed against time using a least-squares linear regression:

$$Y = a + b\,t \tag{2}$$

where Y is the grain yield (t ha^{-1}) of rice or wheat, a is a constant, b is the slope or magnitude of the annual yield change, and t is the time (experimentation period in years).

3. Results

3.1. Effect on Rice and Wheat Yields

The application of RDF either alone or in combination with FYM (IPNS, i.e., 75% RDF + 25% N through FYM in rice and 100% RDF in wheat) significantly ($p < 0.05$) increased the yield of rice over the control (unfertilized) or OF during the initial years (1998–1999 and 1999–2000). The yields of rice under RDF were, however, on par ($p > 0.05$) with the IPNS and IPNS+B or IPNS+C (Table 3). The differences between OF and the control were not significant during the initial years (Table 3). On the other hand, yield reductions under OF as compared with RDF, IPNS, IPNS+B, IPNS+C were in the range of 2.36 to 2.53 t ha^{-1}, during the initial years.

Table 3. Trends in the yields of rice and wheat in a long-term rice-wheat system under different nutrient management options.

Particulars	Annual Yield Changes [a]				Initial Yield [b] (t ha^{-1})
	b-Value	*t*-Statistics	*p*-Value	R^2-Value	
			Rice		
Control	−0.06	−6.98	***	0.83	1.84
RDF	0.01	−4.34	***	0.07	4.62
IPNS	0.07	−3.72	***	0.65	5.01
IPNS+B	0.05	−3.78	***	0.58	4.99
OF	0.09	−5.05	***	0.81	2.90
			Wheat		
Control	−0.06	−7.11	***	0.78	1.58
RDF	−0.05	−4.85	***	0.57	4.42
IPNS	0.07	−3.91	***	0.72	4.54
IPNS+C	0.08	−4.04	***	0.72	4.49
OF	0.14	−5.08	***	0.91	2.33

[a] computed from linear regression; [b] the intercept (a value) of the linear regression; *** significant at $p < 0.001$.

The average rice yields for the terminal years (2016–2017 and 2017–2018) revealed that the treatments receiving conjoint use of fertilizers and FYM, with or without inclusion of a legume, i.e., IPNS, IPNS+B or IPNS+C had significantly ($p < 0.05$) greater yields than those receiving RDF (Table 3). The magnitude of increase in rice yield in IPNS or IPNS+B or IPNS+C over RDF was in the range of 26% to 29% during the terminal years. Rice yield under OF was also significantly ($p < 0.05$) lower as compared with RDF during the terminal years.

Although the residual effects of FYM application to rice on the succeeding wheat yields were not significant ($p < 0.01$) over RDF during the initial years, this varied based on the use of IPNS and different organic and legume combinations during the terminal years (2016–2017 and 2017–2018). Compared with RDF, wheat yields under IPNS, IPNS+B and IPNS+C during the terminal years were greater by 64%, 72% and 73%, respectively (Table 3). Wheat yields under OF also increased with the passage of time and gave significantly ($p < 0.05$) higher yield gains over RDF and the control. Nonetheless, the yields under OF remained 21% to 25% lower than those of IPNS after two decades of experimentation (Table 3).

Further, a comparison of yields during the terminal years vis-à-vis the initial years (2-year average) revealed that OF had the maximum yield gain (63% in rice and 108% in wheat). This yield gain under different IPNS options during the terminal years as compared with the initial years ranged from 20% to 23% in rice and 27% to 34% in wheat in the following order: IPNS > IPNS+C > IPNS+B. After 20 years of RW cropping, changes in the system productivity under the control and RDF showed negative trends (58.1% and 9.43% decline over the initial years) whereas, system productivity gains of 83.4%, 28.2%, 27.2% and 23.6% were noticed under OF, IPNS+C, IPNS+B and IPNS, respectively (Table 4).

Table 4. Changes in rice, wheat and RWS productivity under different long-term nutrient management options during the initial and terminal years.

Treatments	Rice Yield (t ha^{-1})		% Change	Wheat Yield (t ha^{-1})		% Change	RW System Yield (t ha^{-1})		% Change
	Initial Years [a]	Terminal Years [b]		Initial Years	Terminal Years		Initial Years	Terminal Years	
Control	2.01 [c]	0.83 [d]	−58.7	1.81 [d]	0.77 [d]	−57.5	3.82 [c]	1.60 [c]	−58.1
RDF	4.96 [a]	4.88 [b]	−1.6	4.37 [b]	3.57 [c]	−18.35	9.33 [a]	8.45 [b]	−9.4
IPNS	5.09 [a]	6.13 [a]	20.4	4.60 [ab]	5.85 [a]	27.2	9.69 [a]	11.98 [a]	23.6
IPNS+B	5.13 [a]	6.27 [a]	22.2	4.65 [a]	6.18 [a]	32.9	9.78 [a]	12.44 [a]	27.2
IPNS+C	5.09 [a]	6.27 [a]	23.2	4.59 [ab]	6.14 [a]	33.8	9.68 [a]	12.41 [a]	28.2
OF	2.60 [b]	4.23 [c]	62.7	2.15 [c]	4.48 [b]	108.4	4.75 [b]	8.71 [b]	83.4

[a] Average of 1998–1999 and 1999–2000; [b] average of 2016–2017 and 2017–2018.

3.2. Yield Trends in Rice and Wheat

In rice, the trends in yield were positive ($p < 0.001$) in all treatments except for the control (Table 4, Figure 2). The annual increase was 9.2% under OF, 6.6% in IPNS, 5.1% in IPNS+B and 1.4% under RDF, which was statistically significantly different from zero ($p < 0.001$). Skipping fertilizer application caused a significant ($p < 0.001$) annual yield decline (−6.1%) during the study.

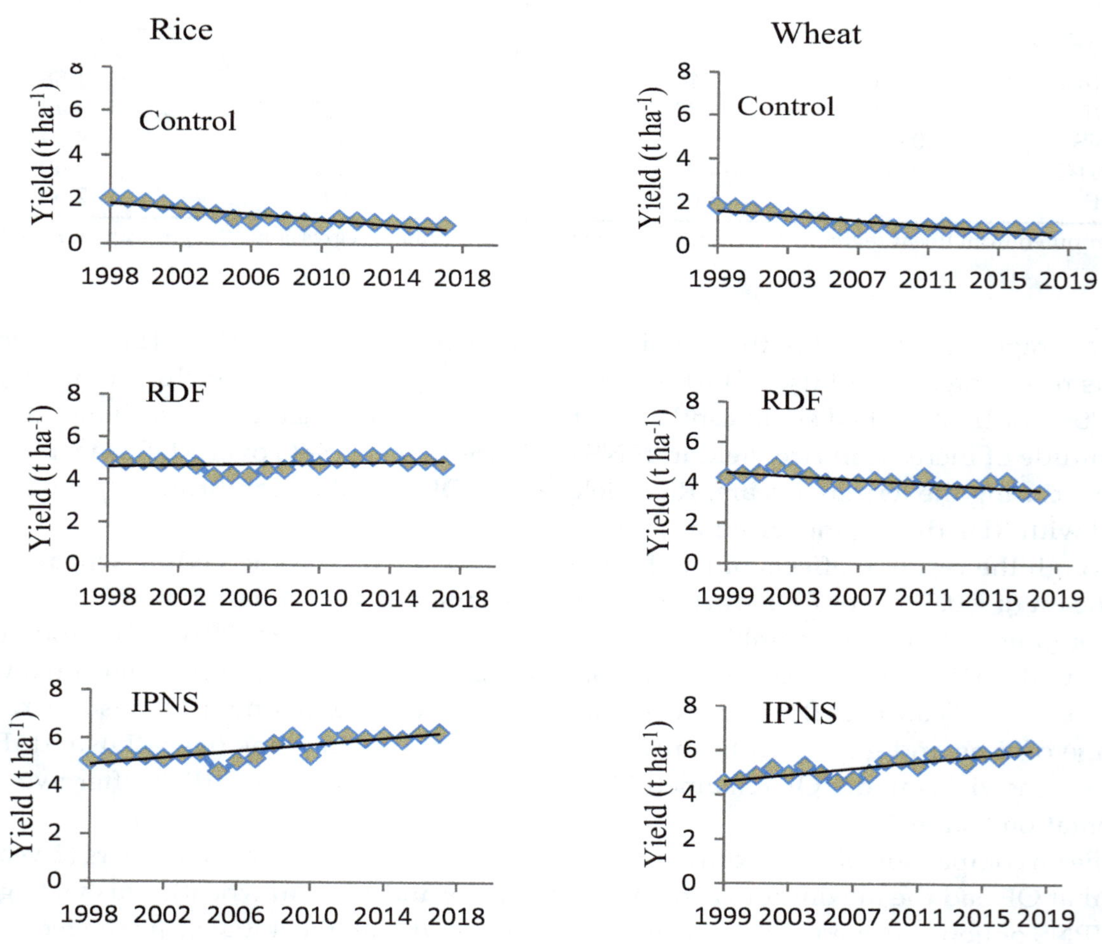

Figure 2. *Cont.*

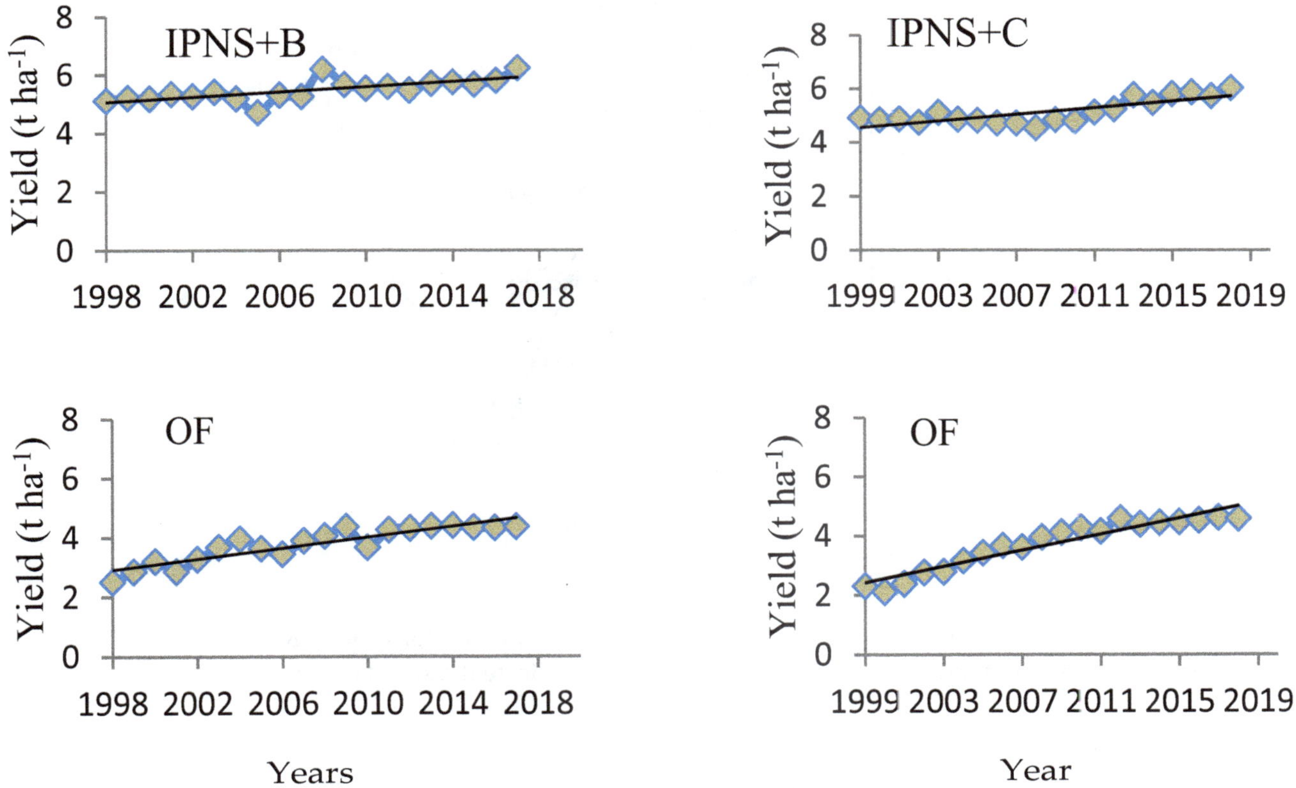

Figure 2. Trends in yields of rice and wheat in a long-term RWS under different nutrient management options. Measured mean yields of the treatments (symbols) and the trend line fitted by linear regression (Table 3) are shown.

The trends in wheat yield in the control as well as RDF were negative ($p < 0.001$), with annual yield declines of −5.5% and −4.5%, respectively (Table 4, Figure 1). In contrast, yield trends were positive and significant ($p < 0.001$) in all other treatments. The maximum annual wheat yield gain (13.7%) was noticed under OF, followed by IPNS+C (7.7%) and IPNS (6.8%).

3.3. Effect on Soil Health

3.3.1. Soil Bulk Density

The soil bulk density (Db) of the surface layer (0–15 cm), was 1.49 ± 0.013 Mg m^{-3} at the onset of the experiment. It, however, increased with an increasing soil profile depth, measuring 1.61 ± 0.010 Mg m^{-3} at a depth of 60–75 cm (Figure 3). At the end of experimentation in 2017–2018, the RDF treatment did not influence the soil Db in the upper soil depths (0–15 cm and 15–30 cm), but the Db values at the 30–45 cm soil depth were greater (1.74 ± 0.012 Mg m^{-3}) compared with the initial Db (1.58 ± 0.016 Mg m^{-3}) at this depth, thereby indicating ta tendency towards sub-surface soil compaction. In the plots having IPNS with legumes as a break crop during every third year, no such compaction was noticed. The Db of soil at a profile depth of 30–45 cm was much smaller (1.36 ± 0.16 Mg m^{-2} and 1.34 ± 0.014 Mg m^{-2} in IPNS+B and IPNS+C plots, respectively) compared with the Db in RDF plots (1.74 ± 0.013 Mg m^{-2}). Although the Db values under IPNS and OF treatments were much smaller (1.42 ± 0.018 and 1.38 ± 0.024) at the 0–15 cm depth as compared with the Db at the onset of the experiment, this did not influence values at the other soil profile depths.

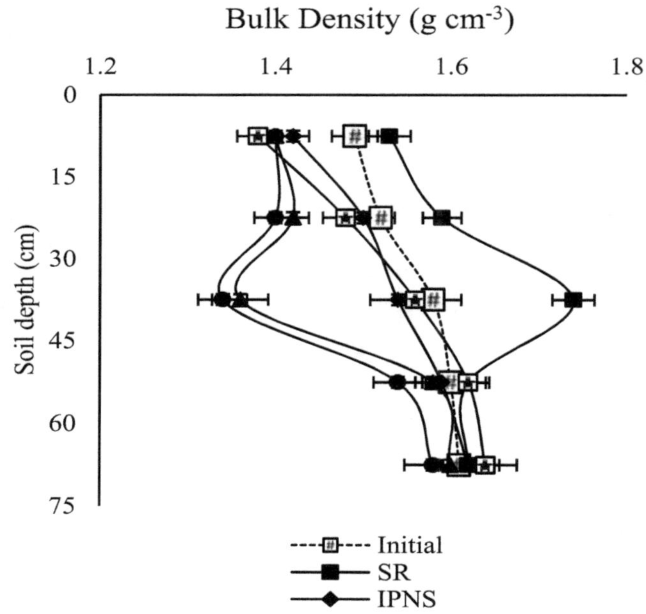

Figure 3. Changes in the bulk density of soil at different profile depths after 20 RWS cycles under the influence of different nutrient management options. Bars indicate the standard error of the mean, $n = 4$.

3.3.2. Soil OC Content

The soil organic carbon (SOC) content measured at profile depths of 0–15 and 15–30 cm during 2002–2003, 2007–2008, 2012–2013 and 2017–2018 varied in accordance with the soil depth and with the passage of time (Table 5). In surface soil (0–15 cm depth), during 2002–2003, the SOC content was highest (5.3 g kg^{-1}) under OF which was on par with IPNS (5.6 g kg^{-1}) and IPNS+C or IPNS+B (5.8 g kg^{-1}), though statistically higher ($p < 0.05$) than RDF (5.2 g kg^{-1}). The SOC content increased ($p < 0.05$) under OF, IPNS and IPNS+B or C over the years but declined in the control plots. The SOC remained unaffected under RDF during the different years of study. During the terminal year (2017–2018), the maximum SOC (6.9 g kg^{-1}) was noticed under OF which was significantly higher ($p < 0.05$) than that under IPNS (6.2 g kg^{-1}) and IPNS+B or IPNS+C (6.4 g kg^{-1}). The RDF and control plots had significantly ($p < 0.05$) lower SOC contents than the different IPNS options.

Table 5. Effects of different nutrient management options on the soil organic carbon (SOC) content (g kg^{-1}) at soil depths of 0–15 and 15–30 cm after 5 RWS cycle intervals.

Treatment	2002–2003	2007–2008	2012–2013	2017–2018	Mean	2002–2003	2007–2008	2012–2013	2017–2018	Mean
	0–15 cm Depth					15–30 cm Depth				
Control	4.7 ^{cA}	4.3 ^{cAB}	3.9 ^{cB}	3.3 ^{dC}	4.1 ^{c}	3.0 ^{cA}	2.8 ^{bA}	2.8 ^{eA}	2.6 ^{dA}	2.8 ^{d}
RDF	5.2 ^{bA}	5.1 ^{bA}	5.3 ^{bA}	5.4 ^{cA}	5.3 ^{b}	3.2 ^{aA}	3.4 ^{aA}	3.5 ^{cA}	3.7 ^{bcA}	3.5 ^{c}
IPNS	5.6 ^{abB}	5.8 ^{aAB}	6.0 ^{aAB}	6.2 ^{bA}	5.9 ^{a}	3.6 ^{aA}	4.0 ^{aB}	4.2 ^{dB}	4.4 ^{cB}	4.1 ^{b}
IPNS+B	5.5 ^{abC}	5.9 ^{aBC}	6.2 ^{aAB}	6.4 ^{bA}	6.0 ^{a}	3.9 ^{b}	4.7 ^{aB}	5.0 ^{aA}	5.3 ^{aAB}	4.7 ^{a}
IPNS+C	5.5 ^{abB}	5.8 ^{aB}	6.4 ^{aA}	6.4 ^{bA}	6.0 ^{a}	4.0 ^{a}	4.5 ^{aB}	5.1 ^{aA}	5.4 ^{aA}	4.8 ^{a}
OF	5.8 ^{aC}	6.4 ^{aBC}	6.9 ^{aB}	7.1 ^{aA}	6.6 ^{a}	3.3 ^{c}	3.8 ^{aA}	4.0 ^{bcA}	4.2 ^{bA}	3.8 ^{bc}

Values followed by different lower case letter(s) within a column are significant at $p < 0.05$. Values followed by different capital letter(s) within a row are significant at $p < 0.05$.

The SOC content also varied with soil depth, and a relatively lower SOC was noticed at the 15–30 cm depth (Table 5). At this soil depth, IPNS combined with legumes had a significant edge over OF in different years of the study. The SOC, measured after 20 annual RW cycles vis-à-vis the onset of the experiment, showed improvements of 6–39% and 19–74% at the 0–15 and 15–30 cm soil depths, respectively, in all treatments except for the control, wherein the SOC declined by 14–35% from its initial value.

3.3.3. Soil Mineral N Content

The nitrate (NO_3)–N content of surface soil (0–15 cm) was greater under IPNS+B and IPNS+C compared with the other nutrient management options; the differences were more prominent during 2012–2013 and 2017–2018 (Figure 4). The NO_3–N content in the surface soil layer after wheat harvest in the IPNS+B treatment was greater than that in the RDF by 103% in 2002–2003, 173% in 2007–2008, 213% in 2012–2013 and 268% in 2017–2018. The corresponding increases with the inclusion of cowpea as a break crop (IPNS+C) were 93%, 159%, 204% and 255%, respectively.

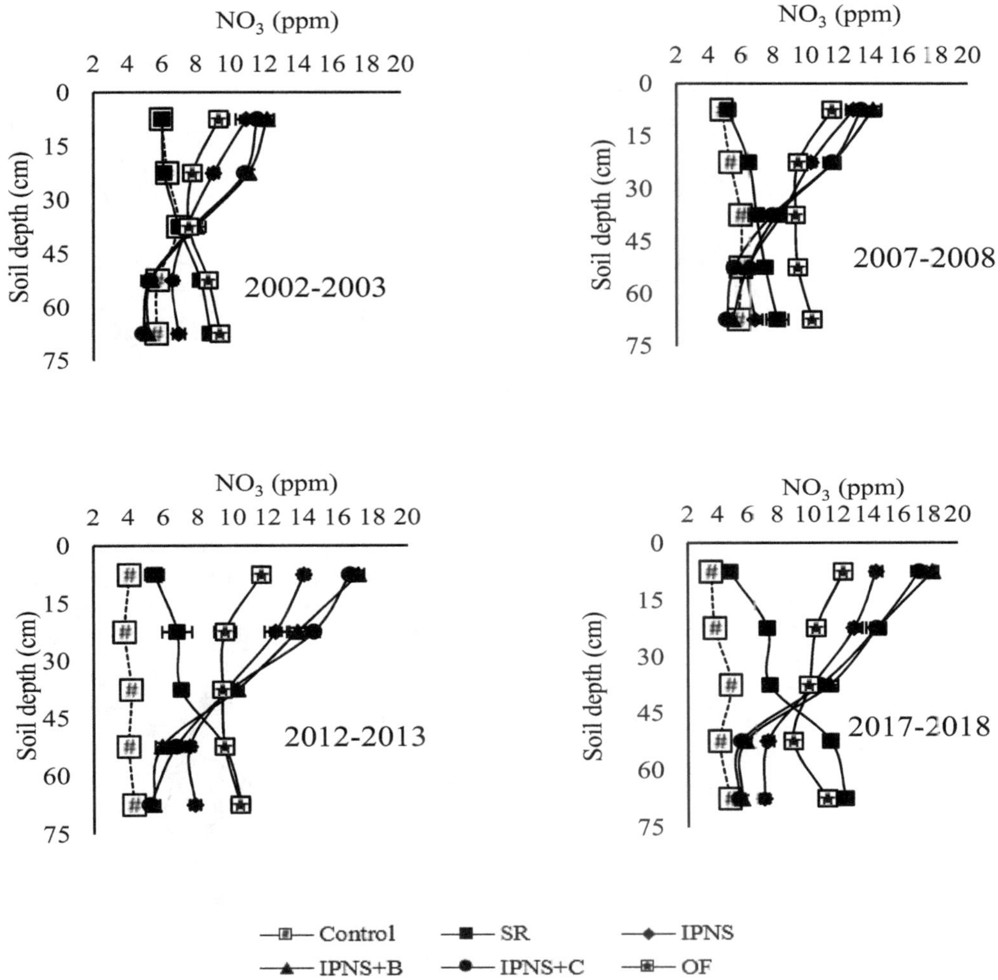

Figure 4. Effects of the nutrient management options on the distribution of nitrate–N in different soil profiles after wheat harvest at five-year intervals of the RWS cycle. Bars indicate the standard error of the mean, $n = 4$.

Compared with the initial NO_3–N content of the surface soil layer (6.9 mg kg^{-1}), the NO_3–N content in IPNS, IPNS+B, IPNS+C, and OF treatments was greater by 7.76, 7.76, 11.5 and 10.6 mg kg^{-1}, respectively, during the terminal year. On the other hand, no application of fertilizer or the application of FYM (control) and RDF led to depletion of NO_3–N by 3.25 mg kg^{-1} and 1.96 mg kg^{-1}, respectively, over the initial status (Figure 4).

Different nutrient management options markedly influenced the distribution of NO_3–N in the soil profile. Whereas the NO_3–N content increased over the initial content up to a profile depth of 75 cm in the RDF and OF plots, a constant decrease in the NO_3–N content was observed under the IPNS + legume options (Figure 4). The inclusion of legumes in every third RW cycle (IPNS+B or IPNS+C) not only favoured a greater NO_3–N content in the surface layer as compared with RDF, but also resulted in a lower NO_3–N content in the deeper soil layers, as was evident from a decrease in the NO_3–N content with an increasing soil profile depth. This advantage (greater retention of NO_3–N in the surface soil) of the inclusion of legumes in RWS as a break crop was, however, more spectacular during the terminal year (2017–2018).

The nutrient supply through fertilizers, organic compounds, or both increased the NH_4–N content as compared with no fertilizer application in the surface soil layer; the magnitude of such increase was greater under OF, followed by IPNS+C, IPNS+B, IPNS and RDF during 2002–2003 (Figure 5).

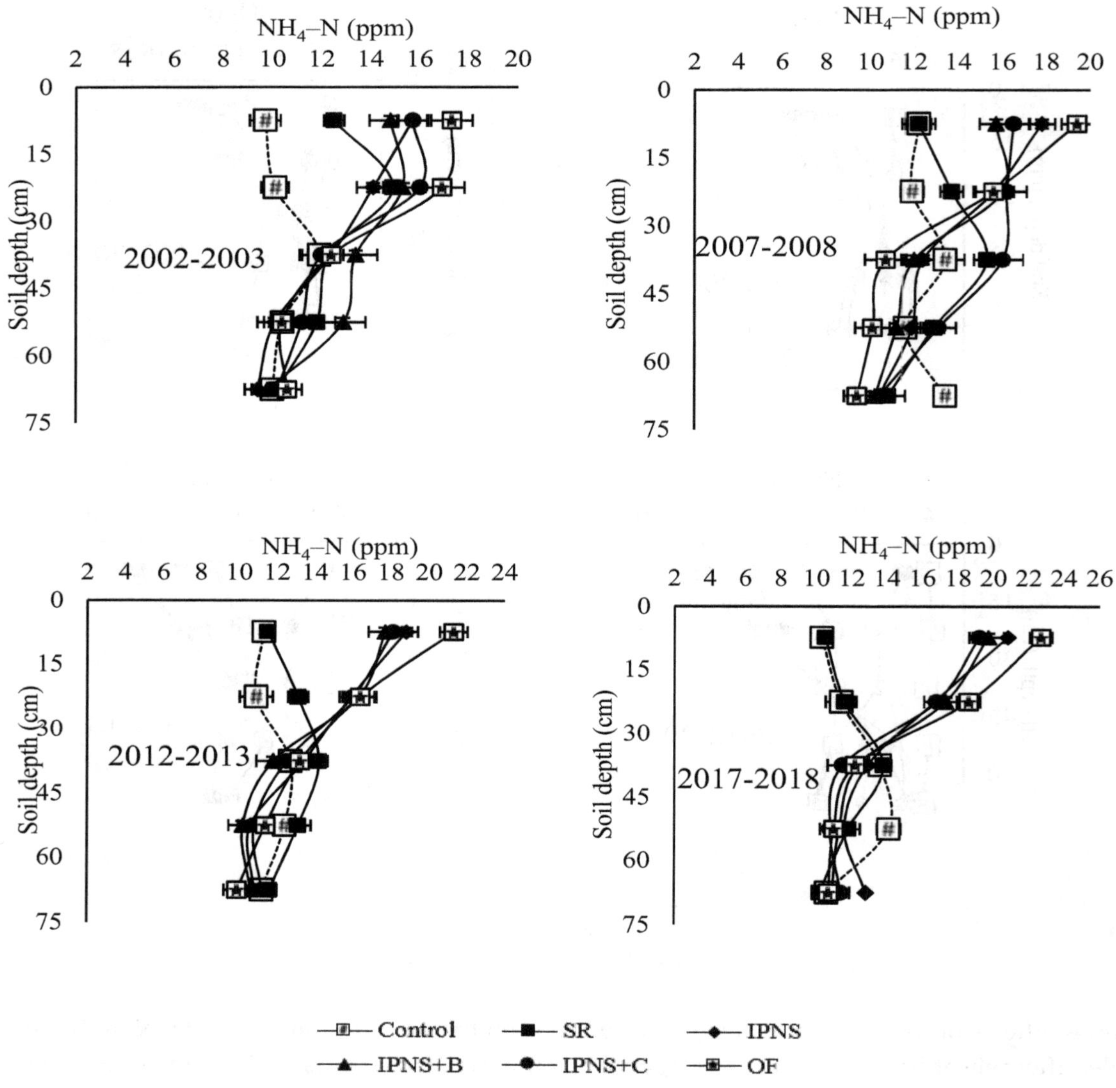

Figure 5. Effect of nutrient management options on the distribution of the ammonium–N content in different soil profiles after wheat harvest at five-year intervals of the RWS cycle. Bars indicate the standard error of the mean, $n = 4$.

Compared with the initial NH_4–N content (11.3 mg kg^{-1}), OF and different IPNS options had a 31–53% higher NH_4–N content in the surface soil layer during 2002–2003 (Figure 5). A similar pattern of increase in NH_4–N content as the consequence of different fertilizer options was also observed after wheat harvest during 2007–2008, 2012–2013 and 2017–2018. The NH_4–N content profile at different depths under OF and IPNS was almost uniformly distributed up to a depth of 0–45 cm, and was relatively greater than the initial NH_4–N concentration. Values beyond a soil depth of 45 cm did not show any consistent trend among different sampling intervals.

3.3.4. Soil Olsen-P Content

The Olsen-P content of the surface soil (0–15 cm) increased compared with the initial content (8.3 mg kg^{-1}) under different nutrient management options and with the passage of time (Table 6). After the 20th RWS cycle (2017–2018), the maximum P build-up in the surface soil was noticed under RDF (21.8 mg kg^{-1}), followed by OF (18.2 mg kg^{-1}), whereas a relatively lower P content (16.0 mg kg^{-1}) was recorded under the IPNS treatments. The effect of fertilizer options was relatively smaller in the sub-surface soil (15–30 cm soil depth). Nonetheless, fertilizer input at the recommended rate increased Olsen-P over the initial content (6.1 mg kg^{-1}) in the sub-surface layer by 122% during the terminal crop cycle (2017–2018). Relatively smaller P accumulations over the initial content in the sub-surface soil (15–30 cm depth) were noticed under IPNS (51%) and OF (11%). On the other hand, a 20–23% decline in Olsen-P over the initial content was noticed under IPNS+B and IPNS+C (Table 6).

Table 6. Effects of different nutrient management options on soil Olsen-P (mg kg^{-1}) at soil depths of 0–15 and 15–30 cm after 5 RWS cycle intervals.

Treatment	2002–2003	2007–2008	2012–2013	2017–2018	Mean	2002–2003	2007–2008	2012–2013	2017–2018	Mean
	0–15 cm Depth					15–30 cm Depth				
Control	8.0 ^{eA}	7.6 ^{dAB}	7.1 ^{eAB}	6.9 ^{dB}	7.4 ^{d}	5.6 ^{cA}	4.9 ^{eB}	4.9 ^{dB}	3.9 ^{eC}	4.8 ^{eB}
RDF	13.0 ^{aD}	15.9 ^{aC}	19.7 ^{aB}	21.9 ^{aA}	17.6 ^{a}	6.3 ^{bD}	8.5 ^{aC}	10.9 ^{aB}	13.6 ^{aA}	9.8 ^{aC}
IPNS	11.7 ^{bD}	13.9 ^{bC}	15.9 ^{bB}	17.7 ^{bA}	14.8 ^{b}	7.3 ^{aC}	7.4 ^{bC}	8.4 ^{bB}	9.2 ^{bA}	8.1 ^{bB}
IPNS+B	8.9 ^{deD}	11.3 ^{cC}	13.7 ^{dB}	15.9 ^{cA}	12.5 ^{c}	5.3 ^{cAB}	5.7 ^{dA}	4.8 ^{dBC}	4.7 ^{dC}	5.1 ^{deB}
IPNS+C	10.0 ^{cdD}	12.1 ^{cC}	14.3 ^{cdB}	16.0 ^{cA}	13.1 ^{c}	6.0 ^{bcA}	6.0 ^{cdA}	5.1 ^{dB}	4.9 ^{dB}	5.5 ^{dA}
OF	12.7 ^{aD}	14.1 ^{aC}	17.2 ^{bcB}	18.2 ^{bA}	15.6 ^{b}	6.5 ^{bAB}	6.4 ^{cB}	6.9 ^{cA}	6.8 ^{cAB}	6.6 ^{cAB}

Values followed by different lower case letter(s) within a column are significant at $p < 0.05$. Values followed by different capital letter(s) within a row are significant at $p < 0.05$.

3.3.5. Soil Exchangeable K Content

Over the years, a consistent increase in soil exchangeable K was noticed under IPNS as well as OF (Table 7). At soil depths of 0–15 and 15–30 cm, the highest exchangeable K contents were recorded under OF (110 and 90 mg kg^{-1}, respectively) and these were on par with the values under the different IPNS options during 2017–2018. When compared with the initial K contents (74.1 and 66.3 mg kg^{-1} at 0–15 and 15–30 cm soil depth, respectively), the exchangeable K content of the soil after 20 RW cycles increased by 48%, 43%, 42% and 36% at the 0–15 cm depth and by 36%, 30%, 29% and 28% at the 15–30 cm depth under OF, IPNS, IPNS+B and IPNS+C, respectively (Table 7). On the contrary, a consistent depletion of exchangeable K was observed in the RDF and unfertilized plots, wherein it remained significantly lower ($p < 0.05$) than that in the other treatments at different soil sampling intervals at both soil profile depths (0–15 and 15–30 cm).

Table 7. Effects of different nutrient management options on the soil exchangeable K content (mg kg^{-1}) in at soil depths of 0–15 and 15–30 cm after 5 RWS cycle intervals.

Treatment	2002–2003	2007–2008	2012–2013	2017–2018	Mean	2002–2003	2007–2008	2012–2013	2017–2018	Mean
	0–15 cm Depth					15–30 cm Depth				
Control	69.1 ^{bA}	67.1 ^{bA}	64.7 ^{bA}	61.9 ^{dA}	65.7 ^{b}	64.7 ^{bAB}	62.5 ^{bA}	60.6 ^{cAB}	58.1 ^{cB}	61.5 ^{b}
RDF	70.5 ^{bA}	65.2 ^{bA}	61.2 ^{bA}	58.9 ^{dA}	64.0 ^{b}	65.8 ^{bA}	62.2 ^{bAB}	59.9 ^{cAB}	56.2 ^{cB}	61.0 ^{b}
IPNS	83.0 ^{aC}	91.3 ^{aAB}	96.5 ^{aAB}	106.4 ^{bcA}	94.3 ^{a}	69.9 ^{abC}	76.0 ^{aB}	79.3 ^{bAB}	86.2 ^{bA}	77.9 ^{a}
IPNS+B	82.3 ^{aC}	92.0 ^{aB}	98.2 ^{aB}	104.5 ^{abA}	94.3 ^{a}	67.5 ^{abC}	75.9 ^{aB}	79.9 ^{abB}	85.7 ^{abA}	77.3 ^{a}
IPNS+C	82.4 ^{aB}	92.6 ^{aA}	96.7 ^{aA}	101.4 ^{cA}	93.3 ^{a}	69.5 ^{abB}	76.8 ^{aA}	80.6 ^{abA}	84.0 ^{bA}	77.7 ^{a}
OF	87.3 ^{aC}	94.6 ^{aB}	99.4 ^{aB}	109.6 ^{aA}	97.7 ^{a}	72.0 ^{aC}	79.8 ^{aB}	84.7 ^{aB}	90.4 ^{aA}	81.7 ^{a}

Values followed by different lower case letter(s) within a column are significant at $p < 0.05$. Values followed by different capital letter(s) within a row are significant at $p < 0.05$.

3.3.6. Soil Available S Content

After 20 RW cycles, the soil available S content increased considerably over the initial content (14.3 and 9.0 mg kg^{-1}) at soil profile depths of 0–15 and 15–30 cm under the OF and IPNS options

(Table 8). Compared with the initial S status, OF had a 114% and 89% higher available S content at the 0–15 and 15–30 cm profile depths during 2017–2018. The increases under IPNS, IPNS+B and IPNS+C were 76% and 70%, 59% and 57% and 62% and 67%, respectively during the terminal years. On the other hand, RDF could not prevent a decline in S content, as depletions of 42% and 31% over the initial soil S content were noticed at soil depths of 0–15 and 15–30 cm, respectively, during 2017–2018 (Table 8). A depletion of available S over the initial S content was also noticed in the control, but the magnitude was smaller than that under the RDF plots.

Table 8. Effects of different nutrient management options on the soil available S content (mg kg^{-1}) at soil depths of 0–15 and 15–30 cm after 5 RWS cycle intervals.

Treatment	2002–2003	2007–2008	2012–2013	2017–2018	Mean	2002–2003	2007–2008	2012–2013	2017–2018	Mean
	0–15 cm Depth					15–30 cm Depth				
Control	11.0 ^{dA}	10.4 ^{eAB}	9.6 ^{dBC}	8.7 ^{dC}	9.9 ^{d}	8.1 ^{cA}	7.8 ^{dA}	7.9 ^{dA}	7.5 ^{dA}	7.8 ^{d}
RDF	10.9 ^{dA}	9.8 ^{eAB}	9.1 ^{dBC}	8.3 ^{dC}	9.5 ^{d}	7.7 ^{cA}	7.1 ^{dAB}	6.9 ^{dB}	6.3 ^{dB}	7.0 ^{d}
IPNS	16.0 ^{bD}	19.2 ^{bA}	22.9 ^{bB}	25.2 ^{bA}	20.8 ^{b}	10.3 ^{bD}	12.9 ^{abC}	14.0 ^{bB}	15.4 ^{bA}	13.2 ^{b}
IPNS+B	14.5 ^{cD}	16.4 ^{dA}	19.7 ^{cB}	22.7 ^{cA}	18.3 ^{c}	9.9 ^{bD}	11.6 ^{cC}	12.6 ^{cB}	14.3 ^{cA}	12.1 ^{c}
IPNS+C	15.1 ^{bcC}	17.8 ^{B}	22.0 ^{bA}	23.1 ^{cA}	19.5 ^{c}	10.4 ^{bD}	12.1 ^{bcC}	13.5 ^{bB}	14.6 ^{bcA}	12.7 ^{bc}
OF	18.2 ^{aD}	22.1 ^{aC}	28.3 ^{aB}	30.6 ^{aA}	24.8 ^{a}	11.5 ^{aD}	13.7 ^{aC}	15.0 ^{aB}	17.2 ^{aA}	14.4 ^{a}

Values followed by different lower case letter(s) within a column are significant at $p < 0.05$. Values followed by different capital letter(s) within a row are significant at $p < 0.05$.

3.3.7. Soil DTPA–Zn Content

The DTPA–Zn content varied among the different nutrient management options, soil profile depths and years of experimentation (Table 9). Whereas DTPA–Zn increased significantly over the initial content (0.54 and 0.46 mg kg^{-1} in the 0–15 and 15–30 soil depths, respectively, under RDF, IPNS, IPNS+C and IPNS+B, a continuous decline in DTPA–Zn was noticed under OF and control over the years at both profile depths. After 20 RW cycles, an additional Zn accumulation of 0.30 to 0.39 mg kg^{-1} at 0–15 cm and 0.14 to 0.18 mg kg^{-1} at the 15–30 cm soil profile depth was noticed over its initial status under RDF and in the different IPNS options. Conversely, declines of 0.18 and 0.12 mg kg^{-1} in DTPA–Zn were noticed under OF treatment at the 0–15 and 15–30 cm profile depths, respectively, during 2017–2018.

Table 9. Effects of different nutrient management options on soil DTPA extractable Zn content (mg kg^{-1}) at soil depths of 0–15 and 15–30 cm after 5 RWS cycle intervals.

Treatment	2002–2003	2007–2008	2012–2013	2017–2018	Mean	2002–2003	2007–2008	2012–2013	2017–2018	Mean
	0–15 cm Depth					15–30 cm Depth				
Control	0.48 ^{bA}	0.43 ^{bAB}	0.40 ^{cBC}	0.37 ^{cC}	0.42 ^{b}	0.41 ^{bA}	0.39 ^{bA}	0.37 ^{cAB}	0.34 ^{bB}	0.38 ^{b}
RDF	0.67 ^{aB}	0.70 ^{aB}	0.83 ^{abA}	0.87 ^{bA}	0.77 ^{a}	0.51 ^{aB}	0.54 ^{aB}	0.60 ^{abA}	0.62 ^{aA}	0.57 ^{a}
IPNS	0.65 ^{aB}	0.70 ^{aB}	0.80 ^{bA}	0.84 ^{bA}	0.75 ^{a}	0.48 ^{aB}	0.51 ^{aB}	0.56 ^{bA}	0.60 ^{aA}	0.54 ^{a}
IPNS+B	0.67 ^{aD}	0.73 ^{aC}	0.87 ^{aB}	0.93 ^{aA}	0.80 ^{a}	0.49 ^{aB}	0.53 ^{aB}	0.58 ^{abA}	0.61 ^{aA}	0.55 ^{a}
IPNS+C	0.69 ^{aB}	0.72 ^{aB}	0.88 ^{aA}	0.88 ^{abA}	0.79 ^{a}	0.50 ^{aC}	0.55 ^{aB}	0.61 ^{aA}	0.64 ^{aA}	0.57 ^{a}
OF	0.51 ^{bA}	0.44 ^{bB}	0.40 ^{cBC}	0.36 ^{cC}	0.43 ^{bB}	0.43 ^{bA}	0.40 ^{bA}	0.35 ^{cB}	0.34 ^{bB}	0.38 ^{b}

Values followed by different lower case letter(s) within a column are significant at $p < 0.05$. Values followed by different capital letter(s) within a row are significant at $p < 0.05$.

3.4. *Effect on Economics*

The average cost of cultivation in rice over the terminal three years was US$644 ha^{-1} under OF, US$481 ha^{-1} under RDF, US$489 ha^{-1} under IPNS, US$464 ha^{-1} under IPNS+B and US$356 ha^{-1} under IPNS+C. The net returns from the rice produce (grain + straw) ranged between US$361 ha^{-1} to US$931 ha^{-1} and were lowest under OF and highest under IPNS.

The cost of wheat was US$530 ha^{-1}, US$368 ha^{-1}, US$375 ha^{-1}, US$303 ha^{-1} and US$351 ha^{-1}, under the different treatments (Figure 6). The net returns for wheat were highest under IPNS (US$1436 ha^{-1}), closely followed by IPNS+C (US$1414 ha^{-1}) and IPNS+B (US$1355 ha^{-1}). Also, the net returns for wheat under OF (US$850 ha^{-1}) were greater than those under RDF (US$763 ha^{-1})

(Figure 6). Apparently, the net returns under OF were greater than those under rice. For the RWS, OF had the highest cost of cultivation (US$1174 ha^{-1}) followed by the IPNS options (US$864) and RDF (US$849 ha^{-1}) and the cultivation costs were lowest under IPNS+C or B (US$707 ha^{-1} to US$767 ha^{-1}). Considering RWS as whole, the economics favoured the integrated use of organics and fertilizers, as the net returns were higher under IPNS (US$2367 ha^{-1}) followed by IPNS+B (US$2260 ha^{-1}) and IPNS+C (US$2233 ha^{-1}). The continuous use of fertilizers (RDF) gave a net return of US$1406 ha^{-1} which was significantly ($p < 0.001$) lower than the IPNS options. The net return for RWS under OF was US$1211 ha^{-1} only.

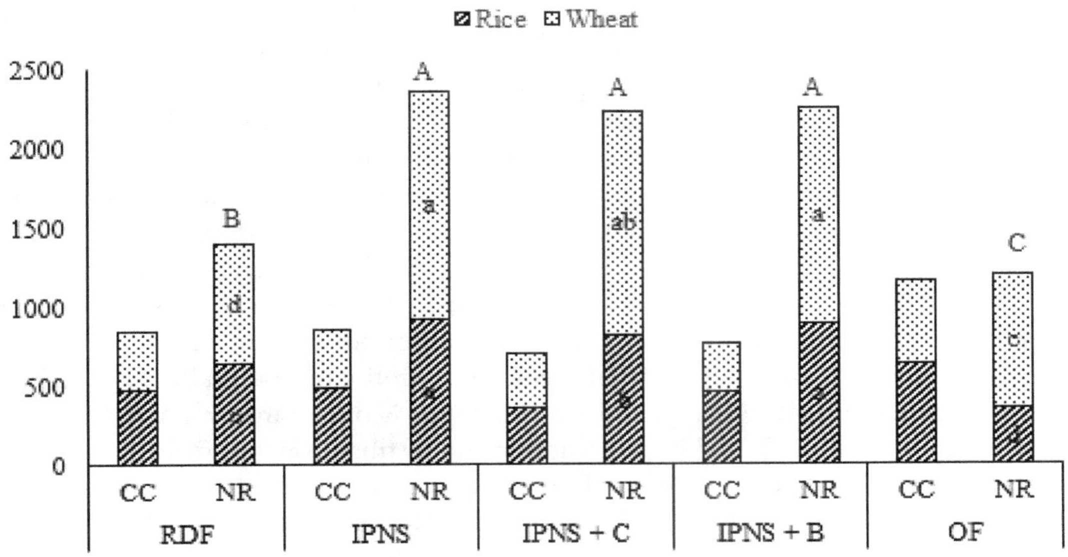

Figure 6. Economics (US$ha^{-1}) of different nutrient management practices in rice and wheat under long-term RWS. CC and NR denote the cost of cultivation and the net returns, respectively. Values followed by different lower case letter(s) within rice or wheat for the net returns are statistically significant ($p < 0.05$). Values followed by different capital letter(s) within RWS for net return are statistically significant ($p < 0.05$).

4. Discussion

4.1. RWS Productivity

In the present investigation, yields of both rice and wheat declined over the years under the RDF plots that received 120 kg N, 26 kg P and 33 kg K to each crop along with 5 kg Zn ha^{-1} to rice. On the other hand, the literature suggests that the soils in the RWS growing area of the IGP are inherently low in organic matter, and suffer from widespread multi-nutrient deficiencies including N, P, K, S, Zn and B [7,17,38,39]. Sustainable high productivity of RWS on these soils is obviously possible with adequate supply of these limiting nutrients through fertilizers and organics. In the present case, the RDF plot exhibited marginal changes in SOC (Table 5), and a decline in mineral N, exchangeable K and available S in the effective root zone (0–45 cm depth) compared with their initial contents at different time intervals (Figures 4 and 5, Tables 7 and 8), thus suggesting a depletion in the native nutrient supply due to continuous RW cropping. Further, changes in soil physico-chemical properties due to continuous flooding and drying under a puddled transplanted rice-conventional till wheat system [40,41] also had adverse effects on the soil nutrient supply [42]. The relatively greater yield reduction in wheat as compared with rice under RDF (Table 3) could, therefore, be explained in two ways: (i) the post-rice drying of soil promoted fixation of P and decreased its availability to subsequent wheat crops; and (ii) the restricted root growth of wheat [7,24,27] caused by increased soil Db (Figure 3) after puddled rice hindered nutrient access in the root zone [7,41]. On the other hand, the superiority of IPNS involving the concurrent use of FYM and fertilizers (75% RDF) vis-à-vis RDF is ascribed to improvements in soil physico-chemical properties due to the organic input and addition of significant quantities of P, K, S and other micro-nutrients through FYM. The manure (FYM) applied to substitute 25% of the recommended N in rice contributed greater amounts of P and K than those curtailed due to the 25% reduction of RDF in the IPNS treatments, as well as supplying an additional 6.7 kg S ha^{-1} annually. Also, the continuous and slow release of nutrients from manure would have enabled their efficient utilization by the crops [22,43].

Among the different IPNS options, the maximum yield gain recorded with the inclusion of legumes as a break crop may be ascribed to a reduction in sub-soil compaction as indicated by the lower Db values beyond the 30 cm soil depth (Figure 3) which improved the root proliferation of both rice and wheat [24,27,44], thus facilitating nutrient absorption from lower soil profiles. The lower NO_3–N concentration below the 30 cm soil depth and its relatively higher content in the surface soil (0–15 cm) under legume inclusion plots supports this contention. The sustainable high yield under the IPNS+B or IPNS+C treatments may also be associated with a reduction in prominent weeds like *Echinochloa* sp. in rice and *Phalaris minor* in wheat under the legume-included plots (data not presented).

The significantly lower yields of rice and wheat under OF plots compared with under RDF and IPNS may be visualized in two possible ways: (i) the FYM equivalent to 120 kg N ha^{-1} applied at the time of sowing/transplanting was unable to supply the nutrients (especially, N) in adequate quantities up to the reproductive stages of crop (data not reported) causing yield reduction; and (ii) the micronutrient content of soil, especially soil Zn, declined over the period (Table 9) as FYM alone could not ensure a sufficient Zn supply. Widespread Zn deficiency is one of major soil fertility constraints in RW growing areas of the IGP [17], and its inclusion in the fertilization schedule is essential to achieve a sustainable high yield. The study, therefore, underlines that the OF nutrient management protocols that ensure adequate supplies of macro- and micro-nutrients (especially Zn) throughout the cropping seasons can hardly sustain high yields of RWS in the IGP [19].

The annual change in the yield of rice under RDF was meagre, whereas the yield of wheat declined significantly over the years. Such yield trends may be ascribed to the high yield level (4.62 t ha^{-1}) in the initial years. An analysis of long-term experiments conducted in the IGP suggested earlier that the magnitude of yield decline in rice-based systems is negatively related to the level of initial yield [45,46]. In the present investigation, the decline in crop productivity under RDF may also explained in terms of greater NO_3–N leaching beyond the root zone (beyond a soil depth of 45 cm) (Figure 4), and depletion of soil exchangeable K and available S contents (Tables 7 and 8). These results are corroborated by earlier reports of other long-term experiments wherein the emergence of multi-nutrient deficiencies [11] led to a yield decline over the passage of time, and the supply of different nutrients in adequate amounts helped to attain yield stability [16,47].

Conversely, initial rice and wheat yield levels under OF plots were lower (2.9 and 2.33 t ha^{-1} respectively) than those under RDF but showed a significant ($p < 0.001$) positive annual yield increase (9.2% $annum^{-1}$) (Table 4). Further, the significantly ($p < 0.05$) higher annual yield increases in both rice and wheat under IPNS and IPNS+B or IPNS+C, even over their initial high yields, is ascribed to the balanced and continuous nutrient supply owing to the improved soil structure (lower Db, Figure 3), SOC and other soil fertility parameters (Tables 5–9 and Figures 2–4). Thus, our study indicates that integrated use of organics and fertilizers can sustain higher productivity of RWS in the IGP [19].

4.2. Soil Health

4.2.1. Bulk Density and Organic Carbon

Continuous RWS under RDF showed an increase in the Db value at soil depths of 30–45 cm (Figure 3). The sub-surface compaction below the puddled zone owing to aggregate disruption [48] is one of the important constraints for sustainable RWS productivity in the IGP [1]. The migration of clay towards a lower profile consequent to puddling increases soil strength rapidly upon drying [41], which restricts the root growth of the subsequent crop [7,24,27]. On the other hand, a decrease in Db, particularly in the sub-soil (30–45 cm depth) under the IPNS+B or IPNS+C plots, consequent to the inclusion of tap rooted legumes and subsequent decomposition of roots increased SOC and helped to improve soil porosity [17,27,49], in turn, facilitating root proliferation. The reduction in Db under IPNS was due to the fibrous nature of added organic matter, which prevents closer packing of soil separates [50]. Also, the use of organic matter input results in better soil aeration and improved soil aggregation, which leads to decreased soil Db [51]. In fact, changes in organic matter not only have a direct bearing on soil Db due to its lower particle density vis-à-vis mineral soil, they also increase soil aggregation and permanent pore development due to the increased soil biological activity [51,52].

The soil organic carbon concentration in the surface layer (0–15 cm) increased significantly under the IPNS options and OF plots (Table 5). In fact, the SOC at any given location largely depends on the annual turnover of organics, root + shoot stubbles and root exudates, and their recycling [17,21,27]. Thus, the use of organics in the IPNS treatments including those with legumes brought further improvement in SOC over sole fertilizers. The relatively higher increase in SOC under OF may be attributed to greater recycling of organic matter compared to under the other nutrient management options. The maintenance of SOC at its initial status may be corroborated with the findings of soil fertility delineation studies in similar agro-ecologies of the IGP (Singh et al., 2015) wherein SOC under balanced fertilization either remained unaffected or even increased depending upon the initial SOC content. The higher SOC in sub-soil under the IPNS treatments involving legumes (cowpea/berseem) is apparently due to the recycling of legume roots in the deeper soil layer, which has created a further conducive environment for the root growth of rice and wheat over the years [24,53].

4.2.2. Soil N, P and K Contents

The NO_3–N content in the upper soil (0–15 cm depth) increased sizeably under the OF and IPNS treatments. The favourable effects of organic manures and legume inclusion on soil N supply have already been documented [14,25], underlining the slow release of nutrients bound in aggregates [43,54] and the addition of N through BNF [55].On the other hand, control (unfertilized) and RDF plots underwent a depletion of NO_3–N by 3.25 mg kg^{-1} and 1.96 mg kg^{-1}, respectively, over the initial status, which suggests excessive nutrient mining due to an inadequate supply of nutrients under high yielding RWS [15,56]. The relatively lower NO_3–N content in the deeper soil profile under IPNS+C or IPNS+B compared with RDF or IPNS was obviously due to greater root proliferation in the deeper soil profile [15,57] leading ultimately to greater NO_3–N uptake by rice and wheat from the deeper layers under treatments involving legumes [24,27,44]. In view of the lower NO_3–N content further down the soil profile, the inclusion of legumes as a break crop at certain interval in RWS could be considered a potential option to overcome the groundwater pollution through curbing NO_3–N leaching in the RWS dominated areas of the IGP [58].

In the present study, the build-up of P highest under RDF (Table 6) was very much expected under regular P addition through fertilizer as cereal crops utilize only a fraction of the applied P [13]. The relatively greater Olsen-P content in soil under OF and IPNS compared with control (unfertilized) may be ascribed to the fact that FYM and legume litter/stubbles supply sizeable amounts of P to the soil. A reduction in the fixation/sorption of applied P in the soil consequent to the enhanced competition of organic molecules with PO_4^{3-} ions for P retention sites [59] under OF/IPNS could be another explanation for these findings. Further, an increased SOC from 5.1 g kg^{-1} to 6.9 g kg^{-1} under OF increased the negative charges that contributed to decreased P sorption by the repulsion of forces between orthophosphate ions and the negatively-charged surface [60]. On the contrary, the Olsen-P content of surface soil (0–15 cm) under the control declined from the initial content by 17% at the end of the 2017–2018 RW cycle. This decline in P content is associated with the lack of addition of P despite continuous removal by rice and wheat crops (data not reported) and a gradual decline in SOC under the control (Table 5). Further, the depletion of soil P content in the soil sub-surface and the relatively lower P content in the surface layer observed with legume inclusion as a break crop is explained by the fact that legumes have higher P demands and better P utilization efficiency [17,57].

The depletion of exchangeable K under RDF despite the regular addition of K through fertilizers can be explained in light of the higher K demands of the constituent crops of RWS. Being exhaustive K feeders, these crops may require 20.5–22.0 kg K t^{-1} of grain production (24). At high productivity levels, the net balance of K in the soil generally remains negative, even with the application of the recommended K content [14,61,62]. Besides, leaching loss of K with percolating water is one of the most significant methods of K removal from the rhizosphere, especially under irrigated ecology. The higher K fixing capacity of the illite-dominant soil minerology of IGP is also one of the major reasons for the depletion of exchangeable K in soil [12,63]. On the other hand, the higher exchangeable K content under OF or IPNS may be ascribed to the release of organic acids during the decomposition which generates negative electric charges in the soil with a preference for di-or-tri-valent cations, such as Al^{3+}, Ca^{2+} and Mg^{2+}, leaving K^+ to be adsorbed by the negatively-charged soil colloids [64]. This phenomenon might have helped to reduce the K fixation and enhance its availability in soil.

4.2.3. Secondary and Micro-Nutrients

The greater S availability under the OF and IPNS options was associated with the greater S addition through FYM in these plots. Since the organic S fraction in the soil is positively related to the organic matter status [65] and is generally considered an important pool of available S [17], lesser deficiencies are expected in soils containing high SOC (Table 5) or those receiving organic manure regularly. On the other hand, the depletion of available S under RDF plots may be explained as (i) the RDF treatments did not include S application, (ii) there was substantial S removal by the crops grown under RDF owing to higher yield levels (data not reported), or (iii) there was reduced $SO_4{}^{2-}$ retention due to higher available P under the RDF plots (Table 6) [17,56]. Since $H_2PO_4{}^-$ is a strong competitor of $SO_4{}^{2-}$ for anion exchange sites, its high availability in the exchange complex can cause concurrent desorption of $SO_4{}^{2-}$ from colloidal surfaces and subsequent leaching with irrigation and rain water [66,67]. The smaller depletion of available S in the control (unfertilized) over the initial S content compared to that in the RDF plots might be due to the smaller S uptake by crops grown under the control.

A gradual accumulation of DTPA–Zn in soil was noticed under RDF and IPNS, whereas a declining trend was noticed under OF. It is pertinent to mention that 5 kg Zn ha^{-1} was applied annually to RDF as well as in the different IPNS options (IPNS, IPNS+B and IPNS+C). The increase in DTPA–Zn reflected the application of zinc in excess of its removal in harvested grain and straw [39]. On the other hand, the addition of Zn in meagre quantities through FYM under OF led to the mining of soil native Zn, resulting in a lower soil Zn content.

4.3. Economics of RWS

The cost of cultivation (CC) for rice, wheat and RWS was maximum under the OF treatments. The higher CC (US$310 to 467) under OF compared with that under the RDF and IPNS options was attributed to the greater cost of FYM and the excess expenditure on hand-weeding. On the other hand, the lower CC of RWS under the IPNS + legume treatments (US$707 and 767) vis-à-vis RDF or IPNS may be ascribed to the lower investment on fertilizer nutrients, irrigation, weed/pest management and laborers under legume fodder grown at every third cycle in place of rice or wheat. Further, the higher economic net returns of RWS under different IPNS options over the RDF and OF treatments was mainly due to improved yields (grain + straw) over time. Although the cost of cultivation under the IPNS+B or IPNS+C treatments was lower than that of the IPNS, the net returns under these treatments remained on par, as the cost of legume fodder produced was much smaller (US$14.3 t^{-1} of fodder) than the cost of grain + straw of rice and wheat. The low returns under OF may be associated with higher CC involved in rice and wheat, and the relatively lower annual productivity as compared with the RDF or IPNS treatments. The economics of RWS thus favour the adoption of IPNS with or without inclusion of legumes.

5. Conclusions

The foregoing results of the present long-term study underline the fact that the sustainable high productivity of RWS in the intensively-cropped IGP cannot be achieved through a single nutrient source, be it fertilizer or organic manure. Conjoint use of both, on the other hand, proved superior in terms of crop productivity, soil health and economic returns. The inclusion of forage legumes as a break crop in the RWS had an added advantage vis-à-vis IPNS alone, as was evident from the improvement in yield and different soil parameters. In view of scarcity of conventional manure (FYM), there is a need to conduct well planned studies involving other locally available organics to develop rational IPNS prescriptions for large-scale adoption. Also, the environmental impacts of continuous use of organic inputs in sizeable amounts under OF need to be studied in a holistic manner.

Author Contributions: In this study, Drs. Vinod K. Singh, Brahma S. Dwivedi, Arvind K Shukla and Rajendra P. Mishra were involve in research conceptualization, data curation, field investigation, project administration and writing original draft for publication. Dr Azad. S. Panwar helped in funding acquisition and general project administration. Drs. Pravin K. Upadhyay, Kapila Shekhawat, Kaushik Majumdar and Jagadish Timsina supported for evolving methodology, formal data analysis, writing and editing of draft at different stages.

Acknowledgments: The authors thank the directors of ICAR—Indian Institute of Farming Systems Research, Modipuram for providing the necessary facilities for field experimentation and laboratory analyses.

References

1. Chauhan, B.S.; Mahajan, G.; Sardana, V.; Timsina, J.; Jat, M.L. Productivity and Sustainability of the Rice-Wheat Cropping System in the Indo-Gangetic Plains of the Indian subcontinent: Problems, Opportunities, and Strategies. *Adv. Agron.* **2012**, *117*, 316–369.
2. Ladha, J.K.; Pathak, H.; Padre, A.T.; Dawe, D.; Gupta, R.K. Productivity trends in intensive rice-wheat cropping systems in Asia. In *Improving the Productivity and Sustainability of Rice-Wheat Systems: Issues and Impacts*; Ladha, J.K., Hill, J., Gupta, R.K., Duxbury, J., Buresh, R.J., Eds.; ASA Special Publication 65; ASA-CSSA-SSSA: Madison, WI, USA, 2003; pp. 45–76.
3. Verma, B.C.; Datta, S.P.; Rattan, R.K.; Singh, A.K. Long-Term Effect of Tillage, Water and Nutrient Management Practices on Mineral Nitrogen, Available Phosphorus and Sulphur Content under Rice-Wheat Cropping System. *J. Indian Soc. Soil Sci.* **2016**, *64*, 71–77. [CrossRef]
4. *FAOSTAT*; Food and Agriculture Organization of the United Nations: Rome, Italy, 2017.
5. Yadvinder-Singh; Gupta, R.K.; Gurpreet-Singh; Jagmohan-Singh; Sidhu, H.S.; Bijay-Singh. Nitrogen and residue management effects on agronomic productivity and nitrogen use efficiency in rice–wheat system in Indian Punjab. *Nutr. Cycl. Agroecosyst.* **2009**, *84*, 141–154. [CrossRef]
6. Das, D.; Dwivedi, B.S.; Singh, V.K.; Datta, S.P.; Meena, M.C.; Chakraborty, D.; Bandyopadhyay, K.K.; Kumar, R.; Mishra, R.P. Long-term effects of fertilizers and organic sources on soil organic carbon fractions under a rice–wheat system in the Indo-Gangetic Plains of north-west India. *Soil Res.* **2017**, *55*, 296–308. [CrossRef]
7. Timsina, J.; Connor, D.J. Productivity and management of rice-wheat cropping systems: Issues and challenges. *Field Crops Res.* **2001**, *69*, 93–132. [CrossRef]
8. Bhatt, R.; Kukal, S.S.; Busari, M.A.; Arora, S.; Yadav, M. Sustainability issues on rice–wheat cropping system. International Soil and Water Conservation Research. *Int. Soil Water Conserv. Res.* **2016**, *4*, 64–74. [CrossRef]
9. Ladha, J.K.; Yadvinder-Singh; Erenstein, O.; Hardy, B. *Integrated Crop and Resource Management in the Rice? Wheat System of South Asia*; International Rice Research Institute: Los Banos, Philippines, 2009.
10. Majumder, B.; Mandal, B.; Bandyopadhyay, P.K. Soil organic carbon pools and productivity in relation to nutrient management in a 20-year old rice-berseem agroecosystem. *Biol. Fertil. Soils* **2008**, *44*, 451–464. [CrossRef]

11. Singh, M.; Reddy, K.S.; Singh, V.P.; Rupa, T.R. phosphorus availability to rice (*Oryza sativa* L.)–wheat (*Triticum aestivum* L.) in a vertical soil after eight years of inorganic and organic fertilizer addition. *Bioresource. Technol.* **2007**, *98*, 1474–1481. [CrossRef]
12. Panaullah, G.M.; Timsina, J.; Saleque, M.A.; Ishaque, M.; Pathan, A.B.M.B.U.; Connor, D.J.; Saha, P.K.; Quayyum, M.A.; Humphreys, E.; Meisner, C.A. Nutrient Uptake and Apparent Balances for Rice-Wheat Sequences. III. Potassium. *J. Plant Nutr.* **2006**, *29*, 173–187. [CrossRef]
13. Saleque, M.A.; Timsina, J.; Panaullah, G.M.; Ishaque, M.; Pathan, A.B.M.B.U.; Connor, D.J.; Saha, P.K.; Quayyum, M.A.; Humphreys, E.; Meisner, C.A. Nutrient Uptake and Apparent Balances for Rice-Wheat Sequences. II. Phosphorus. *J. Plant Nutr.* **2006**, *29*, 157–172. [CrossRef]
14. Singh, V.K.; Sharma, B.B.; Dwivedi, B.S. The impact of diversification of a rice–wheat cropping system on crop productivity and soil fertility. *J. Agric. Sci.* **2002**, *139*, 405–412. [CrossRef]
15. Timsina, J.; Panaullah, G.M.; Saleque, M.A.; Ishaque, M.; Pathan, A.B.M.B.U.; Quayyum, M.A.; Connor, D.J.; Saha, P.K.; Humphreys, E.; Meisner, C.A. Nutrient Uptake and Apparent Balances for Rice-Wheat Sequences. *J. Plant Nutr.* **2006**, *29*, 137–155. [CrossRef]
16. Dwivedi, B.S.; Singh, V.K.; Dwivedi, V. Application of phosphate rock, with or without Aspergillus awamori inoculation, to meet P demands of rice-wheat systems in the Indo-Gangetic plains of India. *Aust. J. Exp. Agric.* **2014**, *44*, 1041–1050. [CrossRef]
17. Singh, V.K.; Dwivedi, B.S.; Shukla, A.K.; Kumar, V.; Gangwar, B.; Rani, M.; Singh, S.K.; Mishra, R.P. Status of available sulphur in soils of north-west Indo-Gangetic Plain and Western Himalayan Region and response of rice and wheat to applied sulphur in farmers' fields. *Agric. Res.* **2015**, *4*, 76–92. [CrossRef]
18. Bijay-Singh. Are Nitrogen Fertilizers Deleterious to Soil Health? *Agron. J.* **2018**, *8*, 48. [CrossRef]
19. Timsina, J. Review Can Organic Sources of Nutrients Increase Crop Yields to Meet Global Food Demand? *Agron. J.* **2018**, *8*, 214. [CrossRef]
20. Bandyopadhyay, K.K.; Misra, A.K.; Ghosh, P.K.; Hati, K.M. Effect of integrated use of farmyard manure and chemical fertilizers on soil physical properties and productivity of soybean. *Soil Tillage Res.* **2010**, *110*, 115–125. [CrossRef]
21. Numbiar, K.K.M. *Soil Fertility and Crop Production under Long-Term Fertilizer Use in India*; ICAR: New Delhi, India, 1994.
22. Dwivedi, B.S.; Singh, V.K.; Meena, M.C.; Dey, A.; Datta, S.P. Integrated nutrient management for enhanced nutrient use efficiency. *Indian J. Fertil.* **2016**, *12*, 62–71.
23. Singh, V.K.; Dwivedi, B.S.; Tiwari, K.N.; Majumdar, K.; Rani, M.; Singh, S.K.; Timsina, J. Optimizing nutrient management strategies for rice-wheat system in the Indo-Gangetic Plains and adjacent region for higher productivity, nutrient use efficiency and profits. *Field Crops Res.* **2014**, *164*, 30–44. [CrossRef]
24. Dwivedi, B.S.; Shukla, A.K.; Singh, V.K.; Yadav, R.L. Improving nitrogen and phosphorus use efficiencies through inclusion of forage cowpea in the rice–wheat system in the Indo-Gangetic plains of India. *Field Crops Res.* **2003**, *84*, 399–418. [CrossRef]
25. Yadav, S.K.; Babu, S.; Yadav, M.K.; Singh, K.; Yadav, G.S.; Suresh, P. A Review of Organic Farming for Sustainable Agriculture in Northern India. *Int. J. Agron.* **2003**, *2013*, 718145. [CrossRef]
26. Wani, S.P.; Rego, T.J.; Rajeswari, S.; Lee, K.K. Effect of legume-based cropping systems on nitrogen mineralization potential of Vertisol. *Plant Soil.* **1995**, *175*, 265–274. [CrossRef]
27. Singh, V.K.; Dwivedi, B.S.; Shukla, A.K.; Chauhan, Y.S.; Yadav, R.L. Diversification of rice with pigeonpea in a rice–wheat cropping system on a Typic Ustochrept: Effect on soil fertility, yield and nutrient use efficiency. *Field Crops Res.* **2005**, *92*, 85–105. [CrossRef]
28. Bremner, J.M.; Keeney, D.R. Steam distillation methods for determination of ammonium, nitrate and nitrite. *Anal. Chem. Acta* **1965**, *32*, 485–495. [CrossRef]
29. Walkley, A.; Black, C.A. An examination of the Degitjareff method for determining soil organic matter, and a proposed modification of the chromic acid titration method. *Soil Sci.* **1934**, *37*, 29–38. [CrossRef]
30. Olsen, S.R.; Cole, C.V.; Watanabe, F.S.; Dean, L.A. *Estimation of Available Phosphorus in Soils by Extraction with Sodium Bicarbonate*; United States Department of Agriculture: Washington, DC, USA, 1954.
31. Washington, D.C.; Helmke, P.A.; Sparks, D.L. Lithium, sodium, potassium, rubidium, and cesium. In *Methods of Soil Analysis. Part 3*; Sparks, D.L., Ed.; Chemical Methods-Soil Science Society of America Book Series No. 5; SSSA and ASA: Madison, WI, USA, 1996; pp. 551–574.

32. Williams, C.H.; Steinbergs, H. Soil sulphur fractions as chemical indices of available sulphur in some Australian soils. *Aust. J. Agric. Res.* **1995**, *10*, 340–352. [CrossRef]
33. Lindsay, W.L.; Norvell, WA. Development of a DTPA Soil Test for Zinc, Iron, Manganese, and Copper. *Soil Sci. Soc. Am. J.* **1978**, *42*, 421–428. [CrossRef]
34. Subbiah, B.V.; Asija, G.L. A rapid procedure for the determination of available nitrogen in soils. *Curr. Sci.* **1956**, *25*, 259–260.
35. Page, A.L.; Millar, R.H.; Keeney, D.R. (Eds.) *Methods of Soil Analysis. Part-2*; American Society of Agronomy and Soil Science Society of America: Madison, WI, USA, 1982.
36. Blake, G.R.; Hartge, K.H. Bulk density. In *Methods of Soil Analysis, Part 1*; Klute, A., Ed.; Physical and Mineralogical Properties, Monograph 9; ASA: Madison, WI, USA, 1986; pp. 363–376.
37. Cochran, W.G.; Cox, G.M. *Experimental Designs*; Wiley: New York, NY, USA, 1957.
38. Dwivedi, B.S.; Singh, V.K.; Shekhawat, K.; Meena, M.C.; Dey, A. Enhancing Use Efficiency of Phosphorus and Potassium under Different Cropping Systems of India. *Indian J. Fertil.* **2017**, *13*, 20–41.
39. Cassman, K.G.; Peng, S.; Olk, D.C.; Ladha, J.K.; Reichardt, W.; Dobermann, A.; Singh, U. Opportunities for increasing nitrogen use effciency from improved resource management in irrigated rice systems. *Field Crops Res.* **1998**, *56*, 7–39. [CrossRef]
40. Dwivedi, B.S.; Singh, V.K.; Shukla Arvind, K.; Meena, M.C. Optimizing dry and wet tillage for rice on a Gangetic alluvial soil: Effect on soil characteristics: Water use efficiency and productivity of the rice-wheat system. *Eur. J. Agron.* **2012**, *43*, 155–165. [CrossRef]
41. Cassman, K.G.; Pingali, P.L. Extrapolating trends from long-term experiments on farmers' felds: The case study of irrigated rice-systems in Asia. In *Agricultural Sustainability: Economic, Environmental and Statistical Considerations*; Barnet, V., Payre, R., Steiner, R., Eds.; Wiley: Chichester, UK, 1995; pp. 63–84.
42. Mahajan, A.; Gupta, R.D. *Integrated Nutrient Management (INM) in a Sustainable Rice-Wheat System*; Springer: Dordrecht, The Netherlands, 2017.
43. Singh, V.K.; Dwivedi, B.S.; Shukla, A.K. Yields, and nitrogen and phosphorus use efficiency as influenced by fertilizer NP additions in wheat (*Triticum aestivum*) under rice (*Oryza sativa*)-wheat and pigeonpea (*Cajanus cajan*)-wheat system on a Typic Ustochrept soil. *Indian J. Agric. Sci.* **2000**, *76*, 92–97.
44. Dawe, D.; Dobermann, A.; Moya, P.; Abdulrachman, S.; Singh, B.; Lal, P.; Li, S.Y.; Lin, B.; Panaullah, G.; Sariam, O.; et al. How widespread are yield declines in long-term rice experiments in Asia? *Field Crops Res.* **2000**, *66*, 175–193. [CrossRef]
45. Yadav, R.L.; Dwivedi, B.S.; Pandey, P.S. Rice–wheat cropping system: Assessment of sustainability under green manuring and chemical fertilizer input. *Field Crops Res.* **2000**, *65*, 15–30. [CrossRef]
46. Swarup, A.; Wanjari, R.H. *Three Decades of All India Coordinated Research Project on Long-Term Fertilizer Experiments to Study Changes in Soil Quality, Crop Productivity and Sustainability*; IISS: Bhopal, India, 2000.
47. Six, J.; Elliott, E.T.; Paustian, K. Soil macroaggregate turnover and microaggregate formation: A mechanism for C sequestration under no-tillage agriculture. *Soil Biol. Biochem.* **2000**, *32*, 2099–2103. [CrossRef]
48. Sharma, P.K.; Ladha, J.K.; Bhushan, L. Soil physical effects of puddling in rice–wheat cropping system. In *Improving the Productivity and Sustainability of Rice–Wheat Systems: Issues and Impacts*; Ladha, J.K., Hill, J., Gupta, R.K., Duxbury, J., Buresh, R.J., Eds.; ASA Special Publication 65; ASA, CSSA and SSSA: Madison, WI, USA, 2003; pp. 97–114.
49. Adams, W.A. The effect of organic matter on the bulk and true densities of some uncultivated podzolic soils. *Eur. J. Soil Sci.* **2006**, *24*, 10–17. [CrossRef]
50. Halvorson, A.D.; Reule, C.A.; Follet, R. Nitrogen Fertilization Effects on Soil Carbon and Nitrogen in a dryland cropping system. *Soil Sci. Soc. Am. J.* **1999**, *63*, 912–917. [CrossRef]
51. Franzluebbers, A.J.; Stuedemann, J.A.; Schomberg, H.H.; Wilkinson, S.R. Soil organic C and N pools under long-term pasture management in the Southern Piedmont USA. *Soil Biol. Biochem.* **2000**, *32*, 469–478. [CrossRef]
52. Oussible, M.; Crookston, R.K.; Larson, W.E. Subsurface Compaction Reduces the Root and Shoot Growth and Grain Yield of Wheat. *Agron. J.* **1992**, *84*, 34–38. [CrossRef]
53. Das, B.; Chakraborty, D.; Singh, V.K.; Aggarwal, P.; Singh, R.; Dwivedi, B.S. Effect of organic inputs on strength and stability of soil aggregates under rice-wheat rotation. *Int. Agrophys.* **2014**, *8*, 163–168. [CrossRef]
54. Awonaike, K.O.; Kumarsinghe, K.S.; Danso, S.K.A. Nitrogen fixation and yield of cowpea as influenced by cultivar and *Bradyrhizobium* strain. *Field Crop Res.* **1990**, *24*, 163–171. [CrossRef]

55. Dwivedi, B.S.; Shukla, A.K.; Singh, V.K.; Yadav, R.L. Response of wheat (*Triticum aestivum*), potato (*Solanum tuberosum*) and mustard (*Brassica juncea*) to potassium applied through muriate of potash of varying particle size. *Indian J. Agric. Sci.* **2001**, *71*, 634–638.
56. Dwivedi, B.S.; Shukla, A.K.; Singh, V.K.; Yadav, R.L. Results of participatory diagnosis of constraints and opportunities (PDCO) based trials from the state of Uttar Pradesh. In *Development of Farmers' Resource-Based Integrated Plant Nutrient Supply Systems: Experience of a FAO–ICAR–IFFCO Collaborative Project and AICRP on Soil Test Crop Response Correlation*; Subba Rao, A., Srivastava, S., Eds.; IISS: Bhopal, India, 2001; pp. 50–75.
57. Singh, V.K.; Dwivedi, B.S.; Shukla, A.K.; Mishra, R.P. Permanent raised bed planting of the pigeonpea-wheat system on a Typic Ustochrept: Effects on soil fertility, yield, and water and nutrient use efficiencies. *Field Crops Res.* **2010**, *116*, 127–139. [CrossRef]
58. Aulakh, M.S.; Singh, B. Nitrogen losses and fertilizer N use efficiency in irrigated porous soils. *Nutr. Cycl. Agroecosyst.* **1997**, *147*, 197–212. [CrossRef]
59. Xie, R.J.; Fyles, J.W.; Mckenzie, A.F.; O'Hollaran, I.P. Ligno-sulphate retention in a clay soil: Casual modeling. *Soil Sci. Soc. Am. J.* **1991**, *55*, 711–716. [CrossRef]
60. Moshi, A.O.; Wild, A.M.; Greenland, D.J. Effect of organic matter on the charge and phosphate adsorption characteristics of Kikuyu red clay from Kenya. *Geoderma* **1974**, *11*, 275–285. [CrossRef]
61. Nayak, A.K.; Gangwar, B.; Shukla, A.K.; Mazumdar, S.P.; Kumar, A.; Raja, R.; Kumar, A.; Kumar, V.; Rai, P.K.; Mohan, U. Long-term effect of different integrated nutrient management on soil organic carbon and its fractions and sustainability of rice–wheat system in Indo Gangetic Plains of India. *Field Crop Res.* **2012**, *127*, 129–139. [CrossRef]
62. Singh, V.K.; Dwivedi, B.S.; Yadvinder-Singh; Singh, S.K.; Mishra, R.P.; Shukla, A.K.; Rathore, S.; Shekhawat, K.; Majumdar, K.; Jat, M.L. Effect of tillage and crop establishment, residue management and K fertilization on yield, K use efficiency and apparent K balance under ricemaize system in north-western India. *Field Crop Res.* **2018**, *224*, 1–12.
63. Tandon, H.L.S.; Sekhon, G.S. *Potassium Research and Agricultural Production in India*; Fertilizer Development and Consultation Organization: New Delhi, India, 1988.
64. Timsina, J.; Singh, V.K.; Majumdar, K. Potassium management in rice-maize systems in South Asia. *J. Plant Nutr. Soil Sci.* **2013**, *176*, 317–330. [CrossRef]
65. Pasricha, N.S.; Sarkar, A.K. Secondary nutrients. In *Fundamentals of Soil Science*, 2nd ed.; Goswami, N.N., Rattan, R.K., Dev, G., Narayanasamy, G., Das, D.K., Pal, D.K., Rao, D.L.N., Eds.; Indian Society of Soil Science: New Delhi, India, 2009; pp. 449–460.
66. Pasricha, N.S.; Aulakh, M.S. *Twenty Years of Sulphur Research and Oilseed Production in Punjab, India*; Sulphur in Agriculture; TSI: Washington, DC, USA, 1991.
67. Tandon, H.L.S. *Biofertilizers and Organic Farming: A Source-Cum-Directory*; Fertilizer Development and Consultation Organisation: New Delhi, India, 2011.

11

Nitrate Assimilation Limits Nitrogen use Efficiency (NUE) in Maize (*Zea mays* L.)

Dale Loussaert *, Josh Clapp, Nick Mongar, Dennis P. O'Neill and Bo Shen

DuPont Pioneer, 7250 NW 62nd Ave, Johnston, IA 50113, USA; joshuaclapp76@gmail.com (J.C.); nick.mongar@pioneer.com (N.M.); dennis.oneill@pioneer.com (D.P.O.); bo.shen@pioneer.com (B.S.)

* Correspondence: dale.loussaert@gmail.com

Abstract: Grain yield in maize responds to N fertility in a linear-plateau fashion with nitrogen use efficiency (NUE) higher under lower N fertilities and less as grain yield plateaus. Field experiments were used to identify plant parameters relative for improved NUE in maize and then experiments were performed under controlled conditions to elucidate metabolism controlling these parameters. Field experiments showed reproductive parameters, including R1 ear-weight, predictive of N response under both high and low NUE conditions. R1 ear-weight could be changed by varying nitrate concentrations early during reproductive development but from V12 onward R1 ear-weight could be changed little by increasing or decreasing nitrate fertility. Ammonia, on the other hand, could rescue R1 ear-weight as late as V15 suggesting nitrate assimilation (NA) limits ear development response to N fertility since bypassing NA can rescue R1 ear-weight. Nitrate reductase activity (NRA (in vitro)) increases linearly with nitrate fertility but in vivo nitrate reductase activity (NRA (in vivo)) follows organic N accumulation, peaking at sufficient levels of nitrate fertility. The bulk of the increase in total plant N at high levels of nitrate fertility is due to increased plant nitrate concentration. Increasing NADH levels by selective co-suppression of ubiquinone oxidoreductase 51 kDa subunit (Complex I) was associated with improved grain yield by increasing ear size, as judged by increased kernel number plant^{-1} (KNP), and increased NRA (in vivo) without a change in NRA (in vitro). These results support NUE is limited in maize by NA but not by nitrate uptake or NRA (in vitro).

Keywords: nitrogen use efficiency (NUE); nitrate assimilation; nitrate reductase activity; maize; nitrate; ammonia; NADH; NADH-dehydrogenase; Complex I

---

1. Introduction

NUE in maize is defined as an incremental increase in grain yield with incremental increases in N fertilizer [1,2]. Grain yield responds to N fertility in a linear-plateau fashion [3]. NUE is greater at lower inputs of N [1,4,5] where grain yield responses approach linearity, and NUE is very low past the inflexion point where little grain yield increases result from increased N fertility. NUE in maize ranges from 50–10% [6–8] but field sources of N are not limited to fertilizer input. Field environments may provide up to 185 kg ha^{-1} of non-fertilizer N [9] with N mineralization and N carryover as sources of non-fertilizer N input. These additional sources of N vary from season to season based on soil temperature and moisture, organic matter, and performance of the previous crop. Precise control of N input in a field environment is difficult and only 50–65% of applied fertilizer is taken up by the plant the same year it is applied [9,10]. Applied N fertilizer can be incorporated into soil organic matter, lost due to ammonia volatility, leached, and lost due to denitrification [11–14]. Precise control of N in field environments required to develop transgenic improvements in NUE is very difficult and requires controlled environments to complement field experimentation to bridge the gap between what is known biochemically, required to make transgenic modifications, and field performance.

Nitrate uptake and assimilation have been studied in controlled environments using the model system, *Arabidopsis thaliana* (L.) Heynh. [8,15,16], but *A. thaliana* is a rather poor model for maize both genetically [17] and physiologically. Not only is photosynthetic carbon metabolism different between maize (C4) and *A. thaliana* (C3) but nitrogen metabolism is dissimilar [18,19].

Recently, Fan et al. [20] showed transgenic expression of a high affinity nitrate transporter significantly improves NUE in rice. It is difficult to conceive that increasing nitrate uptake would have a significant effect on maize, especially under high N fertility, since maize accumulates significant amounts of nitrate under these conditions. As much as 70% of stalk N is nitrate [21]. Cliquet et al. [22] showed that 47% of the N applied after pollination accumulated in the stalk as nitrate. Currently, the main interest in stalk nitrate is in predicting N carryover and overall plant health when soil nitrate is limited by poor fertilization or nitrate leaching [4,23,24] but stalk nitrate is also associated with reduced NUE. Brouder et al. [25] showed that stalk nitrate and agronomic efficiency, defined identically as NUE, were inversely related. Binford et al. [4] showed, using 900 crop years of data, that stalk nitrate is linearly related to grain yield up to the linear regression plateau (LRP) inflection point. This would suggest that accumulation of stalk nitrate is symptomatic of reduced NUE as demonstrated by Varvel et al. [26] and Brouder et al. [25]. Varvel et al. [26] showed stalk nitrate linearly increased with increased N fertility past the LRP inflection point while grain yields were unchanged. With no additional grain yield with increased N fertility, NUE dropped as stalk nitrate dramatically increased. Under normal or high N fertility, maize will concentrate nitrate in the stalk 20–100 times the soil nitrate concentration [24] which would argue that N uptake is not limiting under high N fertility (low NUE).

Linking physiological/biochemical information to relevant field performance is critical in developing transgenic improvements in maize NUE. Though nitrate transporter research is more recent than NA research, NA is likely more relevant in improving NUE in maize. Beevers and Hageman [27] proposed that NRA (in vitro) is the limiting step in N metabolism in plants. Though grain yields in maize can be significantly improved by N fertility, a direct correlation between NRA (in vitro) and grain yield has never been established. Blackmer et al. [28] showed >80% of grain yield was related to spring soil nitrate concentrations. Klepper et al. [29] showed in vivo nitrate reductase activity, NRA (in vivo), or the ability of leaf tissue to generate nitrite from nitrate in the dark, was enhanced by respiratory metabolites. Klepper et al. [29] also demonstrated that NRA (in vitro) could be supported by the addition of glyceraldehyde-3-phosphate and NAD^+ to cell free plant extracts suggesting glyceraldehyde-3-phosphate dehydrogenase (GA3PDH—EC 1.2.1.12) the source of NADH for nitrate reduction. Later, Gowri and Campbell [30] showed that NRA (in vitro) and GA3PDH are coincidentally induced by nitrate in etiolated maize. Though extractable levels of GA3PDH are high enough to provide sufficient NADH to support NRA (in vitro) and NRA (in vitro) and GA3PDH are co-induced by nitrate, this is not a proof of in vivo metabolism. Klepper et al. [29] also noted that NRA (in vitro) was 2.5–20 times higher than the rate of NRA (in vivo) which would suggest that extractable NRA (in vitro), per se, is far in excess of what is required to support NA. Also, since extractable levels of GA3PDH are high enough to support NRA (in vitro) it would follow that GA3PDH is in excess of that required to support NA. Later, Neyra and Hageman [31] suggested malate could be a substrate for generating NADH to support NA. Neither Klepper et al. [29] nor Neyra and Hageman [31] were able to demonstrate enhanced NRA (in vivo), by the addition of malate to the NRA (in vivo) assay medium. Rathnam [32] showed NA, measured by the disappearance of nitrate, in spinach protoplast could be supported by the addition of phospho-3-glyceric acid and oxaloacetic acid (OAA) in the light and/or by glyceraldehyde-3-phosphate and malate in either light or dark showing GA3PDH and/or malate dehydrogenase (MDH—EC 1.1.1.37) capable of supporting NA. Similar to the observation made by Klepper et al. [29], Kaiser et al. [33] also observed higher extractable levels of NRA (in vitro) than NRA (in vivo) and concluded that NADH levels, and not NRA (in vitro), limits NA. Later, increased leaf NADH levels were also reported [33] to be associated with NA.

In order to complete the nitrogen assimilation pathway from nitrate to the formation of glutamate, NADH, ATP, reduced ferredoxin, and α-ketoglutarate (αKG) are required. Mitochondrial or cytoplasmic isocitrate dehydrogenase (ICDH—EC 1.1.1.41, EC 1.1.1.42) produce αKG. Reducing the expression of mitochondrial citrate synthase (CS—EC 2.3.3.1) [34] or ICDH [35] in tomato resulted in elevated levels of lamina nitrate. This suggests a direct link between αKG and nitrate reduction. When isocitrate or αKG moves out of the mitochondria, NADH levels drop in the mitochondria and since NADH does not pass through the mitochondrial membrane it cannot be replenished directly. Reestablishing the mitochondrial NADH concentration may be done by importing malate from the cytoplasm. Assuming malate provides significant reduction power for NA the loss of malate from the cytoplasm results in NA becoming limited by low cytoplasmic NADH. Thus, the pathway becomes self-regulated. Cytoplasmic male sterile (CMS) tobacco [36] and CMS cucumber [37] have improved NUE under lower N fertility. These CMS mutants with defective Complex I (NADH dehydrogenase—EC 1.6.5.3) have reduced capacity to oxidize mitochondrial NADH which results in increased mitochondrial NADH concentrations, reducing the need to import malate to balance the mitochondrial NADH levels due to the transport of αKG out of the mitochondria. Based on these observations, Foyer et al. [38] predicted over-expression of ICDH or co-suppression of Complex I would improve NUE. CMS tobacco [36] and CMS cucumber [37] might be examples of improved NUE through diminished Complex I activity but these have associated deleterious attributes (slow growth, male sterility) which makes these agronomically unsuitable. The mutant NCS2 in maize also has defective Complex I and expresses undesirable traits [39] but has not, specifically, been shown to have improved NUE.

In this report, multi-year field experiments of maize grown at different levels of NUE showed reproductive plant parameters are associated with improved NUE under both low and high NUE conditions. The response of these parameters to different forms of N applied at different times of development were used, under controlled environments, to elucidate key physiological factors related to improved NUE. The accumulation of nitrate in lower internodes was investigated between plants grown under high and low NUE conditions and shown to be inversely related to improved NUE. Finally, increasing cytoplasmic NADH using a transgenic co-suppression of Complex I under the control of a tissue preferred promoter increased kernel number plant^{-1} (KNP) and grain yield under low NUE (high N fertility) conditions. In all, these results suggest that NA limits NUE in maize via reduced availability of cytoplasmic NADH.

2. Materials and Methods

2.1. Field NUE Analyses under Normal and Depleted N Conditions

Field plots depleted of N for a minimum of two years were used in NUE experiments over a period of three years in Johnston, Iowa. Non-depleted plots were plots in previous cropping seasons fertilized with N to obtain maximum economic grain yield of maize. In the depleted plots N treatments consisted of 0, 22, 45, 67 kg ha^{-1} N applied as urea and in the non-depleted plots N treatments were 0, 34, 67, 101, 135, and 168 kg ha^{-1} N applied as urea. All N applications were made at V3. The three DuPont-Pioneer hybrids (33K42, 33W84, and 33T56) used in these studies were selected based on a range of yield responses to N fertility when grown under depleted N conditions in multi-environment trials. The experiments were arranged in a split, split plot experimental design with season as the main plot, N fertility as the subplot and cultivar as the sub, subplot. Sampling dates, R1 and maturity (black layer), were blocked as separate experiments to avoid contamination of the maturity sampling by the R1 sampling. Planting density was 75,000 plants ha^{-1}, each plot consisted of two 5.1 m rows spaced 0.75 m apart. There were five replicates of all treatments. Soluble leaf amino acids at V9 was the first parameter measured. At R1 chlorophyll measurements (Minolta SPAD, Minolta Camera Co., Ramsey, NJ, USA) were made by averaging five samplings taken down the ear leaf of 10 plants in each plot. Chlorophyll measurements were made in a similar manner at R2 and R3. At R1 10 plants were sampled from each plot and the ears removed and dried (70 °C, 72 h). Ear shoots of plants sampled for R1 ear measurements were bagged prior to silk emergence to avoid pollination. Ear length, ear width and ear dry weight were determined. Total fresh weight of the remaining chopped plants was measured and vegetative biomass was determined by weighing and drying (70 °C, 72 h) a subsample. Samples were ground and total N determined. At maturity, 10 plants were sampled, the ears removed and dried (70 °C, 72 h). Ear weight, grain weight, and kernel density were determined of the dried ears. Vegetative biomass and total N were determined similarly as at R1. Statistical analysis was performed as previously described [40].

2.2. *Field Stalk Nitrate Experiments*

The DuPont-Pioneer cultivar 33W84 was grown in field plots in Johnston, IA either fertilized for optimum grain yield (224 kg N ha^{-1}) or in plots depleted for N for at least two years and fertilized with 77 kg N ha^{-1} immediately after planting. Plants were sampled for stalk nitrate at V11, VT, R1, R2 and R3. Leaves associated with each internode were punched (10, 5 mm diameter) and used for metabolic analysis. Leaves, including leaf sheath, were removed and stalk internodes were numbered and cut at the top of each node, dried (70 °C, 72 h), then ground to a fine powder. When physiological measurements were made, plants were cut at ground level and transported to the laboratory in buckets filled with water.

2.3. *Controlled Environment Experiments*

A Conviron PGR15 growth chamber set at 30 °C, 60% RH, 16 h light/25 °C, 50% RH, 8 h dark was used to grow the maize model system plant, Gaspe Flint-3 (GF3) (manufacturer, city, state, country) [41] under controlled environmental conditions. A semi-hydroponics irrigation system [41], similar to that described [42], was used in growth chamber and field hydroponics to attain a high level of control of N input. When ammonia was used as an N source 1 mM NH_4Cl was substituted for KNO_3 but KCl was maintained at 4 mM. GF3 was grown in 1.74 L pots and field hydroponics plants were grown in 15.14 L pots.

2.4. *Controlled Environment Plant Samplings*

When plants were sampled, the Turface™ (Turface/Profile Products, LLC, Buffalo Grove, IL, USA) was washed off the roots and the plant separated into shoot, roots, leaves, midribs, tassel, tillers, husk, and ear. Plant parts were dried (70 °C, 72 h) but when metabolic profiling was performed, plant parts were frozen in dry ice and lyophilized. Individual plant parts were weighed, ground to a fine powder, and a sample (30–60 mg) weighed for extraction. Fresh tissue was also extracted as leaf punches (10, 5 mm diameter). In either case, duplicate samples were extracted by Genogrinder (SPEX SamplePrep, Metuchen, NJ, USAin 500 μL acid and 500 μL base as described by Queval and Noctor [43]. When NAD^+/NADH was quantified a small aliquot of the acid extract was heat-treated for the quantification of NADH and the remaining acid extract was quenched and used for metabolic analysis.

2.5. *NAD^+/NADH Measurements*

The method of Queval and Noctor [43] was used.

2.6. *Total Amino Acids*

An aliquot of each tissue extract was suspended in a total volume of 100 μL water and 50 μL of a solution containing 350 mM Borate buffer pH 9.5, 1% SDS, 0.5% β-mercaptoethanol (ME), and 200 μg o-phthadialdehyde was added to each well. Blank samples were treated similarly but without o-phthadialdehyde. Fluorescence (Excitation (Ex) 360 nm Emission (Em) 520 nm) was determined immediately and each complete sample was corrected with the respective blank sample. Alanine from 0 to 5 nmole in 0.5 nmole increments, were used as standards.

2.7. Total N

N of ground plant samples was converted to $(NO)^n$ equivalents by oxidation using a FlashEA 1112 series combustion analyzer (Thermo Fisher Scientific, Waltham, MA, USA) applying Association of Official Analytical Chemists (AOCS) method Ba 4e-93

2.8. Nitrate Quantification

The method of Miranda et al. [44] was used with slight modification to correct for background anthocyanins that absorb at 540 nm in acid, and for nitrite. An equal volume of 1% sulfanilamide, 0.01% naphthalene ethylene diamine in 2 M H_3PO_4 was added and optical density at 540 nm (OD^{540}) was determined followed by the addition of a sample volume of saturated VCl_2 in 1 M HCl then incubated at 37 °C for 1 h. The OD^{540} was again measured and the previous absorption readings used to correct for anthocyanin and nitrite after correction for differences in path lengths.

2.9. Malate Quantification

An aliquot was suspended in a total volume of 100 μL with water. 10 μL of 10 mM NADP, 5 mM $MgCl_2$, 1 M Tris-HCl pH 7.5 was added to each well followed by 10 μL containing 0.05 units of malic enzyme (EC 1.1.1.40). Blank plates were prepared in the same fashion but 10 μL of water was added instead of malic enzyme. These were incubated for 1 h at room temperature and fluorescence (Ex 345, Em 460) was measured subtracting the blank from the sample plate. Malate standards from 0 to 5 nmoles, in 0.5 nmole increments, were used.

2.10. In Vivo Nitrate Reductase Activity—NRA (In Vivo)

In vivo nitrate reductase activity was measured similar to the method of Reed and Hageman [45]. Leaf punches (10, 5 mm diameter) were submerged in 400 μL 50 mM KH_2PO_4-$KHPO_4$ pH 6.0, 300 mM sorbitol, 0.04% Trition X-100 then vacuum infiltrated. A 100 μL aliquot of a 100 mM KNO_3, 50 mM $KHCO_3$, 300 mM sorbitol, 50 mM KH_2PO_4-$KHPO_4$ pH 6.0, 0.04% Trition X-100 solution was added to each tube so the final assay concentration was 300 mM sorbitol, 50 mM KH_2PO_4-$KHPO_4$ pH 6.0, 0.04% Trition X-100, 20 mM KNO_3, and 10 mM $KHCO_3$. Tubes were incubated in the dark at 30 °C and 50 μL aliquots were remove every 30 min for 2 h. The production of nitrite was determined by adding 100 μL of 1% sulfanilamide, 2 M H_3PO_4, 0.02% naphthalene ethylene diamine, and 50 μL acetonitrile and measuring OD^{540}. The assay was linear for at least 3 h.

2.11. In Vivo Nitrite Reductase Activity—NiRA (In Vivo)

In vivo nitrite reductase activity was measure by the loss of nitrite from the medium. Leaf punches (5, 5 mm diameter) were submerged in 200 μL 50 mM KH_2PO_4-$KHPO_4$ pH 6.0, 300 mM sorbitol, 0.04% Trition X-100 then vacuum infiltrated. A 50 μL aliquot of a 1 mM KNO_2, 50 mM $KHCO_3$, 300 mM sorbitol, 50 mM KH_2PO_4-$KHPO_4$ pH 6.0, 0.04% Trition X-100 solution was added to each tube so the final assay concentration was 300 mM sorbitol, 50 mM KH_2PO_4-$KHPO_4$ pH 6.0, 0.04% Trition X-100, 200 μM KNO_2, and 10 mM $KHCO_3$. Tubes were incubated at 30 °C under a bank of light emitting diodes of photosynthetic quality (Quantum Devices, Model # SL1515-470-670, manufacturer, Barneveld, WI, USA). Duplicate 10 μL aliquots were removed every 30 min for 2 h and treated with 150 μL of 1% sulfanilamide, 2 M H_3PO_4, 0.02% naphthalene ethylene diamine, and 50 μL acetonitrile and OD^{540} determined. The assay was linear for at least 3 h.

2.12. Enzyme (In Vitro) Extraction

Extraction tubes (1.2) mL were filled with 500 μL of a 1% slurry of insoluble polyvinyl polypyrrolidone (PVPP), reduced to dryness by Speedvac (SPD131DDA-115 Thermo Fisher Scientific, Waltham, MA, USA), then stored at room temperature until use. Extraction tubes were filled with 500 μL extraction medium (100 mM Tris-HCl pH 8, 10 mM cysteine, 10 μM leupeptin, 4 °C) and leaf punches (10, 5 mm diameter) were delivered to the tubes. The tubes were arranged in a 96-well plate such that surrounding tubes contained ice and were vacuum infiltrated. Tubes were ground by Genogrinder for 2, 1 min intervals. Plates were centrifuged 4000× *g* for 20 min at 4 °C and a 200 μL aliquot removed and place on a 2 mL bed of Sephadex G-25 previously equilibrated with 50 mM Tris-HCl pH 7.5 with the void volume removed by a 1 min 500× *g* centrifugation. Small molecular weight metabolites were removed from the plant extract without dilution by a second low speed centrifugation with the receiving wells containing concentrated cysteine and leupeptin to maintain 10 mM and 10 μM concentrations, respectively.

2.13. Nitrate Reductase Activity (NRA (In Vitro)) EC 1.6.6.1

Cyclic renewal of NADH was used to avoid excess NADH from interfering with color development. The assay was performed in a 20 μL volume in 384-well plates with the following component concentrations: enzyme extract; 100 mM Tris-HCl pH 7.5; 10 mM cysteine; 10 mM KNO_3; 20 μM NAD^+; 1 mM glucose-6-phosphate; and, 1 unit glucose-6-phosphate dehydrogenase-NAD^+ (EC 1.1.1.388). The reaction was started by the addition of enzyme extract and stopped at 30 min intervals by adding two volumes of 1% sulfanilamide, 2 M H_3PO_4, 0.02% naphthalene ethylene diamine. Activation state of NRA (in vitro) was estimated by including either 5 mM EDTA or 10 mM $MgCl_2$ in the assay. The ratio of $MgCl_2$ to EDTA enzyme activities was used as estimate of the activation state of NRA (in vitro) [46].

Since the concentration of NADH was maintained at 20 μM, inhibition of color development was not statistically ($p \leq 0.05$) significant, relative to the inherent variability of the experiments. If there was interference of color development it would be constant across all samples since the concentration of NADH was maintained constant. In an assay development experiment where the concentration of NAD^+ was varied 0, 20, 50, 100, and 200 μM in the assay media across a range of nitrite concentrations from 0 to 100 μM, in 10 μM increments, NAD^+ (NADH) significantly ($p \leq 0.05$) inhibited color development starting at 50 μM NAD^+ but the absorption of varying NAD^+ concentrations across nitrite concentrations were parallel (equal slopes ($p \leq 0.05$)), demonstrating equal levels of inhibition of color development across nitrite concentrations. This would negate the effect of variable leftover concentrations NADH on the assay since color reduction would be equal at the end of the assay irrespective of the intensity of NRA (in vitro). Also, since NAD(P)H inhibits color development by reducing the diazonium salt formed between sulfanilamide and nitrite to a hydrazone such that the azo-compound is not formed, combining sulfanilamide and naphthalene ethylene diamine increases the chance the diazonium salt reacts with the diamine rather than with NAD(P)H. Finally, the assay was further improved by removing small molecular weight molecules including oxaloacetic acid (OAA) which reduces the level of detectable nitrite over time (not shown), presumably by forming an oxime.

2.14. Nitrite Reductase Activity (NiRA (In Vitro)) EC 1.7.7.1

NiRA (in vitro) was assayed in a 200 μL volume containing enzyme extract, 100 mM KH_2PO_4-$KHPO_4$ pH 6.9, 400 μM KNO_2, 10 mM methyl viologen, 30 mM $Na_2S_2O_4$ and 30 mM $KHCO_3$. The assay was started by addition of $Na_2S_2O_4$ + $KHCO_3$ and incubated at 30 °C. Aliquots (50 μL) were removed after 0, 10 and 20 min and methyl viologen oxidized by shaking. The loss of nitrite was determined after methyl viologen oxidation by adding 150 μL of 1% sulfanilamide, 2 M H_3PO_4, 0.02% naphthalene ethylene diamine and 50 μL acetonitrile and determining OD^{540}.

2.15. Phosphoenolpyruvate Carboxylase (PEP Carboxylase) EC 4.1.1.31

PEP carboxylase activity was measured by the loss of NADH coupled to the conversion of OAA, formed, to malate via malate dehydrogenase (EC 1.1.1.37). The extract was diluted 10× with water and 10 μL was used in a 200 μL assay volume which was 25 mM HEPES pH 7.5, 20% ethylene glycol, 5 mM $MgCl_2$, 10 mM $KHCO_3$, 2 mM phosphoenolpyruvate (PEP), 40 μM NADH, and 1 unit malate dehydrogenase (EC 1.1.1.37). Blanks contained water instead of PEP. The reaction was started by the addition of PEP and fluorescence (Ex 345, Em 440) was measure every minute for 10 min. The linear portion of the reduction in fluorescence was used to determine enzyme activity. Removal of OAA from the enzyme extract was crucial to maintain a constant baseline for this assay.

2.16. Malate Enzyme (ME) EC 1.1.1.40

Malate enzyme activity was determined by measuring NADPH formed as a result of malate conversion to pyruvate and HCO_3^-. The extract was diluted 10× with water and 10 µL was used in a 200 µL assay volume which was 50 mM HEPES pH 7.0, 1 mM $NADP^+$, 5 mM $MgCl_2$ and 2 mM malate. The reaction was started by adding malate; the blanks had water instead of malate. Fluorescence (Ex 345, Em 440) was measured every minute for 10 min. The linear portion of the increase in fluorescence was used to determine enzyme activity.

2.17. Protein

The method of Bradford [47] was used.

2.18. Ear Growth Response to Switching Nitrate Concentrations at Different Stages of Development

The DuPont-Pioneer hybrid, 33W84, was grown in a field semi-hydroponics system previously described. Plants were grown in either 1 or 3 mM KNO_3 as the N source and at V0 (control), V3, V6, V9, V12, V15, and V18 (tassel emergence) plants growing in 1 mM KNO_3 were switched to 3 mM KNO_3 and plants growing in 3 mM KNO_3 switched to 1 mM KNO_3. Ear shoots were bagged as the shoots emerged to avoid seed set. Plants were sampled at R1 and separated into ear and remaining vegetative plant then dried (70 °C, 72 h). There were 10 replicates of each treatment combination. The experiment was repeated a second season but the concentrations of KNO_3 used were 1 and 4 mM KNO_3.

2.19. Response to Nitrate and Ammonia Nutrition during Later Stages of Ear Development

The DuPont-Pioneer hybrid, 33W84, was grown in a field semi-hydroponics system and ear dry weight response to ammonia tested for two consecutive growing seasons. In the first season plants were grown in nutrient medium containing 1 and 3 mM KNO_3 and converted to 1 mM NH_4Cl at V12 and V15 or maintained at 1 and 3 mM KNO_3. The plants were treated similarly in the subsequent year, except the 3 mM KNO_3 treatments were replaced by 4 mM KNO_3 treatments. Ear shoots were bagged as the shoots emerged to avoid seed set. Plants were harvested at R1 and separated into ear and remaining vegetative plant then dried (70 °C, 72 h). There were 10 replicates of each treatment combination.

GF3 was grown semi-hydroponically in either 1 or 2 mM KNO_3 as the sole N source. At 23 days after emergence (DAE), plants growing in 1 mM KNO_3 fertility were switched to 2 mM KNO_3, to 1 mM NH_4Cl, or maintained at 1 mM KNO_3. Plants growing in 2 mM KNO_3 were switched to 1 mM KNO_3, to 1 mM NH_4Cl, or maintained at 2 mM KNO_3. At 30 DAE (R1) plants were separated into ears and remaining plant biomass then dried by lyophilization. Ear weight, remaining plant biomass, and total N were determined.

GF3 was grown semi-hydroponically under controlled environment in 1 mM KNO_3 up to 22 DAE and the plants were switched to 1 mM NH_4Cl, 10 mM KNO_3, or maintained at 1 mM KNO_3. Ear and plant biomass was determined of all treatments at 23, 25, and 27 DAE along with total N.

2.20. Vector Construction, Plant Transformation and Transgene Expression Analysis

A 230-base pair (bp) fragment of maize NADH ubiquinone oxidoreductase (GRMZM2G024484) including the 5′-untranslated region (UTR) and part of the coding sequence was PCR-cloned to make an RNAi construct. An intron from ST-LS1 was added between the two inverted repeats of NADH ubiquinone oxidoreductase. Maize PEP carboxylase gene promoter and sorghum actin gene terminator were used to silence NADH ubiquinone oxidoreductase in mesophyll cells. This cassette was linked to LTP2:DS-RED2:PIN II TERM as a seed marker for transgenic seed as described [48]. The vector construction and maize transformation were carried out as described previously [48,49]. Multiple lines were generated. Single-copy T-DNA integration lines that expressed the transgene were selected for advancement to greenhouse or field test.

2.21. *NADH-Ubiquinone Oxidoreductase 51 kDa Subunit (Complex I) Expression*

RNA was isolated using a Qiagen RNeasy kit followed by Invitrogen Turbo DNA free kit to remove contaminating DNA. Transcript quantification was performed using Biorad's iTaq™ Universal One-Step RT-qPCR (Bio-Rad, Hercules, CA. USA) on a Biorad CFX96 Touch Real-Time PCR Detection System (Bio-Rad, Hercules, CA. USA). NADH-ubiquinone oxidoreductase 51 kDa subunit (Complex I) was quantified using the forward primer CAACTCTGGAACCAAGCTCTAT and the reverse primer GAGCAACTCCTTCAGAGGTATG. Transcriptional corepressor LEUNIG was used as a reference using CATCGACACCTTCCACTCATAC as the forward and TCCGTCAGAGCCAAACATTAC as the reverse primer.

2.22. *Multi-Location Field Yield Trails*

Transgenic Complex I RNAi events and the corresponding null were tested in 5 locations managed for optimal grain yields with four replicates at Johnston IA, Woodland CA, Plainview TX, and Corning AR, and three replicates in Garden City, KS. The field trial and statistical analysis were conducted as described previously [48]. Grain yield was calculated and adjusted to a standard moisture of 155 g kg^{-1}. Yield was predicted using Best Linear Unbiased Predictor (BLUP).

3. Results

3.1. *Field NUE Analyses under Normal and Depleted N Conditions*

During the 2009–2010 growing seasons, experiments were conducted to identify yield parameters responsive to N input and those that were also related to grain yield under conditions of high and low N fertility. These experiments were conducted in fields with corn on corn cropping systems with normal N inputs and in fields depleted of N for at least two years at the start of the experiment. There were significant ($F \leq 0.1$) effects of experimental year and significant ($F \leq 0.05$) effects of N fertility on grain yield but there were no significant ($F \leq 0.1$) effects of cultivar and no significant interactions of experimental year × cultivar, experimental year × N fertility, or cultivar × N fertility, so means were pooled across experimental years and cultivars. Grain yields were linearly related to N fertility in both years under both low N fertility (depleted N) and high N fertility (non-depleted N) ranges. Since grain yields were linearly related to N fertility, NUE was constant in both environments but NUE was 52% efficient (15.132 kg ha^{-1} kg N^{-1}) in the low N fertility range and 23% efficient (4.764 kg ha^{-1} kg N^{-1}) in the high N fertility range. Response parameters measured are shown in Table 1. Leaf total amino acid-V9 was very responsive to N fertility under both low and high ranges of N fertility. However, the relationship to grain yield was much greater under low N range than under high N range (Table 1). R1 biomass was not responsive to N fertility in either low or high N ranges, but R1 biomass was highly related to grain yield in the high N range. Total vegetative N at R1 was responsive to N fertility in the high N range and moderately related to grain yield. Ear length and ear-weight at R1, but not ear-width, were responsive to N and were highly correlated to grain yield in both high and low N ranges. SPAD chlorophyll measured at R1, R2, and R3 all responded to N fertility under both high and low N ranges, but the relationship to grain yield in the high N range was poor. Mature biomass was very responsive to N fertility in the low N range and poorly responsive in the high N range. Plant total N at maturity was non-responsive to N fertility and poorly correlated to grain yield. Ear-weight, KNP, and grain total N were all highly responsive to N fertility and closely related to grain yield under both N fertility ranges. Kernel weight was moderately responsive to N fertility in the high N range but slightly negatively responsive to N fertility in the low N range. Kernel weight was highly correlated to grain yield in the high N fertility range but poorly related to grain yield in the lower N fertility range.

Table 1. Maize parameters responsive to N input and correlated to grain yield.

Parameter	Response to N $p > F$		Relation to Grain Yield r^2	
	High N	Low N	High N	Low N
Total Amino Acids (V9)	<0.01	<0.01	0.77	0.99
Biomass, R1	0.638	0.828	0.96	0.85
Total N, R1	0.018	0.119	0.88	0.91
Ear Length, R1	0.028	<0.01	0.99	0.97
Ear Weight, R1	0.072	<0.01	0.95	0.93
Ear Width, R1	0.476	0.134	0.12	0.78
SPAD, R1	<0.01	<0.01	0.87	0.96
SPAD, R2	<0.01	<0.01	0.62	0.91
SPAD, R3	0.012	0.017	0.77	0.91
Biomass—Maturity	0.302	0.038	0.68	0.84
Plant Total N—Maturity	0.15	0.84	0.23	0.31
Ear Weight—Maturity	0.027	0.01	0.99	0.99
Kernel Weight	0.1	0.048(−)	0.98	0.56(−)
Kernel Number	0.038	0.013	0.99	0.99
Grain Total N	<0.01	0.061	0.97	0.95

During the 2009 and 2010 growing seasons, three DuPont-Pioneer hybrids (33K42, 33W84, 33T56) were grown in fields not depleted for N (High N) and were treated with six N fertilities (0, 34, 67, 101, 134, and 167 kg ha^{-1}). In a similar fashion, 33W84 was planted in an N depleted field (Low N) and treated with four N fertilities (0, 22, 45, and 67 kg ha^{-1}). Various parameters were measured and their response to N-fertility determined along with their relationship to grain yield.

3.2. Stalk Nitrate

Nitrate accumulates in maize lower stalk internodes as early as V11 (11th fully expanded leaf) in high N fertilized plots but little detectable nitrate was found in stalks of plants grown under low N fertility (Figure 1). V11 was the earliest all of the stalk internodes could be sampled and stalk nitrate was maximum at R3 but not significantly different ($p \leq 0.1$) between sampling dates.

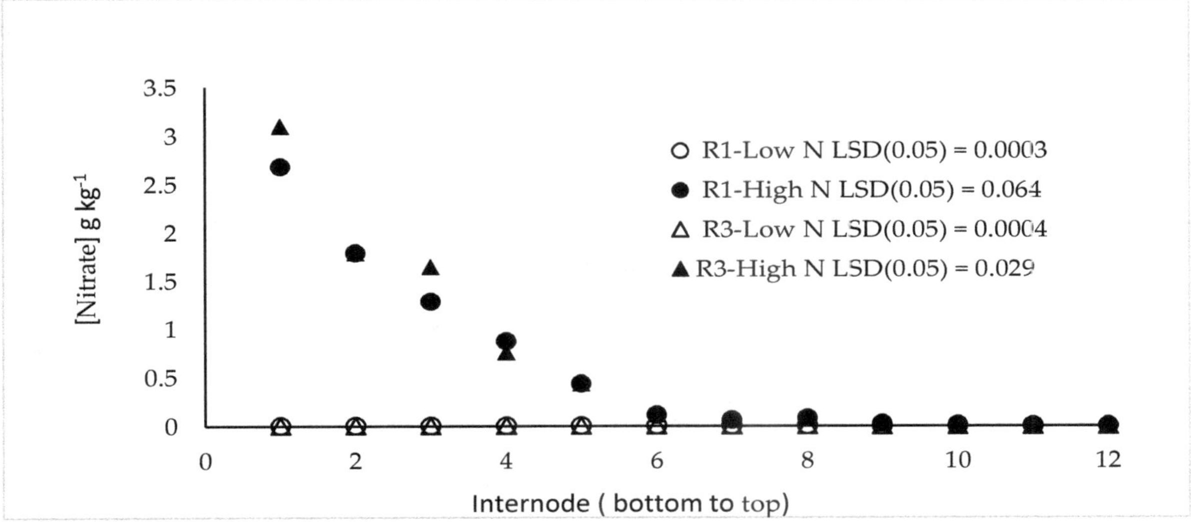

Figure 1. Stalk nitrate, by internode, of plants grown under high N and low N fertility. The DuPont-Pioneer hybrid 33W84 was grown in field plots depleted in N then fertilized with 77 kg ha^{-1} N (Low N), or in plots not N depleted and fertilized with 224 kg ha^{-1} N (High N). At V11, R1, R2, and R3 plants were sampled, leaves and leaf sheaths removed and the stalks separated into segments by internode. Samples were dried, ground to a fine powder and nitrate quantified.

Though generally low, lamina nitrate levels were not different irrespective of N fertility (Table 2). Lower leaves accumulated malate (Table 2) in both high and low N fertility grown plants at all sampling dates from V11 to R3. Lower leaves had less capacity to assimilate nitrate (NRA (in vivo)), (Figure 2) but nitrite assimilation (NiRA (in vivo)) was not significantly different in any leaf. Plants grown in lower N fertility also showed this response in lower leaves but the overall NRA (in vivo) and NiRA (in vivo) capacities were reduced by 70% (not shown) diminishing the ability to make statistically significant ($p \leq 0.1$) inferences. The extractable levels of enzymes followed the same pattern from lower to upper leaves with extractable nitrate reductase activity (NRA (in vitro)—EC 1.6.6.1) significantly less in the lower leaves whereas nitrite reductase activity (NiRA (in vitro)—EC 1.7.7.1) was not different (Figure 3). The ratio of NRA (in vitro) ($MgCl_2$) to NRA (in vitro) (EDTA) showed no signs of change in NRA (in vitro) activation [46]. The data presented is the NRA (in vitro) with EDTA in the assay medium, or maximum in vitro activity. Interestingly, the extractable PEP-carboxylase (EC 4.1.1.31) activity showed the same pattern as NRA (in vitro) with lower activity in lower leaves whereas malic enzyme (EC 1.1.1.40) activity was not different in lower verses upper leaves similarly as NiRA (in vitro) (Figure 3).

Table 2. Leaf metabolites by internode of plants grown under non-depleted N (High N) and depleted N (Low N) conditions of N fertility.

Internode	Leaf Nitrate ($\mu g\ g^{-1}$)		Leaf Malate ($\mu mole\ g^{-1}$)	
	Low N	High N	Low N	High N
1	0.96	1.18	16.28	38.89
2	1.01	1.16	22.43	28.83
3	1.23	1.28	14.73	16.98
4	1.26	1.36	10.91	17.64
5	1.06	1.14	11.67	11.93
6	1.47	1.56	8.89	11.87
7	1.19	1.11	10.27	9.31
8	1.44	1.14	9.49	9.76
9	1.14	1.12	8.41	8.80
10	0.97	1.27	9.68	9.17
11	0.91	1.05	9.68	9.72
12	1.48	1.26	10.07	9.94
LSD (0.05)	1.177		1.336	

DuPont-Pioneer hybrid 33W84 was planted in Johnston, IA under high N fertility (224 kg N ha^{-1}) and in a field depleted of N for at least 2 years (Low N) and fertilized with 77 kg N ha^{-1}. Sampling occurred at V11 and R1. Subtending leaves of each internode were sampled by leaf punching. Samples were dried and metabolites quantified at each sampling date but analyzed together with sampling time included in the replicate term. Leaf 1 is at the base of the plant and leaf 12 is at the top.

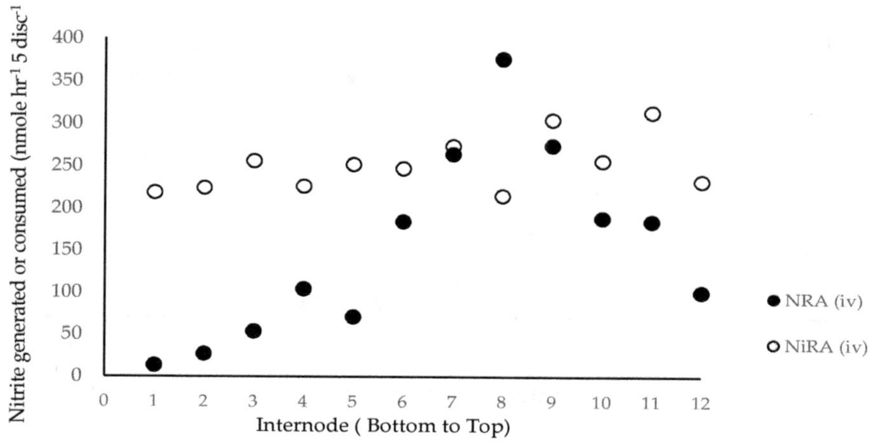

Figure 2. In vivo nitrate assimilation-NRA (in vivo) and in vivo nitrite assimilation-NiR (in vivo) by leaf position. The DuPont-Pioneer hybrid 33W84 was grown in non-depleted field plots and fertilized with 224 Kg N ha^{-1}. At R1 each leaf was sampled for nitrate assimilation (NRA (in vivo)) and nitrite assimilation (NiRA (in vivo)).

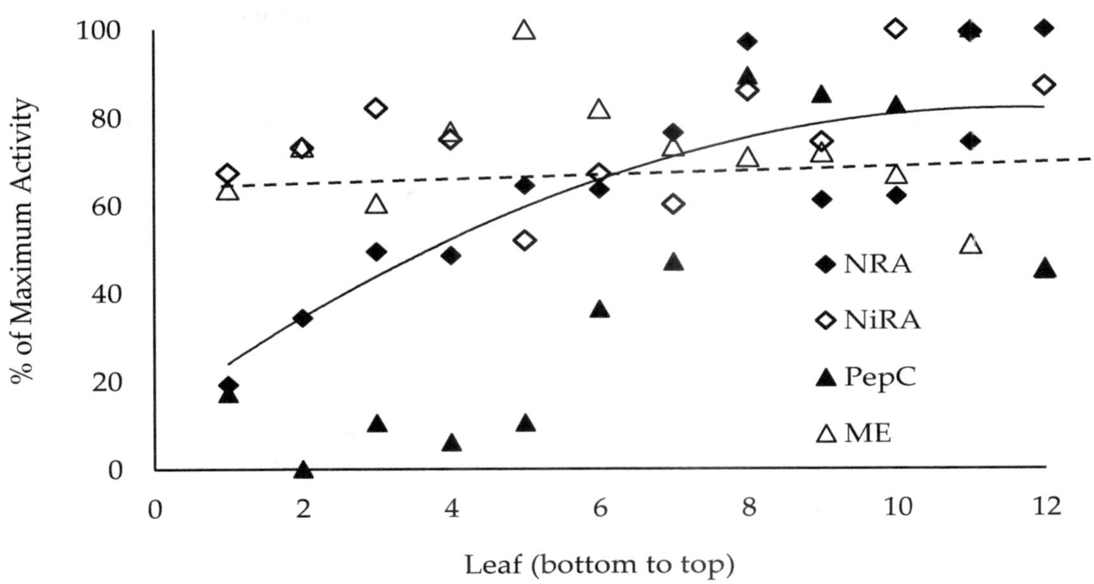

Figure 3. Extractable enzyme activities by internode. The DuPont-Pioneer hybrid 33W84 was grown in non-depleted field plots and fertilized with 224 Kg N ha^{-1}. Cell free extracts were produced from leaf punches (10, 5 mm) of each leaf, small molecular weight molecules removed, and enzymes assayed. Nitrate reductase activity (NRA (in vitro)) had a quadradic fit with leaf position whereas nitrite reductase activity (NiRA (in vitro)) and malic enzyme (ME) did not change with leaf position (slopes not different from 0). PEP-carboxylase activity follows a similar trend as NRA (in vitro) with lower activity in lower leaves.

3.3. Ear Dry Weight Response to Switching Nitrate Concentrations

In these experiments what effect reducing or increasing nitrate concentration has on R1 ear-weight was investigated. These experiments were conducted during two sequential growing seasons. The first season plants were grown in either 1 mM KNO_3 as insufficient levels of N fertility and at specific developmental stages switched to 3 mM KNO_3, considered a luxurious level of N fertility, for the remaining time. Concurrently, plants grown in 3 mM KNO_3 were switched to 1 mM KNO_3 and the effect of reducing nitrate for the remaining time had on R1 ear-weight was determined. In the second year the higher KNO_3 concentration was switched from 3 mM to 4 mM KNO_3. The results were similar for both years so means were summed across years. In Figure 4 plants switched at V0 (planting) would be considered controls; maintained at either 1 or 3–4 mM KNO_3 the entire time. Predictably, reducing nitrate levels at V3 (switching from 3–4 mM KNO_3 to 1 mM KNO_3) resulted in significant loss in R1 ear-weight. Likewise, increasing nitrate levels at V3 (switching from 1 mM KNO_3 to 3–4 mM KNO_3) improved R1 ear-weight but R1 ear-weight could not be completely restored when switched at V3. Recovery of R1 ear-weight was much less when nitrate levels were increased at V6 (switching from 1 mM KNO_3 to 3–4 mM KNO_3) and reducing nitrate at V6 (switching from 3–4 mM KNO_3 to 1 mM KNO_3) was much less effective on reducing R1 ear-weight. Switching at V12, or later, had very little effect on R1 ear-weight, neither increasing (switching from 1 mM KNO_3 to 3–4 mM KNO_3) nor decreasing (switching from 3–4 mM KNO_3 to 1 mM KNO_3) nitrate levels.

3.4. The Effect of Ammonia on Ear Development

Though R1 ear-weight could not be changed by increasing or decreasing nitrate levels after V12, ammonia (1 mM) could significantly change R1 ear-weight as late as V15. These experiments were conducted during sequential growing years. In the first year, plants were grown in either 1 or 3 mM KNO_3 as the insufficient and sufficient N levels, respectively, and in the second year, the 3 mM KNO_3 treatment was replaced with 4 mM KNO_3. There was no significant effect of year so the years were combined in analysis. Switching to 1 mM NH_4Cl significantly increased the ear number at both starting nitrate concentrations (Table 3). There was a significant effect of ammonia on R1 ear-weight when 3–4 mM KNO_3 grown plants were switched to 1 mM NH_4Cl (Table 3) at V12–V15, stages when nitrate was ineffective in influencing R1 ear-weight. There was no significant effect of NH_4Cl on vegetative biomass.

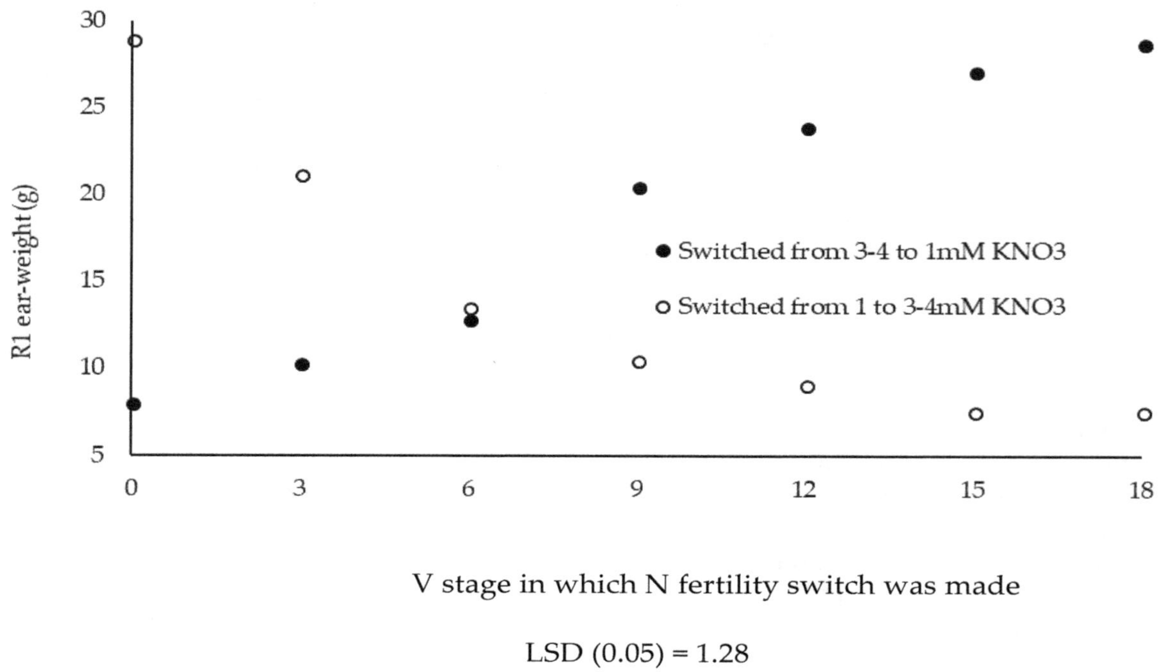

Figure 4. The DuPont-Pioneer 33W84 was grown in field hydroponics during two successive growing seasons. The first season plants were grown in nutrient media containing 1 and 3 mM KNO_3 as the sole N source and the second season grown in nutrient media containing 1 and 4 mM KNO_3 as the sole N source. Starting at emergence (V0), plants growing in 1 mM KNO_3 were switched to either 3 or 4 mM KNO_3 and plants growing in 3 or 4 mM KNO_3 switched to 1 mM KNO_3 every third developmental stage.

Table 3. The effect of V12–15 nutrient replacement with NH_4Cl on R1 ear-weight.

Treatment	Ear Dwt (g)	Plant Biomass (g)	Ear Number (%)
1 mM KNO_3 (No Change)	7.3 c	97.9 b	86 b
Switched from 1 mM KNO_3 to 1 mM NH_4Cl	8.3 c	97.9 b	97 a
3–4 mM KNO_3 (No Change)	28.2 b	152.6 a	93 a
Switched from 3–4 mM KNO_3 to 1 mM NH_4Cl	36.0 a	154.8 a	100 a

DuPont-Pioneer hybrid 33W84 was grown in nutrient media containing 1 or 3 mM KNO_3 as the sole nitrogen source up to V12 and V15 and then switched to media containing 1 mM NH_4Cl. Controls were unchanged. At R1, plants were harvested and ear and plant dry weights were determined along with the proportion of the plants that had a visible ear with emerging silks. The experiment was repeated in the following growing season except the upper KNO_3 concentration was 4 mM. Data were summed across treatment stages (V12 and V15) and experimental seasons. Means with different letters were significantly different by protected LSD (0.05).

Subsequent experiments were performed with Gaspe Flint-3 (GF3) under controlled environmental conditions. In these experiments, switching from insufficient levels of nitrate to sufficient levels of nitrate, and vice-versa, showed a similar R1 ear-weight response (data not shown) as the full season hybrids. GF3 R1 ear-weight was not responsive to changes in nitrate levels 20 days after emergence (DAE), or later. R1 in GF3 occurs from 29–31 DAE so, as in the case of a full season hybrids, GF3 R1 ear-weight was unresponsive to changes in nitrate for a significant amount of time prior to R1, or approximately for the last third of the time required to reach R1. Switching GF3 plants grown on either 1 or 2 mM KNO_3 to 2 or 1 mM KNO_3, respectively, had no significant effect on ear dry weight, vegetative biomass or total vegetative biomass (Table 4) compared to plants maintained at 1 or 2 mM KNO_3 the entire time. Switching to 1 mM NH_4Cl from both nitrate sources significantly increased ear dry weight and ear total N. There were some non-significant ($F \leq 0.1$) reductions in biomass in plants switched to 1 mM NH_4Cl but there were no significant effects of NH_4Cl on vegetative N.

Table 4. R1 ear-weight, ear N, R1 vegetative biomass, and vegetative N of GF3 switch from 1 mM nitrate N nutrition 23DAE.

Treatment	Veg Dwt (g)	Veg N (mg)	Ear Dwt (mg)	Ear N (mg)
1 mM KNO_3 (No Change)	4.11	175.5	193 b	14.4 b
Switched from 1 mM KNO_3 to 2 mM KNO_3	4.02	160.6	212 b	14.7 b
Switched from 1 mM KNO_3 to 1 mM NH_4Cl	3.83	190.7	265 a	22.4 a
Significance level	NS	NS	0.05	0.05
2 mM KNO_3 (No Change)	4.32	197.0	208 b	13.5 b
Switched from 2 mM KNO_3 to 1 mM KNO_3	4.21	187.9	193 b	12.1 b
Switched from 2 mM KNO_3 to 1 mM NH_4Cl	3.79	229.9	297 a	24.7 a
Significance level	NS	NS	0.05	0.05

GF3 was grown in nutrient media containing 1 or 2 mM KNO_3 as the sole nitrogen source up to 23-DAE. Plants grown in 1 mM KNO_3 were switched to 2 mM KNO_3, switched 1 mM NH_4Cl or maintained in 1 mM KNO_3. Likewise, plants grown in 2 mM KNO_3 were switched to 1 mM KNO_3, switched to 1 mM NH_4Cl or maintained in 2 mM KNO_3. Plants were harvested 29-DAE and ear and plant dry weights along with total-N were determined. Means with different letters were significantly different by protected LSD (0.05).

In companion experiments, GF3 was grown in 1 mM KNO_3 and at 22 DAE were either maintained in 1 mM KNO_3, switched to 1 mM NH_4Cl, or switched to 10 mM KNO_3. Switching to ammonia (1 mM) doubled the rate of ear growth within 24 h. Nitrate concentration of 10 mM nitrate was much less affective (Table 5).

Table 5. The effect of ammonia fertility on ear growth rate of GF3 when applied 22DAE.

Treatment	Ear Dry Weight (mg)			
	22 DAE	23 DAE	25 DAE	27 DAE
1 mM KNO_3 (No Change)	34	51 f	110 d	195 c
Switched from 1 mM KNO_3 to 10 mM KNO_3	34	47 f	101 d	284 b
Switched from 1 mM KNO_3 to 1 mM NH_4Cl	34	87 e	244 b	420 a

GF3 was grown in medium containing 1 mM KNO_3 as the sole N source. At 22-DAE plants were either maintained in 1 mM KNO_3 medium, switched to 10 mM KNO_3 or switched to 1 mM NH_4Cl. Ear dry weight was determined at every sampling date. Each mean represents 9 replicates and means with a similar letter are not different by protected LSD (0.05).

3.5. Total-N, Nitrate-N, and Organic-N Accumulation

Nitrogen balance experiments, both quantitatively and qualitatively, in response to nitrate fertility were conducted using GF3 under controlled environment. These experiments would be very difficult to conduct under field conditions, especially, since root:shoot ratios drop with increased N fertility [50] and access to the entire plant is required to quantify whole plant N levels. As in full season maize cultivars, N fertility of 2 mM KNO_3 is sufficient to support R1 ear-weight in GF3 [43]. Above 2 mM KNO_3 no further increase in R1 ear-weight can be observed so N fertilities of 3 and 4 mM KNO_3 would be considered more than sufficient. This range of N fertility was shown to accumulate nitrate in lower leaf midribs and stalk internodes especially when KNO_3 was provided in excess. NRA (in vitro) increases linearly with increased nitrate fertility (Figure 5), whereas NRA (in vivo) does not change above 2 mM KNO_3 fertility (Figure 5). When the accumulation of total plant N is compared to increasing N fertility, organic-N accumulation is not significantly ($F \leq 0.1$) increased from 2 mM KNO_3 or above, similarly as NRA (in vivo). With increased N fertility, the proportion of the total plant N that is nitrate increases (Figure 6). Though there is a numeric difference in organic-N between 2 and 3 mM KNO_3 grown plants, organic-N levels were not significantly changed with increased N fertility.

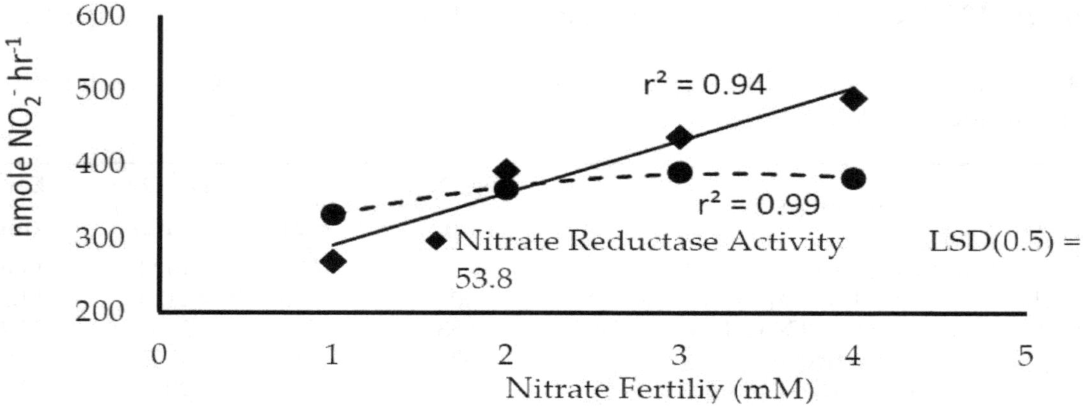

Figure 5. Change in nitrate reductase activity (NRA (in vitro)) and nitrate reductase activity, in vivo, (NRA (in vivo)) with increased nitrate fertility. GF3 was grown under controlled conditions with nitrate fertility varied from 1 to 4 mM KNO_3. The penultimate leaf was sampled 28-DAE for both Nitrate Assimilation (NRA (in vivo) and Nitrate Reductase Activity (NRA (in vitro)).

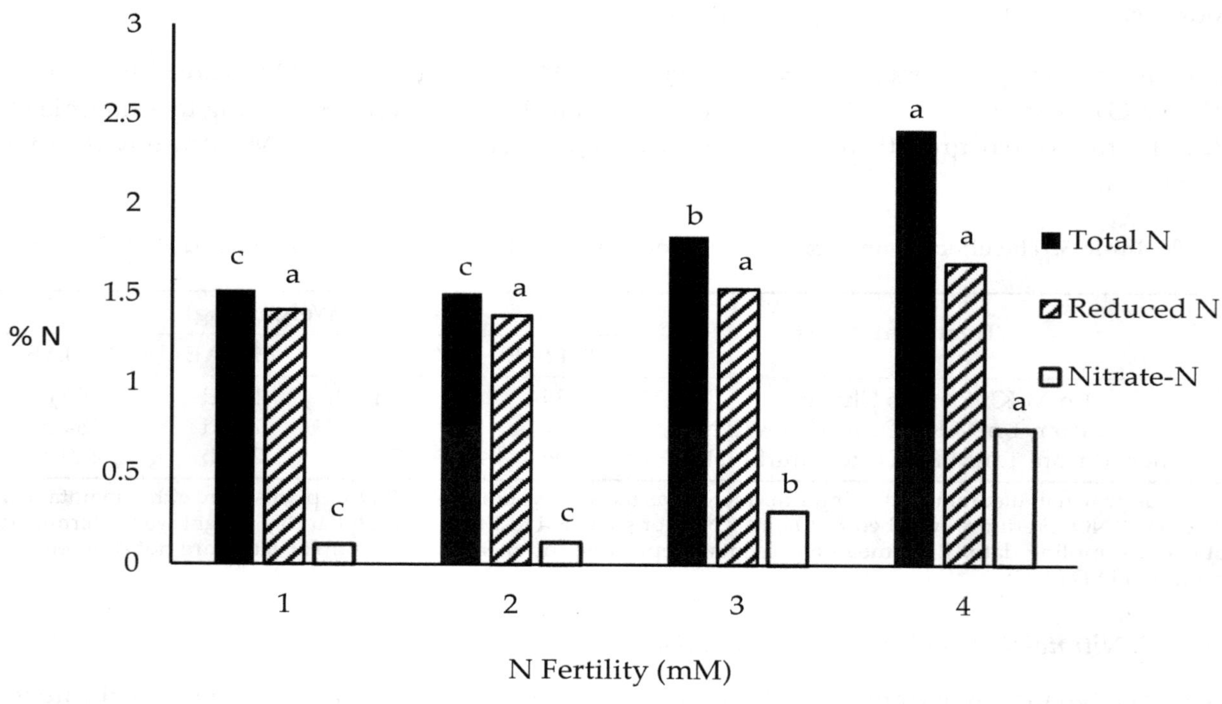

Figure 6. Plant N with increasing nitrate fertility. GF3 was grown under controlled conditions with nitrate fertility varied from 1 to 4 mM KNO_3. At 26-DAE the entire plant was removed from the potting media, dried, and ground to a fine power. Total-N and nitrate-N were determined. Organic-N was determined by subtracting nitrate-N from total-N. Means with similar letters are not different by protected LSD (0.05).

3.6. Transgenic Co-Suppression of NADH-Ubiquinone Oxidoreductase 51 kDa Subunit (Complex I)

Eight events of NADH-ubiquinone oxidoreductase 51 kDa subunit co-suppression driven by PEP-carboxylase promoter were generated and yield tested in multi-location environments managed for maximum yields (high N fertility). Only one event had a statistically significant ($p < 0.1$) grain yield increase of 230.37 kg ha^{-1} across all locations (Table 6), relative to the null hybrid. The positive event was selected for a second-year grain yield test along with physiological/biochemical evaluation. In the second-year field evaluation the positive event showed significant improvements in grain yield that was primarily associated with increased KNP (Table 7). The positive event had significantly lower levels of leaf NADH-ubiquinone oxidoreductase 51 kDa subunit (Complex I) expression, greater extractable NADH, higher NRA (in vivo) but no difference in extractable NRA (in vitro) (Table 7). Complex I expression in root, a non-targeted tissue, was not different between transgene and non-transgene. Though the statistic cannot be used to determine a null effect, since non-significance may be due to higher variability, the numeric difference between transgenic and non-transgenic root expression was small suggesting selective co-suppression of Complex I.

Table 6. NADH-ubiquinone oxidoreductase 51 kDa subunit (Complex I) co-suppression multi-location yield trial (2013).

Event	Grain Yield Increase over Transgenic Null (kg ha^{-1})
1.2	72.18
1.4	162.57
1.5	−23.22
1.8	87.25
1.12	91.64
1.14	6.28
1.18	80.97
1.19	230.37 *

NADH-ubiquinone oxidoreductase 51 kDa subunit (Complex I) inverted repeat co-suppression gene driven by PEP-carboxylase promoter was transformed into an elite inbred. Eight transgenic progenies were produced and crossed pollinated to form a commercial hybrid. Transgenic hybrid seed was separated from null hybrid seed using the color marker associated with the transgene. The hybrid was tested for grain yield in multiple locations managed for maximum yield and a multi-location yield analysis performed. Transgenic null hybrid grain yield across locations—12,384 kg ha^{-1}. *—Best Linear Unbiased Predictor (BLUP) ($p \leq 0.1$).

Table 7. The effect of selective NADH-ubiquinone oxidoreductase 51 kDa subunit (Complex I) co-suppression on maize N metabolism and grain yield.

	Complex I Expression		NADH	Nitrate Metabolism		Yield Parameters		
	Leaf (V9)	Root (V2)	nmole 10 Punches^{-1}	NRA (in vitro) nmole h^{-1} 5 disc^{-1}	NRA (in vivo) nmole h^{-1} 5 disc^{-1}	Kernel Wt (mg)	Kernel Number Plant^{-1}	Grain Yield kg ha^{-1}
Transgene	0.63 b	2.09	1.27 a	578.6	677 a	250.9	559 a	4886.0 a
Null	1.04 a	2.06	1.01 b	603.2	552 b	248.9	526 b	4581.5 b
$P > F$	<0.01	NS	<0.01	NS	<0.01	NS	<0.01	<0.01

Hybrid containing NADH-ubiquinone oxidoreductase 51 kDa subunit (Complex I) co-suppression transgene event, 1.19, was compared to the null (non-transgenic) hybrid. Means with a different letter are significantly (LSD < 0.01) different.

4. Discussion

4.1. Field NUE Analyses under Normal and Depleted N Conditions

Improving NUE under low N fertility is very difficult due to the inherently high NUE under these conditions and the unwillingness of farmers to accept lower grain yields in order to attain higher NUE. Identification of parameters associated with N input, especially under conditions of low NUE (high N fertility), is highly desirable for identifying plant parameters that might lead to increased NUE and higher grain yields. Useful parameters for study are those parameters that respond to N input that are also highly correlated to changes in grain yield. R1-biomass was closely related to grain yield but poorly responsive to N fertility (Table 1) which might suggest R1-biomass and grain yield are related but R1-biomass may not necessarily be vectored by N input. Pearson and Jacobs [51] found no relationship between plant biomass at anthesis and KNP. Andrade et al. [52] and Paponov et al. [53]

proposed increase in KNP was associated with plant growth rate around anthesis. Ciampitti et al. [54] observed a similar relationship between biomass and grain yield, though no direct correlation with N fertility was made. Loussaert et al. [41] showed the developing ear competes for N with the developing tassel, with the tassel having priority; increased R1 ear-weight in male sterile plants was independent of R1-biomass but associated with reduced tassel biomass. Ear length and ear-weight at R1, but not ear-width, were responsive to N and were highly correlated to grain yield in both high and low N ranges. Lemcoff and Loomis [55] showed the effect of N deprivation, verses population stress, was on KNP and not on kernel abortion suggesting the effect of N fertility is primarily on ear development. Poor SPAD chlorophyll relationship to grain yield in high N fertility make these measurements less useful since improvements in NUE are more desirable in high N fertilities. These are similar to results obtained by Blackmer and Schepers [56], in that SPAD measurements were not related to grain yield at higher N fertilities. Grain total N and ear weight would be expected to be highly related to grain yield since ear weight is nearly identical to grain yield and, since there were no significant differences in grain %N, grain total N would also be nearly identical to grain yield. KNP has been routinely associated with improved grain yield in maize [3,51–53] and these data (Table 1) show KNP is highly responsive to N fertility. Somewhat surprising was the effect of N fertility on kernel weight. In the low N range, KNP was reduced such that kernel weight compensated for poor seed set by making larger kernels under conditions which also produced lower grain yields resulting in a negative relationship between kernel weight and grain yield at lower N fertility. Though kernel weight was only moderately responsive to N fertility in the high N range, kernel weight was highly correlated to grain yield in the high N range but not in the lower N range. N fertility, especially in the high N range where NUE is lower, affects ear development that can have a lasting effect on increased KNP. Pearson and Jacobs [51] showed that N fertility manipulations during ear development had a lasting effect on KNP. Kernel weight seems only to be a factor in the higher N fertility when KNP has been optimized and added N may have an effect on increasing kernel weight. Biomass, measured at R1 or maturity, especially in the higher N range, was poorly responsive to N input.

These data would suggest that the prime effect of N under higher N fertility (low NUE) is on reproductive development since under higher N fertility, biomass was not affected by N fertility and SPAD was poorly associated with grain yield. N fertility increases KNP under higher N fertility and this is manifested at R1 by longer, more massive ears. R1 ear-weight, R1 ear-length, KNP, and kernel density are parameters responsive to N fertility and related to grain yield, especially under higher N fertility (low NUE). Ear-weight at R1 is particularly useful since it is a reproductive trait that can be determined early during plant development.

4.2. Stalk Nitrate Accumulation

Another important difference between plants grown under N deficient and N sufficient conditions is the accumulation of nitrate in the lower stalk internodes (Figure 1). Warner and Huffaker [57] and later Espen et al. [58] showed nitrate uptake is not dependent upon NA, so reduction in NRA (in vivo) would have little effect on the uptake of nitrate, but with NRA (in vivo) reduced in the lower leaves (Figure 2), nitrate would accumulate in the lower stalk internodes. Reduced NRA (in vivo) in the lower leaves could be explained by reduced NRA (in vitro) (Figure 3). Maize, being a C4 plant, generates OAA via PEP-carboxylase which is reduced to malate and transported to bundle sheath cells where CO_2 is concentrated. Reduced OAA production and thus reduced malate would be expected with reduced PEP-carboxylase activity in lower leaves (Figure 3). No change in malic enzyme activity between leaves coupled to reduced PEP-carboxylate activity in the lower leaves should result in lower malate levels in the lower leaves unless some other mechanism is responsible for increased malate in lower leaves. If malate is a main source of cytoplasmic NADH supporting NA [31,32], reduced NA capacity might be partially responsible for malate accumulation in lower leaves (Table 2). Thus, accumulation of malate in the lower leaves may be due to reduced NA resulting in the accumulation of reductant, malate.

4.3. *Ear Dry Weight Response to Nitrate and Ammonia*

Soil nitrate concentration sampled when plants are 20–30 cm tall has been shown to be related to maize grain yield [28] and the pre-sidedress nitrate test (PSNT—[59]) is a soil test used to determine whether supplemental N fertilizer is required. According to the PSNT, soil nitrate concentrations greater than 25 mg N L^{-1} (1.78 mM) require no supplemental N fertility, suggesting this level of soil nitrate is sufficient for optimum economic maize yield. R1 ear-weight response to nitrate concentrations from 1 to 4 mM KNO_3 is consistent with the PSNT with 1 mM KNO_3 being insufficient to support R1 ear-weight whereas 2 mM KNO_3 is sufficient and 3 mM KNO_3, and above, being luxurious levels of fertilizer N [41]. In the results shown in Figure 4, R1 ear-weight was established by V12 and increasing or decreasing KNO_3 levels from V12, on, had little effect on R1 ear-weight. This is consistent with Binder et al. [60] who showed delaying side dressing N till V6 resulted in 12% yield decreases. da Silva et al. [61] showed that grain yields could be increased by N side dressing at silking only when sufficient N was applied during vegetative development. In the studies of da Silval et al. [61] NUE of vegetative N application was twice that of N applied at silking and the main yield component improved by N applied at silking was kernel weight not KNP. R1 ear-weight is largely established by V12 and only moderate changes in R1 ear-weight can be obtained by changes in nitrate fertility after V12.

Unlike nitrate applications, moderate levels of ammonia improve ear development applied at a time (V12–V15) when much higher levels of nitrate are ineffective (Table 3). Ammonia improves ear development of plants grown in insufficient (1 mM) levels of nitrate but is even more efficacious in improving R1 ear-weight of plants grown in sufficient (3–4 mM) levels of nitrate (Table 3). R1 ear-weight improvements induced by ammonia were not due to biomass increases since no change in biomass was observed (Table 3). In controlled experiments with GF3, switching N fertility from nitrate to 1 mM NH_4Cl at a time when higher concentrations of nitrate are ineffective (23DAE) increases R1 ear-weight and ear total N without significantly changing vegetative biomass (Table 4). Changes in R1 ear-weight appear to be due to an increase in ear growth rate in response to ammonia (Table 5) with a doubling of the ear dry weight within 24 h of ammonia application. Ten times the concentration of nitrate is not as effective as ammonia at this developmental stage. These results suggest that NA, the conversion of nitrate to ammonia, limits R1 ear-weight and NUE in maize. These data show that bypassing NA by supplying ammonia rescues ear development when nitrate is ineffective. If R1 ear-weight is predictive of improved NUE then improvements in R1 ear-weight by by-passing NA would suggest NA limits NUE in maize.

Few studies have shown differential effects of ammonia, versus nitrate, on ear development. Jung et al. [62] showed that supplemental application of urea or NH_4NO_3 was much more efficacious than KNO_3 when applied 5 to 8 weeks after emergence. KNO_3 applications after 8 weeks were ineffective. Though the developmental stage was not specified in these experiments, a 5–8 week after emergence time frame would be roughly from V8–V15. Pan et al. [63] showed that urea fertilization fortified with a nitrification inhibitor produced more ears than comparable nitrate fertilities. Below and Gentry [64] showed that maize produced more kernels ear^{-1} and more grain yield when supplied with a mixture of nitrate and ammonium than when supplied nitrate alone. Smiciklas and Below [65] showed improved grain yield and increased kernels ear^{-1} were associated with a mixture of ammonia and nitrate verses nitrate alone but related the effect to cytokinin balance. Yasir et al. [66] showed foliar feeding of urea was most effective in improving grain yield, KNP and stover biomass when applied at V12. All of these observations are in agreement with ammonia induced increase in ear growth rate shown in these experiments (Tables 3–5). These results, combined with the observations that maize accumulates nitrate in the stalk under higher N fertilities, Figure 1, [4,23,24], would suggest NA limits ear development since stalk nitrate accumulation at high N fertility would suggest nitrate uptake is not limiting and by-passing NA by providing ammonia stimulates R1 ear-weight, raising NUE. This inference is different from those of Chen et al. [67] and Fan et al. [20] who showed transgenic expression of high affinity nitrate transporters in rice significantly improved grain yield and NUE in rice, implying the results are due to improved nitrate uptake. It is not surprising that maize and rice have different mechanisms limiting NUE since maize is a C4 plant and rice is a C3 plant; rice can grow submerged in water and maize cannot. In the field experiments of Fan et al. [20] rice was grown submerged in water, as is the custom for rice, and fertilized with urea [20]. Under these conditions, the main source of N would be ammonia [68]. Fan et al. [20] showed that when ^{15}N was supplied as $Ca(^{15}NO_3)_2$ transgenic rice expressing OsNrt2.1 had significantly higher rates of ^{15}N uptake but when ^{15}N was supplied as $NH_4{}^{15}NO_3$ there was a significant reduction in ^{15}N uptake in transgenic plants compared to controls. When $^{15}NH_4NO_3$ was used as a substrate not only was the background ^{15}N uptake rate 10–20× higher than when $NH_4{}^{15}NO_3$ was supplied but there was significantly more ^{15}N taken up by the transgenic plants. The rate of ^{15}N uptake when $^{15}NH_4{}^{15}NO_3$ was used as a substrate was not significantly different from ^{15}N uptake from the $^{15}NH_4NO_3$ treatment suggesting the effect of the transgene was on the uptake of ammonia, not nitrate. This would make our current inference of NA limiting NUE in congruence with those of Chen et al. [67] and Fan et al. [20] since improved ammonia uptake would bypass NA, the limiting step, resulting in improved NUE.

4.4. Total N, Organic-N, Nitrate-N, and NA

When GF3 was provided a range of fertilities that spans the range of insufficient to greater than sufficient levels of nitrate, NRA (in vitro) increased linearly with increased nitrate fertility, whereas NRA (in vivo) leveled off at the point in which nitrate fertility reached sufficiency to support R1 ear-weight (Figure 5). Organic-N accumulation mirrored NRA (in vivo) reaching maximum organic-N at 2 mm nitrate fertility (Figure 6). These results show that uptake of nitrate, the predominant form of N in aerobic soils, nor the extractable level of NRA (in vitro) do not limit NUE but the conversion of nitrate-N to organic-N (ammonia) limits NUE, especially, under conditions of high N fertility. Ciampitti et al. [54] concluded NUE in maize is limited by N uptake. The methodology used by Ciampitti et al. [54] was not capable of distinguishing between nitrate-N and organic-N which could significantly bias any inference made. Uncoupling of NRA (in vitro) with NRA (in vivo) (Figure 5) has been previously shown by Klepper et al. [29] and again by Kaiser et al. [33]. NRA (in vitro) does not reflect NA and is in excess of that required for NA which might suggest some other factor other than NRA (in vitro) limits NA. The increase in total N was primarily due to an increase in nitrate-N at higher N fertilities with 30% of the total plant N of plants grown in 4 M KNO_3 being nitrate (Figure 6). With both nitrate and NRA (in vitro) in excess under high nitrate fertility and NiR (in vivo) not limiting (Figure 2), the only other factor affecting NA would be the availability of reductant, NADH, to drive NA. The availability of cytoplasmic NADH as the limiting step in NA as proposed [33,36,38] would be a logical alternative.

4.5. Transgenic NADH-Ubiquinone Oxidoreductase 51 kDa Subunit (Complex I) Co-Suppression

NA is a complex biochemical process both supported and inhibited by respiratory metabolism [38]. The tricarboxylic acid cycle contributes αKG for NA but in the process robs the tricarboxylic acid cycle's ability to generate NADH within the mitochondria, resulting in lower mitochondrial NADH concentrations. NA may become self-regulated when mitochondrial NADH levels are restored by absorbing cytoplasmic malate which lowers the level of reductant available to generate NADH for nitrate reduction. Tobacco [36] and cucumber [37] have cytoplasmic mutants deficient in mitochondrial NADH dehydrogenase (Complex I) which inhibits the oxidation of NADH, artificially increasing the concentration of NADH inside the mitochondria. These mutants have improved capacity to utilize nitrate, but because the trait is constitutively expressed, also have undesirable traits associated with reduced capacity to generate reduction power in non-photosynthetic tissues. NCS2, a Complex I mutant in maize, has striped leaves, reduced growth rate and other deleterious traits [39]. A constitutive co-suppression of Complex I would likely be similarly compromised. Since NA occurs exclusively in the mesophyll cells in maize leaves, selective Complex I co-suppression in the mesophyll cells might minimize deleterious effects not related to NA. Also, since mitochondrial respiration shifts toward providing carbon skeletons for NA in the light when nitrate is the N source [69], generation of leaf mitochondrial energy might not be important in the light. The promoter of PEP-carboxylase, preferentially expressed in the mesophyll cells [70], was selected to drive Complex I co-suppression with the aim to maximize the expression of Complex I co-suppression in mesophyll cells while minimizing non-targeted effects.

One transgenic event showed positive grain yield responses when maintained in high N fertility (low NUE) in two years, one year in a multi-environment experiment (Tables 6 and 7). In the second-year field evaluation, the positive event showed significant improvements in grain yield that was primarily associated with increased KNP (Table 7). The positive event had significantly lower levels of leaf NADH-ubiquinone oxidoreductase 51 kDa subunit (Complex I) expression, greater extractable NADH, higher NRA (in vivo) but no difference in extractable NRA (in vitro) (Table 7). NADH-ubiquinone oxidoreductase 51 kDa subunit (Complex I) expression in root, a non-targeted tissue, was not different between transgene and non-transgene. These results are expected if NA is limited by NADH rather than by NRA (in vitro). Since the transgenic hybrid showed higher NADH levels associated with improved NRA (in vivo) without increase in NRA (in vitro) it would follow that increases in KNP associated with increased grain yields under high N fertility could have been brokered by improved NA during ear development. It may be suggested that increased reductant (NADH) limits NA, which limits NUE. A disturbing issue is the low frequency of positive events. The difference in transgene and non-transgene expression of NADH-ubiquinone oxidoreductase 51 kDa subunit (Complex I) was greater with plant maturity, with the greatest difference in leaf expression at R2. This is similar to PEP-carboxylase expression shown by Cho et al. [70], increasing with plant maturity with maximum expression at R3. The PEP-carboxylase promoter may be sufficiently tissue specific but not have sufficient strength to drive gene co-suppression during the critical time of ear development between V9–V15. If maximum expression of the promoter gene (PEP-carboxylase) is R2-R3 [70] not all of the transgene events may have had sufficient expression levels from V9–V15 when ear development would require maximum NADH to support NA. These data demonstrate the potential of increasing leaf NADH levels to improve maize NUE.

5. Conclusions

R1 ear-weight is a good estimate of maize NUE under both high N fertility (low NUE) and low N fertility (high NUE) conditions. The assimilation of nitrate to ammonia limits ear development and NUE in maize. This is concluded by observation that nitrate cannot rescue ear development at later stages of reproductive growth but ammonia can. Nitrate uptake is not limiting under high N fertility and accumulates in the stalk to much higher concentrations than soil nitrate concentrations. Stalk nitrate accumulation is associated with reduced leaf NRA (in vivo) and increased leaf malate concentrations. At higher nitrate fertilities, NRA (in vitro) increases linearly but NRA (in vivo) levels off at higher nitrate fertilities. Organic-N follows NRA (in vivo), not NRA (in vitro) with increasing nitrate fertility, with the bulk of increased total-N due to nitrate accumulation at higher nitrate fertilities. Improved NUE in maize is associated with improved NA which may be brokered by increased NADH levels available for nitrate reduction.

Author Contributions: Conceptualization, D.L., N.M.; Methodology, D.L.; Validation, B.S.; Investigation, N.M., J.C., D.P.O.; Writing-Original Draft Preparation, D.L.; Writing-Review & Editing, D.L.; Funding Acquisition, B.S.

Acknowledgments: David Sevenich for total N analyses. Renee Lafitte for supervising the multi-environment field yield trial.

References

1. Uriberlarrea, M.; Moose, S.P.; Below, F.E. Divergent selection for grain protein affects nitrogen use in maize hybrids. *Field Crops Res.* **2007**, *100*, 82–90. [CrossRef]
2. Ciampitti, I.A.; Vyn, T.J. A comprehensive study of plant density consequences on nitrogen uptake dynamics of maize plants from vegetative to reproductive stages. *Field Crops Res.* **2011**, *12*, 2–18. [CrossRef]
3. Uhart, S.A.; Andrade, F.H. Nitrogen Deficiency in Maize: II. Carbon-Nitrogen Interaction Effects on Kernel Number and Grain Yield. *Crop Sci.* **1995**, *35*, 1384–1389. [CrossRef]
4. Binford, G.D.; Blackmer, A.M.; El-Hout, N.M. Tissue Test for Excess Nitrogen during Corn Production. *Agron. J.* **1990**, *82*, 124–129. [CrossRef]
5. Rutto, E.; Bossenkemper, J.P.; Kelly, J.; Chim, B.K.; Raun, W.R. Maize grain yield response to the distance nitrogen is placed away from the row. *Exp. Agric.* **2013**, *49*, 3–18. [CrossRef]
6. Raun, W.R.; Johnson, G.V. Improving nitrogen use efficiency for cereal production. *Agron. J.* **1999**, *91*, 357–363. [CrossRef]
7. Xu, G.; Fan, X.; Miller, A.J. Plant nitrogen assimilation and use efficiency. *Annu. Rev. Plant Biol.* **2012**, *63*, 153–182. [CrossRef] [PubMed]
8. McAllister, C.H.; Beatty, P.H.; Good, A.G. Engineering nitrogen use efficiency crop plants: The current status. *Plant Biotechnol. J.* **2012**, *10*, 1011–1025. [CrossRef] [PubMed]
9. Sawyer, J.; Nafziger, E.; Randall, G.; Bundy, L.; Rehm, G.; Joern, B. Concepts and Rational for Regional Nitrogen Rate Guidelines for Corn. Iowa State University Extension: Ames, IA, USA, 2006. Available online: http://publications.iowa.gov/id/eprint/3847 (accessed on 2 June 2010).
10. Kohl, D.H.; Shearer, G.B.; Commoner, B. Variation of ^{15}N in Corn and Soil Following Application of Fertilizer Nitrogen. *Soil Sci. Soc. Am. J.* **1973**, *37*, 888–892. [CrossRef]
11. Sogbedji, J.M.; van Es, H.M.; Yang, C.L.; Geohring, L.D.; Magdoff, F.R. Nitrate Leaching and Nitrogen Budget as Affected by Maize Nitrogen Rate and Soil Type. *J. Environ. Qual.* **2000**, *29*, 1813–1820. [CrossRef]
12. Nelson, K.A.; Scharf, P.C.; Stevens, W.E.; Burdick, B.A. Rescue Nitrogen Applications for Corn. *Soil Sci. Soc. Am. J.* **2010**, *75*, 143–151. [CrossRef]
13. Wang, Z.-H.; Liu, X.-J.; Ju, X.-T.; Zhang, F.-S.; Malhi, S.S. Ammonia Volatilization Loss from Surface-Broadcast Urea: Comparison of Vented- and Closed-Chamber Methods and Loss in Winter Wheat–Summer Maize Rotation in North China Plain. *Commun. Soil Sci. Plant Anal.* **2011**, *35*, 2917–2939. [CrossRef]
14. Pelstera, D.E.; Larouche, F.; Rochette, P. Nitrogen fertilization but not soil tillage affects nitrous oxide emissions from a clay loam soil under a maize–soybean rotation. *Soil Tillage Res.* **2011**, *115–116*, 16–26. [CrossRef]
15. Good, A.G.; Shrawat, A.K.; Muench, D.G. Can less yield more? Is reducing nutrient input into the environment compatible with maintaining crop production? *Trends Plant Sci.* **2004**, *9*, 597–605. [CrossRef] [PubMed]
16. Kant, S.; Bi, Y.-M.; Rothstein, S.J. Understanding plant response to nitrogen limitation for the improvement of crop nitrogen use efficiency. *J Exp. Bot.* **2011**, *62*, 1499–1509. [CrossRef] [PubMed]
17. Brendel, V.; Kurtz, S.; Walbot, V. Comparative genomics of *Arabidopsis* and maize: Prospects and limitations. *Genome Biol.* **2002**, *3*. [CrossRef]
18. Basra, A.S.; Dhawan, A.K.; Goyal, S.S. DCMU inhibits in vivo nitrate reduction in illuminated barley (C3) leaves but not in maize (C4): A new mechanism for the role of light? *Planta* **2002**, *215*, 855–861. [CrossRef] [PubMed]
19. Bloom, A.J.; Burger, M.; Asensio, J.S.R.; Cousins, A.B. Carbon Dioxide Enrichment Inhibits Nitrate Assimilation in Wheat and Arabidopsis. *Science* **2010**, *328*, 899–903. [CrossRef] [PubMed]
20. Fan, X.; Tang, Z.; Tan, Y.; Zhang, Y.; Luo, B.; Yang, M.; Lian, X.; Shen, Q.; Miller, A.J.; Xu, G. Overexpression of a pH-sensitive nitrate transporter in rice increases crop yields. *Proc. Natl. Acad. Sci. USA* **2017**, *113*, 7118–7123. [CrossRef] [PubMed]
21. Rizzi, E.; Balconi, B.C.; Manusardi, C.; Gentinetta, E.; Motto, M. Genetic variation for traits relating to nitrogen content of maize stalks. *Euphytica* **1991**, *52*, 91–98.
22. Cliquet, J.-B.; Deléens, E.; Bousser, A.; Martin, M.; Lescure, J.-C.; Prioul, J.-L.; Mariotti, A.; Morot-Gaudry, J.-F. Estimation of Carbon and Nitrogen Allocation during Stalk Elongation by ^{13}C and ^{15}N Tracing in *Zea mays* L. *Plant Physiol.* **1990**, *92*, 79–87. [CrossRef] [PubMed]

23. Binford, G.D.; Blackmer, A.M.; Meese, E.G. Optimal Concentrations of Nitrate in Cornstalks at Maturity. *Agron. J.* **1992**, *84*, 881–887. [CrossRef]
24. Sawyer, J. Corn Stalk Nitrate Interpretation. 2010. Available online: http://www.extension.iastate.edu/CropNews/2010/0914sawyer.htm (accessed on 5 September 2012).
25. Brouder, S.M.; Mengel, D.B.; Hofmann, B.S. Diagnostic Efficiency of the Blacklayer Stalk Nitrate and Grain Nitrogen Tests for Corn. *Agron. J.* **2000**, *92*, 1236–1247. [CrossRef]
26. Varvel, G.E.; Schepers, J.S.; Francis, D.D. Chlorophyll Meter and Stalk Nitrate Techniques as Complementary Indices for Residual Nitrogen. *J. Prod. Agric.* **1997**, *10*, 147–151. [CrossRef]
27. Beevers, L.; Hageman, R.H. Nitrate Reduction in Higher Plants. *Annu. Rev. Plant Physiol.* **1969**, *20*, 495–522. [CrossRef]
28. Blackmer, A.M.; Pottker, D.; Cerrato, M.E.; Webb, J. Correlations between Soil Nitrate Concentrations in Late Spring and Corn Yields in Iowa. *J. Prod. Agric.* **1988**, *2*, 103–109. [CrossRef]
29. Klepper, L.; Flesher, D.; Hageman, R.H. Generation of Reduced Nicotinamide Adenine Dinucleotide for Nitrate Reduction in Green Leaves. *Plant Physiol.* **1971**, *48*, 580–590. [CrossRef] [PubMed]
30. Gowri, G.; Campbell, W.H. Communication cDNA Clones for Corn Leaf NADH:Nitrate Reductase and Chloroplast NAD(P)$^+$:Glyceraldehyde-3-Phosphate Dehydrogenase; Characterization of the Clones and Analysis of the Expression of the Genes in Leaves as Influenced by Nitrate in the Light and Dark. *Plant Physiol.* **1989**, *90*, 792–798. [PubMed]
31. Neyra, C.A.; Hageman, R.H. Relationships between Carbon Dioxide, Malate, and Nitrate Accumulation and Reduction in Corn (*Zea mays* L.) Seedlings. *Plant Physiol.* **1976**, *58*, 726–730. [CrossRef] [PubMed]
32. Rathnam, C.K.M. Malate and Dihydroxacetone Phosphate-dependent Nitrate Reduction in Spinach Leaf Protoplasts. *Plant Physiol.* **1978**, *62*, 220–223. [CrossRef] [PubMed]
33. Kaiser, W.M.; Kandlbinder, A.; Stoimenova, M.; Glaab, J. Discrepancy between nitrate reduction rates in intact leaves and nitrate reductase activity in leaf extracts: What limits nitrate reduction in situ? *Planta* **2000**, *210*, 801–807. [CrossRef] [PubMed]
34. Sienkiewicz-Porzucek, A.; Nunes-Nesi, A.; Sulpice, R.; Lisec, J.; Centeno, D.C.; Carillo, P.; Leissea, A.; Urbanczyk-Wochniak, E.; Fernie, A.R. Mild Reductions in Mitochondrial Citrate Synthase Activity Result in a Compromised Nitrate Assimilation and Reduced Leaf Pigmentation But Have No Effect on Photosynthetic Performance or Growth. *Plant Physiol.* **2008**, *147*, 115–127. [CrossRef] [PubMed]
35. Sienkiewicz-Porzuceka, A.; Sulpicea, R.; Osorioa, S.; Krahnerta, I.; Leissea, A.; Urbanczyk-Wochniaka, E.; Hodges, M.; Ferniea, A.R.; Nunes-Nesia, A. Mild Reductions in Mitochondrial NAD Dependent Isocitrate Dehydrogenase Activity Result in Altered Nitrate Assimilation and Pigmentation But Do Not Impact Growth. *Mol. Plant* **2010**, *3*, 156–173. [CrossRef] [PubMed]
36. Hager, J.; Pellny, T.K.; Mauve, C.; Lelarge-Troverie, C.; DePaepe, R.; Foyer, C.H.; Noctor, G. Conditional modulation of NAD levels and metabolite profiles in *Nicotiana sylvestris* by mitochondrial electron transport and carbon/nitrogen supply. *Planta* **2010**, *231*, 1145–1157. [CrossRef] [PubMed]
37. Szal, B.; Jastrzebska, A.; Kulka, M.; Lesniak, K.; Podgórska, A.; Pärnik, T.; Ivanova, H.; Keerberg, O.; Gardeström, P.; Rychter, A.M. Influence of mitochondrial genome rearrangement on cucumber leaf carbon and nitrogen metabolism. *Planta* **2010**, *232*, 1371–1382. [CrossRef] [PubMed]
38. Foyer, C.H.; Noctor, G.; Hodges, M. Respiration and nitrogen assimilation: Targeting mitochondria-associated metabolism as a means to enhance nitrogen use efficiency. *J. Exp. Bot.* **2011**, *62*, 1467–1482. [CrossRef] [PubMed]
39. Marienfeld, J.R.; Newton, K.J. The maize NCS2 abnormal growth mutant has a chimeric nad4-nad7 mitochondrial gene and is associated with reduced complex I function. *Genetics* **1994**, *138*, 855–863. [PubMed]
40. Loussaert, D. Microcomputer-based experiment management system: II. Data analysis. *Agron. J.* **1992**, *84*, 256–259. [CrossRef]
41. Loussaert, D.; DeBruin, J.; San Martin, J.P.; Schussler, J.; Pape, R.; Clapp, J.; Mongar, N.; Fox, T.; Albertsen, M.; Trimnell, M.; et al. Genetic Male Sterility (Ms44) Increases Maize Grain Yield. *Crop Sci.* **2017**, *57*, 2718–2728. [CrossRef]
42. Tollenaar, M.; Migus, W. Dry matter accumulation of maize grown hydroponically under controlled-environment and field conditions. *Can. J. Plant Sci.* **1984**, *64*, 475–485. [CrossRef]

43. Queval, G.; Noctor, G. A plate reader method for the measurement of NAD, NADP, glutathione, and ascorbate in tissue extracts: Application to redox profiling during *Arabidopsis* rosette development. *Anal. Biochem.* **2007**, *363*, 58–69. [CrossRef] [PubMed]
44. Miranda, K.M.; Espey, M.G.; Wink, D.A. A rapid, simple spectrophotometric method for simultaneous detection of nitrate and nitrite. *Nitric Oxide* **2001**, *5*, 62–71. [CrossRef] [PubMed]
45. Reed, A.J.; Hageman, R.H. The relationship between nitrate uptake, nitrate flux and nitrate reductase in four maize (*Zea mays* L.) genotypes. I. Genotypic variation. *Plant Physiol.* **1980**, *66*, 1179–1183. [CrossRef] [PubMed]
46. Kaiser, W.M.; Huber, S.C. Correlation between apparent activation state of nitrate reductase (NR), NR hysteresis and degradation of NR protein. *J. Exp. Bot.* **1997**, *48*, 1367–1374. [CrossRef]
47. Bradford, M.M. A rapid and sensitive method for the quantitation of microgram quantities of protein utilizing the principle of protein-dye binding. *Anal. Biochem.* **1976**, *72*, 248–254. [CrossRef]
48. Fox, T.; DeBruin, J.; Haug-Collet, K.; Trimnell, M.; Clapp, J.; Leonard, A.; Li, B.; Scolaro, E.; Collinson, S.; Glassman, K.; et al. A single point mutation in Ms44 results in dominant male sterility and improves nitrogen use efficiency in maize. *Plant Biotechnol. J.* **2017**, *15*, 942–952. [CrossRef] [PubMed]
49. Unger, E.; Betz, S.; Xu, R.; Cigan, A.M. Selection and orientation of adjacent genes influences DAM-mediated male sterility in transformed maize. *Transgenic Res.* **2001**, *10*, 409–422. [CrossRef] [PubMed]
50. Bonifas, K.D.; Walters, D.T.; Cassman, K.G.; Lindquist, J.L. Nitrogen supply affects root:shoot ratio in corn and velvetleaf (*Abutilon theophrasti*). *Weed Sci.* **2005**, *53*, 670–675. [CrossRef]
51. Pearson, C.J.; Jacobs, B.C. Yield components and nitrogen partitioning of maize in response to nitrogen before and after anthesis. *Aust. J. Agric. Res.* **1987**, *38*, 1001–1009. [CrossRef]
52. Andrade, F.H.; Echarte, L.; Rizzalli, R.; Della Maggiora, A.; Casanovas, M. Kernel Number Prediction in Maize under Nitrogen or Water Stress. *Crop Sci.* **2002**, *42*, 1173–1179. [CrossRef]
53. Paponov, J.A.; Sambo, P.; Presterl, T.; Geiger, H.H.; Engels, C. Kernel set in maize genotype differing in nitrogen use efficiency in response to resource availability around flowering. *Plant Soil* **2005**, *272*, 101–110. [CrossRef]
54. Ciampitti, I.A.; Zhang, H.; Friedemannc, P.; Vyn, T.J. Potential Physiological Frameworks for Mid-Season Field Phenotyping of Final Plant Nitrogen Uptake, Nitrogen Use Efficiency, and Grain Yield in Maize. *Crop Sci.* **2012**, *52*, 2728–2742. [CrossRef]
55. Lemcoff, J.H.; Loomis, R.S. Nitrogen and density influences on silk emergence, endosperm development, and grain yield in maize (*Zea mays* L.). *Field Crop Res.* **1994**, *38*, 63–72. [CrossRef]
56. Blackmer, T.M.; Schepers, J.S. Aerial Photography to Detect Nitrogen Stress in Corn. *J. Plant Physiol.* **1996**, *148*, 440–444. [CrossRef]
57. Warner, R.L.; Huffaker, R.C. Nitrate Transport Is Independent of NADH and NAD(P)H Nitrate Reductases in Barley Seedlings. *Plant Physiol.* **1989**, *91*, 947–953. [CrossRef] [PubMed]
58. Espen, L.; Nocito, F.F.; Cocucci, M. Effect of NO_3^- transport and reduction on intracellular pH: An in vivo NMR study in maize roots. *J. Exp. Bot.* **2004**, *55*, 2053–2061. [CrossRef] [PubMed]
59. Magdoff, F.R. Understanding the Magdoff pre-sidedress nitrate test for corn. *J. Prod. Agric.* **1991**, *4*, 297–305. [CrossRef]
60. Binder, D.L.; Sander, D.H.; Walters, D.T. Maize Response to Time of Nitrogen Application as Affected by Level of Nitrogen Deficiency. *Agron. J.* **2000**, *92*, 1228–1236. [CrossRef]
61. Da SilvaI, F.P.R.; StriederI, M.L.; da Silva CoserI, R.P.; RamboI, L.; SangoiII, L.; ArgentaIII, G.; ForsthoferIII, E.L.; da SilvaI, A.A. Grain yield and kernel crude protein content increases of maize hybrids with late nitrogen side-dressing. *Sci. Agric.* **2005**, *62*, 487–492. [CrossRef]
62. Jung, P.E., Jr.; Peterson, L.A.; Schader, L.E. Response of Irrigated Corn to Time, Rate and Source of Applied N on Sandy Soils. *Agron. J.* **1972**, *64*, 668–670. [CrossRef]
63. Pan, W.L.; Kamprath, E.J.; Moll, R.H.; Jackson, W.A. Prolificacy in Corn: Its Effects on Nitrate and Ammonium Uptake and Utilization. *Soil Sci. Soc. Am. J.* **1984**, *48*, 1101–1106. [CrossRef]
64. Below, F.E.; Gentry, L.E. Maize Productivity as Influenced by Mixed Nitrogen Supplied before or after Anthesis. *Crop Sci.* **1992**, *32*, 163–168. [CrossRef]
65. Smiciklas, K.D.; Below, F.E. Role of cytokinin in enhanced productivity of maize supplied with NH_4^+ and NO_3^-. *Plant Soil* **1992**, *142*, 307–313. [CrossRef]

66. Yasir, A.M.; Khlil, S.K.; Jan, M.T.; Khan, A.Z. Phenology, growth, and grain yield of maize as influenced by foliar applied urea at different growth stages. *J. Plant Nutr.* **2010**, *33*, 71–79.
67. Chen, J.; Fan, X.; Qian, K.; Zhang, Y.; Song, M.; Liu, Y.; Xu, G.; Fan, X. pOsNAR2.1:OsNAR2.1 expression enhances nitrogen uptake efficiency and grain yield in transgenic rice plants. *Plant Biotechnol. J.* **2017**, *15*, 1273–1283. [CrossRef] [PubMed]
68. Yu, T.-R. Soil and plants. In *Physical Chemistry of Paddy Soils*; Yu, T.-R., Ed.; Science Press: Beijing, China, 1985; pp. 197–217.
69. Gardeström, P.; Igamberdiev, A.U.; Raghavendra, A.S. Mitochondrial Functions in the Light and Significance to Carbon-Nitrogen Interactions. In *Photosynthetic Nitrogen Assimilation and Associated Carbon and Respiratory Metabolism*; Foyer, C.H., Noctor, G., Eds.; Springer: Dordrecht, The Netherlands, 2002; Chapter 12; pp. 151–172.
70. Cho, Y.; Fernandes, J.; Kim, S.-H.; Walbot, V. Gene-expression profile comparisons distinguish seven organs of maize. *Genome Biol.* **2002**, *3*. [CrossRef]

12

Optimizing Potassium Application for Hybrid Rice (*Oryza sativa* L.) in Coastal Saline Soils of West Bengal, India

Hirak Banerjee [1], Krishnendu Ray [2], Sudarshan Kumar Dutta [3,*], Kaushik Majumdar [4], Talatam Satyanarayana [5] and Jagadish Timsina [6]

1 Bidhan Chandra Krishi Viswavidyalaya, Regional Research Station (CSZ), Kakdwip 743347, South 24 Parganas, West Bengal, India; hirak.bckv@gmail.com
2 Ramakrishna Mission Vivekananda Educational and Research Institute, Sasya Shyamala Krishi Vigyan Kendra, Narendrapur, Kolkata 700103, West Bengal, India; krishnenduray.bckv@gmail.com
3 International Plant Nutrition Institute, South Asia (East India and Bangladesh) Program, 36 Gorakshabasi Road, Kolkata 700028, West Bengal, India
4 International Plant Nutrition Institute, Asia and Africa Programs, Palm Drive, B-1602, Golf Course Extension Road, Sector-66, Gurgaon 122001, Haryana, India; kmajumdar@ipni.net
5 International Plant Nutrition Institute, South Asia Program, 354, Sector 21, Gurgaon 122016, Haryana, India; tsatya@ipni.net
6 Soils and Environment Research Group, Faculty of Veterinary and Agricultural Sciences, University of Melbourne, Melbourne, VIC 3010, Australia; timsinaj@hotmail.com
* Correspondence: sdutta@ipni.net

Abstract: The present study assesses the response of hybrid rice (variety Arize 6444) to potassium (K) application during rainy (wet) seasons of 2016 and 2017 in coastal saline soils of West Bengal, India. The study was conducted at the Regional Research Farm, Bidhan Chandra Krishi Viswavidyalaya, Kakdwip, West Bengal. The soil is clayeywith acidic pH (5.91), saline (Electrical conductivity/EC 1.53 dS $m^{-1})$ and of high K fertility (366 kg ha^{-1}). The experimental plots were laid out in a randomized complete block design with five (5) K treatments (0, 30, 60, 90, and 120 kg K_2O ha^{-1}) with four replications. Plant height, dry matter (DM) in different plant parts, number of tillers, and grain yield were measured in each treatment for the determination of optimum K dose. The study revealed that the stem, leaf, and grain dry matter production at 60 days after transplanting (DAT) and harvest were significantly ($p \leq 0.05$) higher at 90 kg K_2O ha^{-1} application. The number of tillers $hill^{-1}$ was also higher ($p \leq 0.05$) in plants fertilized with 90 kg K_2O ha^{-1} over K omission. At harvest, grain K concentration improved ($p \leq 0.05$) with K fertilization at 90 kg K_2O ha^{-1}, 116% more than the zero-K. Omission of K application from the best treatment (90 kg K_2O ha^{-1}) reduced grain yield by 3.5 t ha^{-1} even though the available K content was high. Potassium uptake restriction due to higher Mg content in the soil may have caused reduced uptake of K leading to yield losses. The present study also showed higher profits with 90 kg K_2O ha^{-1} with higher net returns (US$ 452 ha^{-1}) and benefit:cost ratio (1.75) over other treatments from hybrid rice (var. Arize 6444). From the regression equation, the economic optimum level of K (K_{opt}) was derived as 101.5 kg K_2O ha^{-1} that could improve productivity of hybrid rice during the wet season in coastal saline soils of West Bengal.

Keywords: hybrid rice; K use efficiency; potassium; saline tract

1. Introduction

With growing population and urbanization, the total area under rice cultivation in India as well as in the state of West Bengal is decreasing at a rapid rate. However, the demand for rice in the future is bound to increase with the growing population [1]. Predominantly medium and low lands in the eastern part of India leave no other option for farmers but to go for rice cultivation in the rainy (*khairf*/wet) season. Rice is grown in about 1 million hectares under the coastal ecosystem of West Bengal, accounting for nearly 17% of the net ricearea in the state. The farmers usually grow low yielding, long duration (145–150 days) traditional rice cultivars during the *kharif* (rainy) season, while the landsremainfallow during the *rabi* (winter) season [2].

Salinity is a major yield-reducing factor in thecoastal ecosystems for rice [3]. Rice experiences osmotic stress in saline soils that result in reduced osmotic potential and water stress [4]. Salinity reduces dry matter, grain yield, and the harvest index, and thus affects bothvegetative and reproductive stages of the crop [5]. The productivity of rice is not satisfactory in the coastal areas due to aberrant climatic conditionsand non-availability of hybrid varieties [6]. Water stagnation (medium-deep, 25–50 cm) for most of the crop growing season, flash floods (complete submergence for 1–2 weeks), and water and soil salinity [7] are typical deterrents to high productivity. The South 24-Parganas district ranks 12th amongst all rice growing districts of West Bengal in rice productivity [8]. Considering the demand–supply gap of rice, adoption of superior rice varieties could be effective for narrowing yield gap, and for breaking the yield ceiling in the coastal ecosystem.

Large scale adoption of hybrid rice during wet season in the coastal area could boost rice yield by about 15–20% with the same level of input [6,9]. The hybrid seeds are also tolerant to biotic and abiotic stresses [10]. Along with the improved seeds, a good nutrient management plan is also needed to make the higher production sustainable.

Hybrid rice varieties were already introduced to the region to augment the rice production. However, lack of awareness about the recommended package of practices was the main barrier to its adoption [11]. Consequently, an inadequate nutrient management strategy failed to produce the desired result. The use of nutrient rates suitable for high yielding varieties (HYVs) did not produce the higher yields expected from hybrid rice varieties [12]. Hybrid varieties with higher yield potential require larger quantities of nutrients compared to HYVs. Studies in West Bengal have shown that integrated and adequate use of chemical fertilizers and organic sources of fertilizers can result in improved soil fertility, higher nutrient-use efficiency, and better crop growth, translating into higher yield and profit [13,14].

Potassium is a key nutrient required for optimum yield of hybrid rice. Hybrid rice with high yield potential per unit area and time requires higher amount of K than HYVs. On an average, the crop accumulates 27–36 kg K ha^{-1} to produce a ton of grain, with an equal amount of straw during wet season [15]. A K management strategy based only on attainable yield potential of HYVs are not sufficient to supply the requirement of hybrid rice [9]. Hybrid rice was reported to absorb 79% of total K requirement from soil and remainder from the fertilizer, and utilized 28.1% of the applied K [16]. For improved K use efficiency (KUE) of hybrid rice, appropriate K fertilization strategies should be adopted to fine tune the supply–demand balance of crop and soil. A regular application of K to the crop increased total tillers, dry matter accumulation, effective tillers, number and weight of filled grains and KUE, and enhanced the grain yield besides improving soil properties [17,18]. It was also reported that greater K uptake improved carbohydrate metabolism in plants [18], and adequate K increases the translocation of N to the grain during grain filling period, increasing efficiency of N. Potassium helps to overcome stresses common in this region, and high K uptake in the panicle, especially at the early developmental stage, mitigates the negative effect of sodium uptake by the panicles [3].

For exploiting the full heterotic potential of hybrids, it is necessary to assess the performance of promising rice hybrids at graded levels of K when other management practices are optimum. A quantitative understanding of the crop response to K fertilizer is crucial to optimize K input for higher productivity. Reports on yield performance of superior hybrid rice cultivars under proper K management in the coastal saline soils are limited. The present study evaluates the effects of K fertilization on growth, yield and K-acquisition pattern to estimate appropriate K rates for achieving optimum yield and KUE in hybrid rice cultivation.

2. Materials and Methods

2.1. Field Site

The field study was undertaken during two consecutives rainy (wet) seasons of 2016 and 2017 at the Research Farm of Bidhan Chandra Krishi Viswavidyalaya (BCKV), Kakdwip, South 24-Parganas, West Bengal (22°40′ N latitude, 88°18′ E longitude and 7 m above mean sea level). The maximum and minimum air temperatures fluctuated from 24.9 to 32.1 °C and from 12.3 to 20.4°C during rainy season of 2016, and from 23.2 to 32.9°C and from 12.1 to 19.3 °C during the same period of 2017, respectively. The maximum and minimum air relative humidity was between 85% and 86% (max) and 49% and 63% (min) during rainy season of 2016 and between 87% and 89% (max) and 34% and 58% (min) during rainy season of 2017. The rainfall during the experimental period (July to November) was 474 mm and 654 mm during 2016 and 2017, respectively. The long-term average values of weather parameters showed that average maximum and minimum temperatures fluctuated from 25.3 °C to 35.7 °C and from 10.5 °C to 24.3 °C during rainy season, respectively. The average maximum and minimum relative humidity prevailed between 84% and 85% (max) and 50% and 65% (min) during rainy season, respectively. The long-term average of total rainfall during July to November was 524 mm. Hence, overall weather conditions during the experimental periods were congenial for growth and development of hybrid rice.

Surface soil samples from the field site were collected and analyzed using established procedures mentioned in Table 1.

Table 1. Initial physico-chemical properties of the experimental soil (0–30 cm depth).

Parameter	Values	Methodology	Citation
pH	5.91	Soil–water suspension (1:2.5)	[19]
Electrical conductivity/EC (dS m^{-1})	1.53	Soil–water suspension (1:2.5)	[19]
Sand (%)	16.8		
Silt (%)	28.0	Hydrometer method	[20]
Clay (%)	55.2		
Available N (kg ha^{-1})	155	Hot alkaline permanganate	[21]
Available P (kg ha^{-1})	106	0.5 M $NaHCO_3$	[22]
Available K (kg ha^{-1})	366	Neutral N NH_4OAc	[23]

2.2. Treatment Arrangements and Cultural Practices

The experimental plots were laid out in a randomized complete block design with five rates of K (0, 30, 60, 90, and 120 kg K_2O ha^{-1}) that were replicated four times. The individual plot size was 5 × 5 m. Pre-germinated seeds of hybrid rice var. Arize 6444 with 125 days duration (produced and marketed by Bayer Crop Science Company and notified by Govt. of India) were sown at 20 g m^{-2} of nursery area on the second week of July. In the main field, total 3–4 ploughings were done followed by one laddering. For the next 5–7 days, standing water (5–7 cm) was maintained to control all weeds and full decomposition of stubbles. Then another 2–3 ploughings were given followed by laddering in order to make leveled land. Seedlings of 25 days old were manually transplanted in the first week of August on puddled and leveled land at 1 seedling $hill^{-1}$ with a spacing of 20 × 20 cm. Urea, single super phosphate (SSP) and muriate of potash (MOP) were manually applied as sources of N, P and K, respectively. Based on the soil test recommendations provided by Chinsurah Rice Research Station, Government of West Bengal, 80 kg N ha^{-1} and 40 kg P ha^{-1} were applied in all plots. About 25% of total N, entire amount of P, and 75% of the K as per treatments were applied as basal after draining out the standing water but before final puddling. Rest of the N was top-dressed in threeequal splits, each at an interval of threeweeks, i.e., after transplanting, panicle initiation (PI), and panicle emergence stages. The remaining 25% of the K fertilizer was also applied at the PI stage. Post-emergence application of Bispyribac sodium 10% soluble concentrates/SC (Nominee Gold) at 200 mL ha^{-1} at 15 days after transplanting (DAT) followed by one hand weeding (HW) at 42 DAT were done to promote early crop growth by controlling weeds. Other agronomic management practices were followed based on recommended standards for the coastal region of West Bengal [24] and were applied uniformly across all treatments. The crop was harvested in the first week of November when 80% of the grains in the panicles were ripe, and later dried, winnowed, and weighed for yield estimation.

2.3. Field Measurements

At harvest maturity, plants from 25 hills (1 m^2 area) were harvested and grains were separated to estimate grain yield, while the moisture content of grain was adjusted to 0.15 g H_2O g^{-1}. Growth attributes and yield components were measured from randomly harvested 10 plants. The plants were then partitioned into leaf, stem and panicles. Production of stem, leaf and grain dry matter (DM) was recorded at 60 DAT and harvest. Panicles were hand-threshed, and the filled and unfilled seeds were separated by submerging them into water. Light weight unfilled seeds floating on the water surface were removed, while filled seeds settled down. Filled seeds were further sun-dried to 14% moisture. Dry weight of different plant parts, after oven-drying at 70 °C until constant weight, was determined to estimate above-ground biomass. The grain weight was determined from filled grains per panicle.

2.4. Potassium Determination and Performance Indicators

For K analysis, plant samples (stem, leaf, and grain) were digested with tri-acid mixture (HNO_3:H_2SO_4:$HClO_4$, 9:1:4), and the K concentration (K%) in plant was determined using a flame photometer [25].

Total K uptake was first calculated by the following formula [26].

$$\text{K Uptake}\left(\text{kg ha}^{-1}\right) = \frac{\text{K \%} \times \text{Dry matter}\left(\text{kg ha}^{-1}\right)}{100} \tag{1}$$

To measure the re-translocation efficiency of absorbed K from vegetative plant parts to grain and also to measure K partitioning in plant, K harvest index (KHI) was estimated as per the following formula [27].

$$\text{KHI (\%)} = \frac{\text{Uptake of K in grain}\left(\text{kg ha}^{-1}\right)}{\text{Uptake of K in total above ground biomass}\left(\text{kg ha}^{-1}\right)} \times 100 \tag{2}$$

K mobilization efficiency index (KMEI) of applied K was calculated using the following expressions [28].

$$\text{KMEI (\%)} = \frac{\text{K concentration in grain}\left(\text{mg kg}^{-1}\right)}{\text{K concentration in stem + leaf}\left(\text{mg kg}^{-1}\right)} \times 100 \tag{3}$$

KUE was calculated using following formulae [29].

$$\text{Partial factor productivity of K }(\text{PFP}_\text{K}) = \frac{\text{Grain yield}\left(\text{kg ha}^{-1}\right)}{\text{Applied fertilizer K}\left(\text{kg ha}^{-1}\right)} \tag{4}$$

$$\text{Agronomic efficiency of K }(\text{AE}_\text{K}) = \frac{Y_K - Y_C}{K_a} \tag{5}$$

$$\text{Physiological efficiency of K }(\text{PE}_\text{K}) = \frac{Y_K - Y_C}{U_K - U_C} \tag{6}$$

where, Y and U refer to grain yield and K uptake by hybrid rice and subscripts K and C refer to K fertilized and control plots, respectively. K_a refers to applied fertilizer K. All values are in kg ha^{-1}.

2.4.1. Dose–Response Curve

Dose–response curves were drawn to evaluate the changes in grain yieldwith increasing dose of K fertilization. The curve was drawn by fitting the following quadratic response model to grain yield data

$$y = a + bK + cK^2 \tag{7}$$

where a, b, and c are the regression co-efficient of the quadratic equation; y is the grain yield (kg ha^{-1}); K is the applied fertilizer K (kg ha^{-1}).

The economic optimum for K (K_{opt}) was estimated from the above quadratic regression by using the formula [29]

$$K_{opt}\left(\text{kg ha}^{-1}\right) = \frac{[q - (p \times b)]}{2 \times p \times c} \tag{8}$$

where, q is price per kg of K, and p is price per kg of grain.

2.4.2. Sustainable Yield Index (SYI)

In this study, the SYI for each treatment was calculated from the ratio of minimum assured grain yield to maximum observed yield [28].

$$\text{Sustainable yield index (SYI)} = \frac{(Y_a - \sigma)}{Y_m} \tag{9}$$

where, Y_a = mean yield achieved with the treatment, σ = standard deviation of yield, Y_m = maximum yield achieved with the treatment. The SYI is a quantitative measure to judge sustainability of an agricultural practice [30]. Under SYI concept, low value of σ suggests sustainability of anagricultural system. Conversely, if the σ value is large then SYI will be low indicating unsustainable agricultural practice. The value of SYI varies between zero and unity. The best technology is one where σ is zero and mean = maximum observed yield (Y_m), indicating SYI = 1, hence, the practice gives consistently maximum yield over the years.

2.5. Potassium Balance

The post-harvest pH, EC, and available K contents were assessed for each of the treatments and the extent of change of these values as compared to the initial soil values were estimated. The K balance sheet was prepared according to nutrient balance sheet model [31]. Expected balance was derived from the equation

$$\text{Expected K balance} = \text{Total K available} - \text{Total K uptake} \tag{10}$$

where total K available is the sum of initial K content and K applied through fertilizer. Total K uptake is the sum of K uptake in stem, leaf, and grain at harvest. Actual K balance represents the values derived from post-harvest soil analysis. Finally, net K gain (+) or loss (−) in post-harvest soil was estimated by subtracting actual balance from expected balance.

2.6. Economic Analysis

Common cost of hybrid rice cultivation was derived from summing all expenditure, excluding fertilizer cost. The cost of fertilizer (treatment cost) was estimated on the basis of average retail price in West Bengal. The gross returns from the crop were calculated based on minimum procurement price of the Government of West Bengal. The following equations were used for calculating gross returns, net returns and benefit:cost ratio [28].

$$GR = Y_t \times P_t \quad (11)$$

$$NR = GR - C_p \quad (12)$$

$$BCR = \frac{GR}{C_p} \quad (13)$$

where GR is gross returns (US$ ha^{-1}); NR is net returns (US$ ha^{-1}); BCR is benefit: cost ratio; Y_t is grain yield (t ha^{-1}); P_t is the minimum support price of grain (US$ t^{-1}) and C_p is total cost of production. All economic data were converted into US$ using an exchange rate of 1 US$ = INR 64.56.

2.7. Statistical Analysis

Experimental data were subjected to analysis of variance (ANOVA) as randomized complete block design and the mean values were adjudged by Tukey's HSD (honest significant difference) test method using SPSS (Version 23.0, IBM SPSS Statistics for Windows, IBM Corporation, Armonk, NY, USA) software. Bartlett's chi-square test was performed to test the homogeneity of variance over both the years and pooled values of observations are given to explain the results logically. The Excel software (version 2007, Microsoft Inc., Redmond, WA, USA) was used to draw graphs and figures.

3. Results

3.1. Grain and Straw Yield and Yield Components of Hybrid Rice

Both grain and straw yield of hybrid rice (var. Arize 6444) were significantly ($p \leq 0.05$) influenced by K application levels (Table 2); grain yield increased with the increase in K level from 0 to 90 kg K_2O ha^{-1}. The grain and straw yielddid not vary significantly between 90 kg and 120 kg K_2O ha^{-1}. The grain yield obtained with 90 kg K_2O ha^{-1} was 85% more than the yield obtained with control (zero-K). Figure 1 depicts the regression between applied fertilizer K and grain yield of hybrid rice. Straw yield also increased with increase in K from 0 to 90 kg K_2O ha^{-1}. Significantly ($p \leq 0.05$) higher straw yield was obtained with 90 kg K_2O ha^{-1}, 80.9% more than with K omission. Zero K resulted in the largest reduction in grain and straw yield.

Table 2. Yield components and yield of hybrid rice (var. Arize 6444) as influenced by potassium fertilization (means for 2016 and 2017).

Levels of K (kg ha^{-1})	No. of Panicles Hill^{-1}	No. of Filled Seeds Panicle^{-1}	No. of Unfilled Seeds Panicle^{-1}	Panicle Length (cm)	Panicle Weight (g)	1000-Seed Weight (g)	Grain Yield (t ha^{-1})	Straw Yield (t ha^{-1})
K_0	7 a	115.5 b	47.0 c	25.95 a	2.88 c	22.8 a	4.08 c	4.29 d
K_{30}	8 a	102.5 b	50.5 c	26.60 a	2.97 bc	23.5 a	5.74 b	5.52 c
K_{60}	9 a	143.0 a	58.0 b	27.65 a	3.63 a	23.6 a	7.04 ab	7.02 b
K_{90}	10 a	146.0 a	76.0 a	27.90 a	3.64 a	25.7 a	7.56 a	7.76 a
K_{120}	10 a	117.0 b	70.0 a	27.40 a	3.35 ab	24.5 a	7.49 a	6.97 b
LSD ($p \leq$ 0.05)	NS	14.38	6.35	NS	0.38	NS	1.62	0.61

NS, non-significant; Means followed by a different letter are significantly different at $p \leq 0.05$ by Tukey's HSD (honest significant difference) test; LSD, Least Significant Difference.

3.2. Growth Attributes of Hybrid Rice

The growth attributes (except plant height) of hybrid rice were influenced ($p \leq 0.05$) by different K levels (Table 3). Dry weights of stem and leaf were higher ($p \leq 0.05$) at 90 kg K_2O ha^{-1} compared to the treatments receiving less K, and were statistically at par with 120 kg K_2O ha^{-1}. Similar trend was observed for leaf dry matter at harvest although no significant difference was observed for the stem dry matter. The grain dry weight was also higher ($p \leq 0.05$) with 120 kg K_2O ha^{-1} compared to the lower application rates. The number of tillers hill^{-1} increased significantly ($p \leq 0.05$) with 120 kg K_2O ha^{-1}, accounting for 64.3% more tillers than that with K omission.

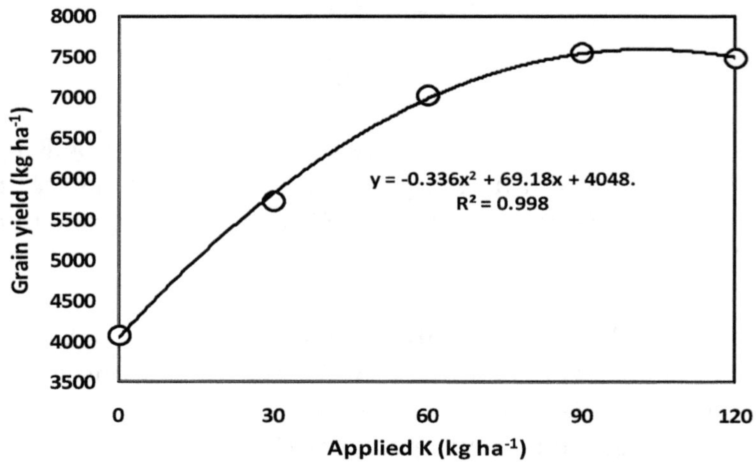

Figure 1. Dose–response curve showing the regression between applied K (kg ha^{-1}) and grain yield (kg ha^{-1}) (Optimum dose of potassium/K_{opt} = 101.5 kg ha^{-1}).

Table 3. Growth attributes of hybrid rice (var. Arize 6444) as influenced by potassium fertilization (means for 2016 and 2017).

Levels of K (kg ha^{-1})	Plant Height (cm)	Dry Weight (g plant^{-1})					No. of Tillers Hill^{-1}
		At 60 DAT		At Harvest			
		Stem	Leaf	Stem	Leaf	Grain	
K_0	84 ^{a}	24.15 ^{bc}	1.38 ^{c}	10.14 ^{a}	23.57 ^{c}	24.80 ^{d}	14 ^{c}
K_{30}	87 ^{a}	17.69 ^{c}	1.25 ^{c}	10.65 ^{a}	26.32 ^{bc}	27.79 ^{cd}	16 ^{bc}
K_{60}	87 ^{a}	27.06 ^{b}	1.88 ^{b}	11.47 ^{a}	27.33 ^{bc}	30.78 ^{bc}	20 ^{ab}
K_{90}	89 ^{a}	32.78 ^{a}	2.71 ^{a}	11.75 ^{a}	30.82 ^{ab}	32.91 ^{ab}	21 ^{a}
K_{120}	89 ^{a}	33.86 ^{a}	2.54 ^{a}	12.88 ^{a}	32.96 ^{a}	33.29 ^{ab}	23 ^{a}
LSD ($p \leq 0.05$)	NS	7.48	0.19	NS	4.43	3.90	4.00

NS, non-significant; DAT, days after transplanting; Means followed by a different letter within a column are significantly different at $p \leq 0.05$ by Tukey's HSD (honest significant difference) test; LSD, Least Significant Difference.

K fertilization significantlyaffected ($p \leq 0.05$) yield components of hybrid rice, such as, number of filled seeds panicle^{-1}, number of unfilled seeds panicle^{-1} and panicle weight, while the effect was non-significant ($p \geq 0.05$) for number of panicles hill^{-1}, panicle length and 1000-seed weight (Table 3). K application at 90 kg K_2O ha^{-1} produced higher number of filled seeds panicle^{-1} accounting for 26.4% more than that with K omission. Panicle weight increased with increase in K dose up to 60 kg ha^{-1}, remained at par at K dose of 90 kg ha^{-1}, and decreased with further increase in K rate. Plants in the control plot produced the lowest panicle weight.

3.3. K Accumulation in Different Plant Parts

The K concentration (%) in stem at peak growth stage, i.e., 60 DAT, increased steadily upto 90 kg K_2O ha^{-1} although the difference is statistically non-significant ($p \geq 0.05$) (Table 4). The leaf K concentration also increased up to 60 kg K_2O ha^{-1}, although insignificantly. A similar trend of K concentration was observed for stem, leaf, and grain at harvest (Table 4). As expected, the K accumulation in grain was lower than in stem or leaf. The relationships between grain yield and stem (R^2 = 0.77) as well as grain K concentration (R^2 = 0.89) were significant ($p \leq 0.05$), while between grain yield and leaf K concentration was non-significant (R^2 = 0.32) (Figure 2).

Table 4. Potassium concentration (%) in different plant parts, mobilization, and utilization by hybrid rice (var. Arize 6444) as influenced by potassium fertilization (means for 2016 and 2017).

Levels of K ($kg\ ha^{-1}$)	Potassium Concentration (%)					KHI (%)	KMEI (%)
	At 60 DAT		At Harvest				
	Stem	Leaf	Stem	Leaf	Grain		
K_0	3.68 ^{a}	2.52 ^{a}	4.48 ^{a}	1.57 ^{ab}	0.043 ^{d}	1.29 ^{c}	0.71 ^{c}
K_{30}	3.73 ^{a}	2.72 ^{a}	4.49 ^{a}	1.27 ^{c}	0.055 ^{cd}	1.87 ^{b}	0.96 ^{bc}
K_{60}	3.83 ^{a}	2.82 ^{a}	4.52 ^{a}	1.58 ^{ab}	0.068 ^{bc}	2.21 ^{b}	1.12 ^{b}
K_{90}	4.02 ^{a}	2.82 ^{a}	4.56 ^{a}	1.70 ^{a}	0.073 ^{b}	2.31 ^{b}	1.18 ^{b}
K_{120}	4.02 ^{a}	2.87 ^{a}	4.62 ^{a}	1.42 ^{bc}	0.093 ^{a}	2.96 ^{a}	1.54 ^{a}
LSD ($p \leq 0.05$)	NS	NS	NS	0.23	0.01	0.59	0.29

NS, non-significant; DAT, days after transplanting; KHI, potassium harvest index; KMEI, potassium mobilization efficiency index; Means followed by a different letter within a column are significantly different at $p \leq 0.05$ by Tukey's HSD (honest significant difference) test; LSD, Least Significant Difference.

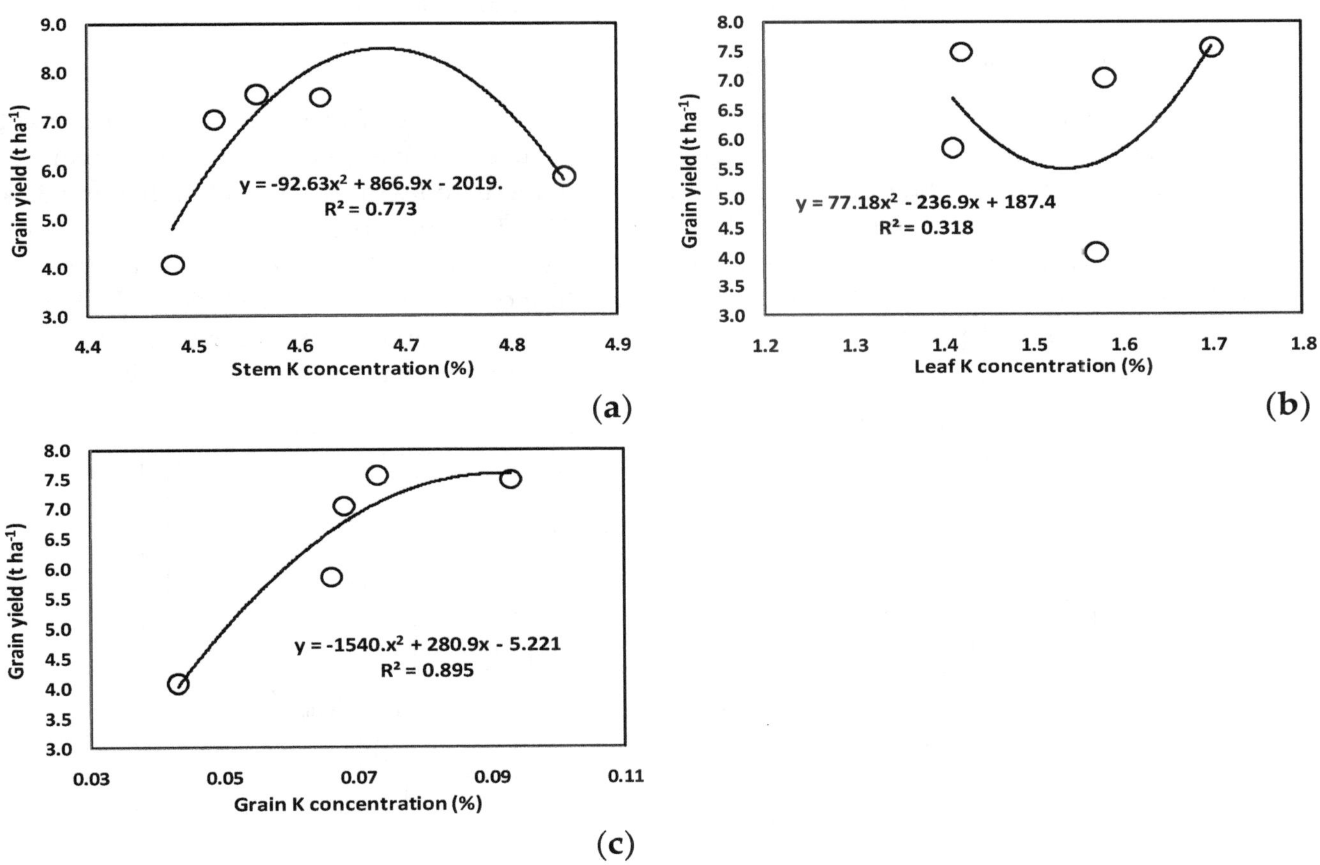

Figure 2. Relationship between grain yield ($t\ ha^{-1}$) and stem (**a**), leaf (**b**), and grain (**c**) K concentration (%).

3.4. *K-Use Efficiency, Sustainable Yield Index, and Partial Factor Productivity*

Potassium harvest index (KHI) in hybrid rice ranged from 1.29 to 2.96% at different K rates (Table 4). Potassium fertilization had a significant ($p \leq 0.05$) influence on KHI of the tested hybrid, and it was maximum with 120 kg K_2O ha^{-1}, accounting for 130% higher KHI over the control. Theapplication rates of 30, 60, and 90 kg K_2O ha^{-1} resulted in statistically at par KHI values. Potassium mobilization efficiency index (KMEI) in the tested hybrid rice cultivar ranged from 0.71 to 1.54% (Table 4), with significant ($p \leq 0.05$) variation over control. Agronomic efficiency and physiological efficiency decreased with every incremental dose of K (Table 5). The KUE indices were highest at 30 kg K ha^{-1} and lowestat the highest K dose (120 kg K_2O ha^{-1}).

Table 5. Potassium use efficiency, sustainable yield index, and partial factor productivity of hybrid rice as influenced by potassium fertilization (means for 2016 and 2017).

Levels of K (kg ha^{-1})	Agronomic Efficiency (kg Grain kg^{-1} K)	Physiological Efficiency (kg Grain kg^{-1} K)	Sustainable Yield Index	Partial Factor Productivity (kg Grain kg^{-1} K)
K_{30}	55.33	48.64	0.83	191.33
K_{60}	49.33	48.14	0.71	117.33
K_{90}	38.66	53.35	0.65	84.00
K_{120}	28.42	53.35	0.64	62.42

Overall trend shows that the SYI varied from 0.64 to 0.83 due to changes in K rates from 30 to 120 kg K_2O ha^{-1} (Table 5). The highest value of SYI (0.83) was obtained with 30 kg K_2O ha^{-1}. Partial factor productivity (PFP_K) decreased as the K ratesincreased from 30 to 120 kg K_2O ha^{-1}. As expected, the highest PFP_K (191.33 kg grain kg^{-1} of K fertilizer) was recorded in plots with 30 kg K_2O ha^{-1} and the lowest (62.42 kg grain kg^{-1} of K-fertilizer) was found at 120 kg K_2O ha^{-1}.

3.5. Post-Harvest Soil Status and K Balance Sheet

Significant changes in pH, EC, and available K in post-harvest soil was observed compared to their initial values (Figure 3). K application significantly reduced the soil pH (7.8 to 12.9% lower than the initial value) (Figure 3a). The EC of the post-harvest soil increased with K application beyond 30 kg K_2O ha^{-1} (6.5 to 34.6% higher than initial value) (Figure 3b). The available K content in post-harvest soil samples improved significantly ($p \leq 0.05$) over the initial value by 24 to 91% when the crop received K fertilization (Figure 3c). For better understanding of K gain or loss, we followed a balance sheet approach (Table 6). A positive balance of K in all plots was observed, irrespective of K rates. The actual net gain (+) of K in post-harvest soil was higher compared to the expected net gain.

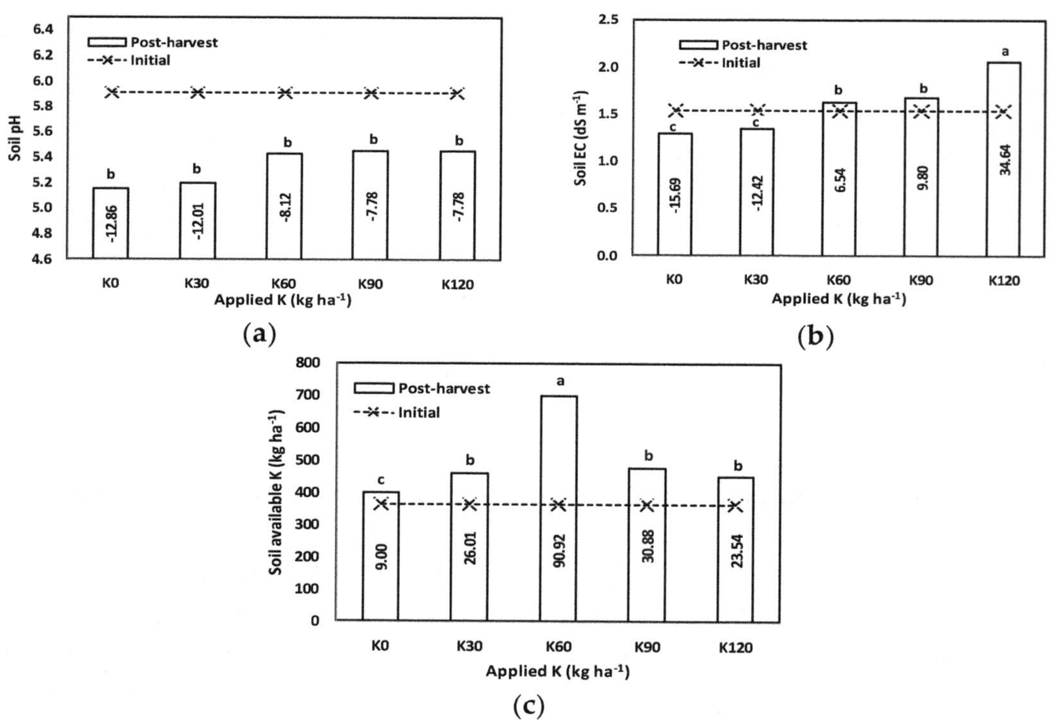

Figure 3. Soil pH (**a**), EC (dS m^{-1}) (**b**), and available K (kg ha^{-1}) (**c**) at different levels of K (kg ha^{-1}) (means followed by same letter are statistically at par, otherwise significantly different at $p \leq 0.05$ by Tukey's HSD (honest significant difference) test; data within bars are percent increase or decrease (−) in any parameter from its initial value).

Table 6. Potassium balance in soil after harvesting of hybrid rice (var. Arize 6444) as influenced by potassium fertilization (means for 2016 and 2017).

Levels of K (kg ha^{-1})	Initial Soil K Status (kg ha^{-1}) (a)	K Added through Fertilizer (kg ha^{-1}) (b)	Total K (kg ha^{-1}) (c = a + b)	Crop K Uptake (kg ha^{-1}) (d)	Expected Balance (kg ha^{-1}) (e = c − d)	Actual K Balance * (kg ha^{-1}) (f)	Net K Gain (+) or Loss (−) (kg ha^{-1}) (f − e)
K_0	343.6	0	343.6	208.7	134.9	398.8	263.9
K_{30}	343.6	30	373.6	243.2	130.4	451.0	320.6
K_{60}	343.6	60	403.6	270.2	133.4	461.0	327.6
K_{90}	343.6	90	433.6	273.9	159.6	498.8	339.2
K_{120}	343.6	120	463.6	272.7	190.9	698.5	507.6

Crop K uptake was calculated by multiplying total plant K concentration (%) with dry matter (kg ha^{-1}); * Represents data on available K content in post-harvest soil.

3.6. Economic Returns

Higher profitability from hybrid rice was achieved with 90 kg K_2O ha^{-1} due to higher net returns (US$ 452 ha^{-1}) and BCRs (1.75) over other treatments (Table 7). Both higher (120 kg ha^{-1}) and lower K rates (30 kg K_2O ha^{-1}) resulted in decline in net return and BCR. Both net return and BCR showed a sharp decline in the K omission treatment.

Table 7. Economic analysis of hybrid rice (var. Arize 6444) cultivation as influenced by potassium fertilization (means for 2016 and 2017).

Levels of K (kg ha^{-1})	Common Cost (US$ ha^{-1})	Treatment Cost * (US$ ha^{-1})	Total Cost (US$ ha^{-1})	Gross Return (US$ ha^{-1})	Net Return (US$ ha^{-1})	BCR
K_0	517	47	564	569	12	1.01
K_{30}	517	60	577	800	416	1.39
K_{60}	517	73	590	981	476	1.66
K_{90}	517	86	603	1054	452	1.75
K_{120}	517	99	616	1044	208	1.69

BCR, benefit:cost ratio; * Treatment cost varies only due to difference in fertilizer levels; 1 US$ = 65.00.

4. Discussion

The cultivation of hybrid rice using K rates applicable to high yielding varieties (HYVs) has failed to realize higher yields under any given conditions [12]. Thus, improved K management (particularly timing and splitting) for superior cultivarsis required to increase yield and KUE of hybrid rice. An uninterrupted supply of K during the entire crop growth period was found to be more effective as it increased growth parameters, yield attributes, yield, and KUE of rice hybrids [17]. The present study quantified the impact of K fertilization on growth, development, and yield of var. Arize 6444. In this study, leaf and grain dry matter (DM) at harvest and number of tillers hill^{-1} were highest at 120 kg K_2O ha^{-1}, followed by 90 kg K_2O ha^{-1}. This finding is supported by a previous study indicating improved growth parameters due to increase in K level [32]. Higher uptake of K in above-ground biomass with high dose of K might have resulted in larger canopy of the hybrid cultivar [33]. Another study has shown that macro and micronutrient uptake increased considerably under higher K application rates. Hence, increased apparent recovery of N, P, S, and Ca with higher K application rates could improve growth of rice hybrids [12]. Other investigators opined that significant ($p \geq 0.05$) increase in total photosynthetic rate and net assimilation rate at heading and maturing stages with high K application might have led to greater DM production [34]. On the other hand, plants in control plots had lower DM accumulation. This result is in conformity with the findings of other investigators who observed increased total tillers, dry matter accumulation, and effective tillers with 75 kg K_2O ha^{-1} [17].

We observed significant ($p \leq 0.05$) beneficial effect of K fertilization on yield components of hybrid rice, mainly on number of filled and unfilled grains panicle^{-1} and panicle weight, and K application at 90 kg ha^{-1} had the best result. In line with our results, other investigators found that high levels of K significantly increased ($p \geq 0.05$) panicle weight due to higher number of filled grains panicle^{-1} [35,36]. Potassium appeared to help enhance grain weight by delaying abscisic acid (ABA) peak by fourto fivedays, thereby delaying maturation and increasing carbohydrate translocation to the seeds [37]. The same study also reported that K application markedly reduced the number of unfilled spikelets due to its promoting effect on cytokinin synthesis, which resulted in less zygote degeneration.

The present study highlights that the number of filled grains panicle^{-1} and panicle weight were strongly associated with grain yield. This finding again confirms that productivity of crop is collectively determined by vegetative growth and its yield components. Better vegetative growth coupled with higher yield components might have resulted in higher grain and straw yield of hybrid rice [9,13,14,38]. In addition, limited supply of assimilates to the grains might be responsible for the reduced panicle weight at lower doses of K. The panicle length and test weight had non-significant contribution to grain yield. Several investigators identified the number of grains panicle^{-1} as the key yield components for higher grain yield in hybrids, rather than panicle number and panicle length [39]. Similar to the present study, other researchers also did not find any relation between test weight and grain yield of hybrid rice grown in the wet season [1].

The regression between K and grain yield indicated that the relationship was highly significant in the observed quadratic relationship (Figure 1). From the regression equation, the economic optimum level of K (K_{opt}) was derived as 101.5 kg K_2O ha^{-1}. Other investigators also used a similar approach for determining K_{opt} and found 80 kg K_2O ha^{-1} as the best dose for hybrid rice in summer season at red and lateritic soils of Orissa [40]. Results are in accordance with earlier reports [17], wherein higher yield response of hybrid rice to K nutrition was found at 75 kg ha^{-1}. However, in the salt affected soil the K requirement is higher as the soil salinity caused a reduction in growth and yield of rice and higher K is required for alleviation [41]. A study in a coastal saline soil of Bangladesh reported that the required K dose is more than 100 kg ha^{-1} to achieve hybrid rice yield of 4 t ha^{-1} [41], which is similar to the present study.

The K content of rice plant is an important index for obtaining higher grain yield of hybrid rice [42]. The regression study with grain yield as a dependent variable showed that the stem (R^2 = 0.77) and grain K concentrations (R^2 = 0.89) had positive significant relationship with yield. Strategies to increase stem and grain K concentration can thereby enhance grain yield. However, leaf K concentration was not a good indicator of the response of hybrid rice to K fertilization. Information regarding increase of both grain yield and KUE, as well as their relationship with K accumulation and utilization characteristics is limited. Only few earlier studies revealed higher K accumulation during the period from primary branch initiation to spikelet initiation [43]. About 45–67% of total K accumulation was observed from panicle initiation (PI) to heading stage. In addition, 50% of the K in leaf sheath might had been translocated to spikelets after heading which accounted for two-thirds of grain K concentration [43]. In the present study, Arize 6444 had high K accumulation both in vegetative (particularly leaf) and reproductive part (grain) at harvest with 90 and 120 kg K_2O ha^{-1} that might have contributed significantly to high yield. Additionally, the higher K accumulation in its stems both at 60 DAT and harvest helped increase lodging resistance [44]. The present work has also shown that K omission treatment decreased K uptake, transfer and efficiency.

The KUE indices (agronomic efficiency and physiological efficiency) for the tested hybrid cultivar decreased with the increase in K from 30 to 120 kg ha^{-1}, being maximum and minimum at 30 and 120 kg K_2O ha^{-1}, respectively. There was not much difference in SYI of hybrid rice due to variation in K fertilization. The fertilizer productivity, estimated as PFP, decreased with the increasing level of K fertilization. This indicates poor rate of K utilization at higher application rates. This might be due to the curvilinear return to the conversion of K to seed as yield approaches the ceiling at higher K-levels [24,45]. The fertilizer productivity decreased to a greater degree passing from 30 to 60 kg K_2O ha^{-1} (PFP 117.33 kg grain kg^{-1} of K-fertilizer) than passing from 60 to 90 (PFP 84.00 kg grain kg^{-1} K) or 90 to 120 (PFP 62.42 kg grain kg^{-1} K). The results point to the advantage of using an optimum K rate that creates the right balance between efficiency and effectivity of applied K fertilizer.

The K balance in the post-harvest soil was positive for all the treatments. Interestingly, the net K balance was positive even for the K omission plots. Thepositive balance was driven by the high initial K level of the experimental soil (366 kg K_2O ha^{-1}). A recent study [46] from the same area revealed that the soils of the area are rich in illitic clays and are expected to release non-exchangeable K (NEK)effectively when under stress. The same study, through sequential extraction with boiling 1M HNO_3, showed that the NEK release could be triggered at a relatively high activity coefficient (AR^K) of K [46]. Such NEK release could augment the already high native K level in the soil and may explain the positive net K balance observed in our study. Crop response to K fertilizer application in soils with such high initial K level is not expected. However, every incremental dose of K over the zero-K treatment produced significant yield increases in this study, and the best yield was achieved at 90 kg K_2O ha^{-1}. This suggest that hybrid rice was unable to use the native K till K fertilizer application increased the activity of K to a sufficiently high level. Such K uptake restriction may happen when soils have high magnesium concentration relative to K concentration. The Mg-induced K deficiency in crops is well documented and can occur in soils with an available K/Mg concentration below 0.32 [47]. A recent assessment of $K^+/(Ca^{2+} + Mg^{2+})$ ratios in soils from two similar locations nearby our experimental sites indicated values less than 0.17 [46]. Such low values of the said ratio in our study area may create restriction to K uptake leading to hybrid rice yield losses when K fertilizer was not applied.

Application of 90 kg K_2O ha^{-1} has been found to be economically effective, mainly because of increased grain yield realized at this K application rate. However, the economic optimum level of K_{opt} was derived from regression study as 101.5 kg K_2O ha^{-1}. On the contrary, the lowest net returns and BCR were found in K-omitted plots because of lower grain yield, resulting from a smaller number of filled grains $panicle^{-1}$, short length of panicle and reduced panicle weight. This result corroborates the earlier studies that hybrid rice cultivars responded to K application appreciably, although economic efficiency of K declined at higher levels of application [9,32,33].

5. Conclusions

In the main rice growing areas of the coastal saline belt of West Bengal (India), productivity, and profitability of hybrid rice (var. Arize 6444) could be increased through site-specific K management practice. Our research confirms that adequate K application is necessary to obtain high grain yield in hybrid rice even though the soils of the coastal areas are high in illitic clay and high available K content. This explains why a K management strategy appropriate for HYVs may not work for hybrid rice. Our results also confirm that the present recommended level (40 kg K_2O ha^{-1}) is inadequate to support attainable yield potentials of hybrid rice. A rate of 101.5 kg K_2O ha^{-1} (through muriate of potash/MOP) may be recommended to achieve higher productivity of hybrid rice during wet season in coastal region of West Bengal (India).

Author Contributions: H.B. and S.K.D. conceived and designed the experiment; H.B. and S.K.D. conducted the experiments; H.B. recorded the data; H.B. and K.R. analyzed the data; H.B. and K.R. wrote the paper; S.K.D., K.M., T.S., and J.T. reviewed, edited, and approved the final manuscript.

Acknowledgments: Authors pay special thanks to Bidhan Chandra KrishiViswavidyalaya (BCKV), India, for giving the opportunity to conduct field experiment in their Research Farm. We are grateful to Pravat Kumar Maity, Ajay Kar, and GurupadaKhanra for managing the field experiments at Kakdwip, West Bengal. We also thank BiswanathHalder who carried out the chemical analysis for plant and soil samples at BCKV.

References

1. Banerjee, H.; Pal, S. Effect of planting geometry and different levels of nitrogen on hybrid rice. *Oryza* **2011**, *48*, 274–275.
2. Banerjee, H.; Chatterjee, S.; Sarkar, S.; Gantait, S.; Samanta, S. Evaluation of rapeseed-mustard cultivars under late sown condition in coastal ecosystem of West Bengal. *J. Appl. Natl. Sci.* **2017**, *9*, 940–949. [CrossRef]
3. Asch, F.; Dingkuhn, M.; Wittstock, C.; Doerffling, K. Sodium and potassium uptake of rice panicles as affected by salinity andseason in relation to yield and yield components. *Plant Soil* **1999**, *207*, 133–145. [CrossRef]

4. Castillo, E.G.; Tuong, T.P.; Ismail, A.M.; Inubushi, K. Response to Salinity in Rice: Comparative Effects of Osmotic and Ionic Stresses. *Plant Prod. Sci.* **2007**, *10*, 159–170. [CrossRef]
5. Aslam, M.; Qureshi, R.H.; Ahmad, N. A rapid screening technique for salt tolerance in rice (*Oryza sativa* L.). *Plant Soil* **1993**, *150*, 99–107. [CrossRef]
6. Mondal, P.; Pal, S.; Alipatra, A.; Mandal, J.; Banerjee, H. Comparative study on growth and yield of promising rice cultivars during wet and dry season. *Plant Arch.* **2012**, *12*, 659–662.
7. Islam, M.R.; Sarker, M.R.A.; Sharma, N.; Rahman, M.A.; Collard, B.C.Y.; Gregorio, G.B.; Ismail, A.M. Assessment of adaptability of recently released salt tolerant rice varieties in coastal regions of South Bangladesh. *Field Crop. Res.* **2016**, *190*, 34–43. [CrossRef]
8. Adhikari, B.; Bag, M.K.; Bhowmick, M.K.; Kundu, C. Status Paper on Rice in West Bengal. In *Rice Knowledge Management Portal (RKMP)*; Directorate of Rice Research: Hyderabad, India, 2011.
9. Banerjee, H.; Pal, S. Response of hybrid rice to nutrient management during wet season. *Oryza* **2012**, *49*, 108–111.
10. Khandkar, V.; Gandhi, V.P. Post-adoption experience of hybrid rice in India: Farmers' satisfaction and willingness to grow. *Agric. Econ. Res. Rev.* **2018**, *31*, 95–104. [CrossRef]
11. Singh, A.; Kumar, B.; Baghel, R.; Singh, R. Sustainability of hybrid rice technology vis a vis inbred rice in Uttar Pradesh. *Ind. Res. J. Ext. Educ.* **2009**, *9*, 22–25.
12. Pattanayak, S.K.; Mukhi, S.K.; Majumdar, K. Potassium unlocks the potential for hybrid rice. *Better Crop.* **2008**, *92*, 8–9.
13. Mondal, S.; Mallikarjun, M.; Ghosh, M.; Ghosh, D.C.; Timsina, J. Effect of Integrated Nutrient Management on Growth and Productivity of Hybrid Rice. *J. Agric. Sci. Technol.* **2015**, 297–308.
14. Mondal, S.; Mallikarjun, M.; Ghosh, M.; Ghosh, D.C.; Timsina, J. Influence of integrated nutrient management (INM) on nutrient use efficiency, soil fertility and productivity of hybrid rice. *Arch. Agron. Soil Sci.* **2016**. [CrossRef]
15. Fageria, N.K.; Sant'ana, E.P.; Morais, O.P.; de Morais, O.P. Response of promising upland rice genotypes to soil fertility. *Pesqui. Agropecu. Bras.* **1995**, *30*, 1155–1161.
16. Shi, H.; Zhang, X.H.; Pan, X.H.; Guo, J.Y.; Zhang, P. Studies on the characteristics of potassium absorption and yield formation on early hybrid rice. *Acta Agric. Univ. JiangXiensis* **1990**, *12*, 54–59.
17. Meena, S.L.; Singh, S.; Shivay, Y.S. Response of hybrid rice (*Oryza sativa*) to nitrogen and potassium application. *Ind. J. Agron.* **2002**, *47*, 207–211.
18. Prasad, K.; Chauhan, R.P.S. Efficiency of potassium under rainfed lowland rice. *Ann. Plant Soil Res.* **1999**, *1*, 52–57.
19. Jackson, M.L. *Soil Chemical Analysis*; Prentice Hall of India Pvt. Ltd.: New Delhi, India, 1967.
20. Dewis, J.; Freitas, F. Physical and Chemical Methods of Soil and Water Analysis. In *Soil Bulletin*; FAO: Rome, Italy, 1984.
21. Subbiah, B.; Asija, G.L. A rapid procedure for the estimation of available N in soils. *Curr. Sci.* **1956**, *25*, 259–260.
22. Olsen, S.R.; Cole, C.V.; Watanale, F.S.; Dean, L.A. *Estimation of Available Phosphorus in Phosphorus in Soils by Extraction with Sodium Bicarbonate*; United States Department of Agriculture: Washington, DC, USA, 1954.
23. Hanway, J.J.; Heidel, H. Soil analysis methods as used in Iowa State College Soil Testing Laboratory. *Iowa Agric.* **1952**, *57*, 1–13.
24. Sarangi, S.K.; Maji, B.; Singh, S.; Sharma, D.K.; Burman, D.; Mandal, S.; Haefele, S.M. Using improved variety and management enhances rice productivity in stagnant flood-affected tropical coastal zones. *Field Crop. Res.* **2016**, *190*, 70–81. [CrossRef]
25. Jackson, M.L. *Soil Chemical Analysis*; Prentice Hall of India Pvt. Ltd.: New Delhi, India, 1973; pp. 38–56.
26. Sharma, N.K.; Singh, R.J.; Kumar, K. Dry matter accumulation and nutrient uptake by wheat (*Triticumaestivum* L.) under poplar (*Populusdeltoides*) based agroforestry system. *ISRN Agron.* **2012**, *2012*, 359673.
27. Fageria, N.K. Nitrogen harvest index and its association with crop yields. *J. Plant Nutr.* **2014**, *37*, 795–810. [CrossRef]
28. Sarkar, S.; Banerjee, H.; Ray, K.; Ghosh, D. Boron fertilization effects in processing grade potato on an inceptisol of West Bengal, India. *J. Plant Nutr.* **2018**, *41*, 1456–1470. [CrossRef]

29. Mozumder, M.; Banerjee, H.; Ray, K.; Paul, T. Evaluation of potato (*Solanumtuberosum*) cultivars for productivity, N requirement and eco-friendly indices under different nitrogen levels. *Ind. J. Agron.* **2014**, *59*, 327–335.
30. Vittal, K.P.R.; MaruthiSankar, G.R.; Singh, H.P.; Samra, J.S. Sustainability of Practices of Dryland Agriculture: Methodology and Assessment. Available online: http://www.crida.in/AICRPDA/Sustainability.pdf (accessed on 4 October 2018).
31. Rana, R.; Banerjee, H.; Dutta, S.K.; Ray, K.; Majumdar, K.; Sarkar, S. Management practices of macronutrients for potato for smallholder farming system at alluvial soil (Entisols) of India. *Arch. Agron. Soil Sci.* **2017**, *63*, 1963–1976. [CrossRef]
32. Maiti, S.; Saha, M.; Banerjee, H.; Pal, S. Integrated nutrient management under hybrid rice—Hybrid rice (*Oryza sativa*) cropping sequence. *Ind. J. Agron.* **2006**, *51*, 157–159.
33. Banerjee, H.; Pal, S. Integrated nutrient management for rice-rice cropping system. *Oryza* **2009**, *46*, 32–36.
34. Fan, M.; Ge, D.Z. Potassium nutrition in hybrid rice. *Int. Rice Res. Newsl.* **1987**, *12*, 21.
35. Thakur, R.B.; Pandey, S.K.; Singh, H. Contribution of production factors on yield of midland rice. *Oryza* **1994**, *31*, 271–293.
36. Bhowmick, N.; Nayak, R.L. Response of hybrid rice (*Oryza sativa* L.) varieties to nitrogen, phosphorus and potassium fertilizers during dry (*boro*) season in West Bengal. *Ind. J. Agron.* **2000**, *45*, 323–326.
37. Yuan, L.; Huang, J.G. Effects of potassium on the variation of plant hormones in developing seeds of hybrid rice. *J. South West Agric. Univ.* **1993**, *15*, 38–41.
38. Dwivedi, A.P.; Dixit, R.S.; Singh, G.R. Effect of nitrogen, phosphorus and potassium levels on growth, yield and quality of hybrid rice (*Oryza sativa* L.). *Oryza* **2006**, *43*, 64–66.
39. Om, H.; Katyal, S.K.; Dhiman, S.D. Effect of time of transplanting and rice (*Oryza sativa*) hybrids on growth and yield. *Ind. J. Agron.* **1997**, *42*, 261–264.
40. Das, L.K.; Panda, S.C. Economics of hybrid rice cultivation as influenced by nitrogen and potassium levels. *J. Res. Orissa Univ. Agric. Technol.* **2002**, *20*, 50–52.
41. Kibria, M.G.; Farhad; Hoque, M.A. Alleviation of soil salinity in rice by potassium and zinc fertilization. *Int. J. Exp. Agric.* **2015**, *5*, 15–21.
42. Li, M.; Zhang, H.; Yang, X.; Ge, M.; Ma, Q.; Wei, H.; Dai, Q.; Huo, Z.; Xu, K.; Luo, D. Accumulation, and utilization of nitrogen, phosphorus and potassium of irrigated rice cultivars with high productivities and high N use efficiencies. *Field Crop. Res.* **2014**, *161*, 55–63. [CrossRef]
43. Huang, Y.M.; Li, Y.Z.; Zheng, J.S.; Zhung, Z.L. Studies on the NPK accumulation and transfer of high yielding colony of hybrid rice. *J. Fujian Acad. Agric. Sci.* **1997**, *12*, 1–6.
44. Yang, C.; Yang, L.; Yan, T.; Ou-Yang, Z. Effects of nutrient and water regimes on lodging resistance of rice. *Chin. J. Appl. Ecol.* **2004**, *15*, 646–650. (In Chinese with English abstract)
45. Premi, O.P.; Rathore, S.S.; Shekhawat, K.; Kandpal, B.K.; Chauhan, J.S. Sustainability of fallow-Indian mustard (*Brassica juncea*) system as influenced by green manure, mustard straw cycling and fertilizer application. *Ind. J. Agron.* **2012**, *57*, 229–234.
46. Sarkar, G.K.; Chattopadhyay, A.P.; Sanyal, S.K. Release pattern of non-exchangeable potassium reserves in Alfisols, Inceptisols and Entisols of West Bengal, India. *Geoderma* **2013**, *207–208*, 8–14. [CrossRef]
47. Hannan, J.M. Potassium-Magnesium Antagonism in High Magnesium Vineyard Soils. Available online: https://lib.dr.iastate.edu/etd/12096 (accessed on 24 May 2018).

13

Can Organic Sources of Nutrients Increase Crop Yields to Meet Global Food Demand?

Jagadish Timsina [1,2]

[1] Soil and Environment Research Group, Faculty of Veterinary & Agricultural Sciences, University of Melbourne, Victoria 3010, Australia; jtimsina@unimelb.edu.au

[2] Agriculture & Forestry University, Rampur, Chitwan 44209, Nepal

Abstract: Meeting global demand of safe and healthy food for the ever-increasing population now and into the future is currently a crucial challenge. Increasing crop production by preserving environment and mitigating climate change should thus be the main goal of today's agriculture. Conventional farming is characterized by use of high-yielding varieties, irrigation water, chemical fertilizers and synthetic pesticides to increase yields. However, due to either over- or misuse of chemical fertilizers or pesticides in many agro-ecosystems, such farming is often blamed for land degradation and environmental pollution and for adversely affecting the health of humans, plants, animals and aquatic ecosystems. Of all inputs required for increased agricultural production, nutrients are considered to be the most important ones. Organic farming, with use of organic sources of nutrients, is proposed as a sustainable strategy for producing safe, healthy and cheaper food and for restoring soil fertility and mitigating climate change. However, there are several myths and controversies surrounding the use of organic versus inorganic sources of nutrients. The objectives of this paper are: (i) to clarify some of the myths or misconceptions about organic versus inorganic sources of nutrients and (ii) to propose alternative solutions to increase on-farm biomass production for use as organic inputs for improving soil fertility and increasing crop yields. Common myths identified by this review include that organic materials/fertilizers can: (i) supply all required macro- and micro-nutrients for plants; (ii) improve physical, chemical and microbiological properties of soils; (iii) be applied universally on all soils; (iv) always produce quality products; (v) be cheaper and affordable; and (vi) build-up of large amount of soil organic matter. Other related myths are: "legumes can use entire amount of N_2 fixed from atmosphere" and "bio-fertilizers increase nutrient content of soil." Common myths regarding chemical fertilizers are that they: (i) are not easily available and affordable, (ii) degrade land, (iii) pollute environment and (iv) adversely affect health of humans, animals and agro-ecosystems. The review reveals that, except in some cases where higher yields (and higher profits) can be found from organic farming, their yields are generally 20–50% lower than that from conventional farming. The paper demonstrates that considering the current organic sources of nutrients in the developing countries, organic nutrients alone are not enough to increase crop yields to meet global food demand and that nutrients from inorganic and organic sources should preferably be applied at 75:25 ratio. The review identifies a new and alternative concept of Evergreen Agriculture (an extension of Agroforestry System), which has potential to supply organic nutrients in much higher amounts, improve on-farm soil fertility and meet nutrient demand of high-yielding crops, sequester carbon and mitigate greenhouse gas emissions, provide fodder for livestock and fuelwood for farmers and has potential to meet global food demand. Evergreen Agriculture has been widely adapted by tens of millions of farmers in several African countries and the review proposes for evaluation and scaling-up of such technology in Asian and Latin American countries too.

Keywords: organic farming; conventional farming; organic nutrients; chemical fertilizers; global food demand; agroforestry system; evergreen agriculture

---

1. Introduction

Providing enough, safe and healthy food to their citizens by avoiding environmental degradation under current and the projected climate change are the most important issues that all countries are facing in the world. Global food production increased by 70% from 1970 to 1995 in developing countries, largely due to the green revolution technologies (also called conventional agriculture) which uses high-yielding inputs such as improved and high-yielding varieties (HYVs), irrigation, chemical fertilizers and synthetic pesticides [1,2] and the production has been increasing after that period too. As per FAO's revised projection, global food production should be 60% higher in 2050 than in 2005/2007 to feed the projected global population of 10 billion [3,4]. To close this gap, total crop production needs to be increased even more from 2006 to 2050 than it did in the same number of years from 1962 to 2006 [5]. Though in the past green revolution technologies have increased crop yields and produced food to meet caloric requirements of the global population [6], there are also increasing concerns about the environmental costs, such as increased soil erosion, surface and groundwater contamination, greenhouse gas emissions, increased pest resistance and reduced biodiversity and so forth, with use of such technologies [7,8]. These concerns suggest that more sustainable methods of food production are essential to meet the food requirements of ever-increasing population now and into the future but at the same time such methods must maintain natural resource base by avoiding land degradation and mitigating climate change. The challenge now is to fine-tune the existing technologies or develop alternative technologies that can increase crop yields to meet global food demand of increasing population but without compromising with the natural resources or the environment.

Over the past 2–3 decades organic agriculture has been advocated as an alternative form of farming to produce food sustainably by reducing the impact of agriculture on the environment [9–12]. All these authors believe that a widespread shift from conventional to organic farming could feed the world with safe and healthy food now and into the future and also could avoid environmental degradation. Their claims however have been widely criticized by many authors [13–18], as they all argue that organic farming without the use of synthetic fertilizers, pesticides, or genetically-engineered crops simply cannot feed the projected 10 billion people for 2050 and that extra lands and water would be required for organic farming to produce similar amount of food to that from conventional farming. Kirchmann et al. [2,19,20] warned that expansion of areas for organic farming into forests or natural lands to feed the projected global population would lead to loss of biodiversity or natural habitats, increase of greenhouse gas emissions and depletion of ecosystem services. Ammann [21] and Ronald and Adamchak [22] however proposed a mid-way or a balanced view suggesting that a combination of high technology and organic techniques (i.e., a hybrid of organic and conventional farming) may provide more realistic and sustainable solutions.

Although many production factors (nutrients, water, pest and diseases, labour, prices of inputs and outputs, etc.) contribute to crop yield, it seems from various debates and arguments surrounding the use of conventional and organic farming that availability of required amounts of plant nutrients and the practicality of their use to produce enough food to feed 10 billion people remain the central issues of all these debates. While role of nutrients, whether organic or inorganic, for increased crop production is universally and unequivocally recognised, there seem to be several myths or misconceptions of using organic farming and/or organic sources of nutrients. Some sectors of the society, particularly those activists or advocates influenced by International Non-governmental Organizations (INGOs) or Non-governmental Organizations (NGOs), some researchers and extensions workers and even the government policy makers in many developing countries claim that use of the chemical fertilizers adversely affects soil quality and decreases the soil and crop productivity, whereas the use of organic farming or organic nutrients unquestionably and universally increases soil and crop productivity. These claims however have very little scientific basis and any decline in soil or crop productivity due to the use of chemical fertilizers could be due to their either over- or misuse.

This paper focuses on the discussion on nutrient sources for crops grown under organic and conventional farming and tries to argue whether existing sources of organic materials can supply enough nutrients to increase crop yields so as to meet the food demand of the growing population. The specific objectives are: (i) to clarify some of the myths or misconceptions regarding organic nutrients/fertilizers and chemical fertilizers by providing scientific facts and realities so that the applications of appropriate amounts of inorganic or organic fertilizers either alone or in their combination can be advised to farmers and (ii) to propose alternative solutions to increase on-farm biomass production for use as organic inputs for maintaining or improving on-farm soil fertility and increasing crop yields. Such clarifications and alternative solutions could help planners and policy makers of any country to develop policies and programs to promote for the rationale use of inorganic and/or organic nutrient inputs to achieve food security and get rid of poverty. The paper is organised into the following sections: (i) Differences between organic and conventional agriculture (iii) Sources of inorganic and organic nutrients for crops (ii) Myths and realities of use of organic materials/fertilizers and chemical fertilizers (iii) Nutrient requirements and supply for organically- and conventionally-grown crops (iv) Need for site-specific nutrient management (v) Alternative approaches to increase on-farm soil fertility and nutrient supply (vi) Conclusions and research and policy implications.

2. Organic Agriculture: Concepts, Principles and Global Performance against Conventional Agriculture

FAO has defined organic agriculture as a holistic production management system which promotes and enhances agro-ecosystem health, including biodiversity, biological cycles and soil biological activity. It emphasizes the use of management practices in preference to the use of off-farm inputs, considering that regional conditions require locally adapted systems (http://www.fao.org/docrep/meeting/x0075e.html). This is accomplished by using, where possible, agronomic, biological and mechanical methods, as opposed to using synthetic materials, to fulfil any specific function within the system. International Federation of Organic Agriculture Movement (IFOAM) has defined organic agriculture as a production system that sustains the health of soils, ecosystems and people and relies on ecological processes, biodiversity and cycles adapted to local conditions, rather than the use of external inputs with adverse effects and such agriculture combines tradition, innovation and science to benefit the shared environment and promote fair relationships and a good quality of life for all involved [23]. IFOAM stresses that organic agriculture is based on the principles of health, ecology, fairness and care. It can sustain and enhance the health of soil, plant, animal, human and planet; sustain living ecological systems and cycles; build on relationships that ensure fairness about the common environment and life opportunities; and manage in a precautionary and responsible manner to protect the health and well-being of current and future generations and the environment. National Research Council (NRC) from USA [24] has also identified organic methods as one of several innovative systems that can meet production, environmental and socio-economic objectives and sustainability goals. Despite the heavy emphasis on, and importance given to organic agriculture by various national and international entities it is now practised only on about 50.9 Mha (1.1% of total agricultural land) by 2.4 million farmers globally, with about 87 countries having some sort of organic regulations or certifications [25].

Organic farming is a form of agriculture that deliberately follows a set of management practices, which exclude the use of chemical fertilizers and other chemical inputs such as synthetic pesticides and genetically-engineered crops. Organic agriculture uses organic materials to supply nutrients and to control pests and diseases. There are contrasting results from both developed and developing countries regarding the performance of organic versus conventional agriculture. In one of the earliest reviews, Stanhill [26], using data from developed countries and mostly from prior to 1985, reported 9% lower yield for organic crops compared to conventional ones. Penning de Vries et al. [27], based on results from a crop simulation model, concluded that organic agriculture can only produce enough food to feed 9 billion people assuming moderate amounts of diet and with animal proteins. Lotter [28] reported that large scale conversion from conventional to organic agriculture is possible only if meat consumption is reduced. Subsequently, Badgley et al. [9], comparing yield data between organic and conventional agriculture from 293 studies reported about 8% lower yield for organic agriculture. Kirchmann et al. [19,20], however reported that organic crop yields are 25–50% lower than conventional ones, which were mainly attributed to lower nutrient availability, poorer weed control and limited possibility to improve the soil nutrient status.

Ponti et al. [29], using a meta-dataset of 362 studies globally, concluded that individual yields of organic crops, on an average, were 20% lower than conventional crops. In their study, the organic yield gaps varied significantly between crop groups and regions and the gaps increased as yields of conventional crops increased. The results of Ponti et al. however are not surprising because potential yield of any crop is climate-derived and not limited by water or nutrients, with pests and diseases fully controlled [30]. Yield potential of any crop can be increased with increasing amounts of nutrients and water and by fully controlling weeds, pests and diseases so that crop will not be stressed by any biotic and abiotic factors [30–32]. On the other hand, even the well-managed organic crops generally do not receive adequate amounts of nutrients and pests and diseases are not fully controlled. Also from a meta-analysis of 316 yield comparisons in 66 studies, Seufert et al. [12] concluded that organic crops in developed countries yielded 20% lower but when developed and developing countries were combined, they yielded 25% lower than their conventional counterparts. However, they also found that for certain crops and for certain growing conditions and management practices, yields of organic crops matched their conventional counterparts. Results from Badgely et al. [9], Ponti et al. [29] and Seufert et al. [12] suggest that adoption of organic agriculture under conditions in which it performs well might close the yield gap between organic and conventional crops.

The results and conclusions of these three studies [9,12,29] were heavily disputed by Cassman [33], Connor [16], Goulding et al. [15] and Dobermann [34], who all argued that their yield data and assumptions made on nutrient availability from organic sources were quite unrealistic. Connor [16] and Dobermann [34] questioned about the analysis methods of Badgely et al. owing to their reliance on yield ratios as in many cases they represented large differences in crop management. Due to flaws and criteria for design and evaluation of comparisons between organic and conventional agriculture, Kirchmann et al. [2] proposed three stringency criteria to ensure scientific quality of data: requirements of similar initial soil fertility, comparable crop production type and quantification of off-farm organic nutrient inputs. Based on the review of above studies, crop yields from organic farming are generally about 20–50% lower although in some cases their yields and economic returns are higher than from conventional farming.

3. Sources of Organic Nutrients

Several terms (e.g., organic farming, natural farming, alternative farming, regenerative farming, low-input agriculture, sustainable agriculture, etc.) are used in literature, sometime interchangeably, to describe organic farming. Likewise, many other terms (e.g., organic materials, organic fertilizers, organic nutrients and bio-fertilizers, etc.) are used to describe sources of nutrients. Crop plants require nutrients derived from organic as well as inorganic sources. Common inorganic sources of nutrients include fossil-fuel derived chemical fertilizers while organic sources include decomposed or undecomposed plant and animal materials. Many different types of chemical fertilizers in various formulations are available in markets in all countries. Chemical fertilizers, especially the nitrogenous fertilizers, are prone to losses from soil-plant systems. Hence, smart and innovative fertilizers such as controlled-release or slow-release fertilizers (e.g., poly-coated urea), deep placement (e.g., urea super granules), or nitrification inhibitors are being developed and used to reduce losses and increase the efficiency of fertilizers [35]. Precision nutrient management such as site-specific nutrient management (SSNM) can help reduce and/or optimize the fertilizer use considering the field, soil or site history and characteristics and resulting nutrient needs. This will be discussed in Section 6.

Some of the organic materials used as organic sources of fertilizers in Asia include (i) agricultural wastes such as crop residues (including rice and wheat straw, maize stover, legume leaves and residues, etc.), rice hulls, wheat chaffs, weeds and grasses in farms, homesteads and farmsteads, biochars, biogas slurry, oilcakes and so forth, (ii) biodegradable wastes, including kitchen and market wastes, fruits and vegetables peelings and biosolids and so forth, (iii) farmyard manure (FYM) and litters such as cattle manure, poultry manure, composts, vermicomposts and so forth, from on-farm and off-farm sources and (iv) forest and grasslands wastes, such as tree leaves, branches and twigs, shrubs and herbs underneath trees, roadside and community grasses and weeds and so forth. Other common sources of organic nutrients include growing food and non-food legumes as intercrops or rotational crops for current or residual N contribution, surface or residue recycling and in situ or ex-situ N_2-fixing green manure crops and so forth [36,37]. Organic fertilizers refer only to decomposed or partially-decomposed plant or animal materials used as a source of nutrients for crops. These also refer to small-sized pellets or granules (for example, granules made from cattle or poultry manure) developed from processing of organic materials. Finally, bio-fertilizers refer to microbial amendments of organisms such as *Rhizobia* or *Azospirilium*, bacteria promoted to stimulate biological N_2 fixation, or *Trichoderma*, a fungus promoted to hasten decomposition of organic materials [36].

In recent years, many kinds and formulations of organic fertilizers and bio-fertilizers are produced in many countries in South and SE Asia as well as imported from other countries and are floated in the markets as organic fertilizers. Some of such fertilizers, for example in Nepal, are bio-organic fertilizers, *Jaibik* Superphosphate (P), *Jaibik dhulo* (N), *Jhol mal*, HB 101, Bonsoon Super *Prangarik Mal*, Green Gold Super *Prangarik Mal* (Nepal), Chao Nang granules, Super Green plus, Super Green plant, Super Green mix and Premium Azosp, Premium Phospofix and Premium Azotoplus (India) [38]. Likewise, some common organic fertilizers in the Philippines are coco-composts, vermicomposts, Kalikasan Organic, Norfarco Bioorganic, Bio-green Compost, Foundation LCF Organic, Green Harvest Organic, Bio-earth Organic and so forth [36].

4. Organic versus Inorganic Materials/Fertilizers: Myths and Realities

Organic materials are widely used, albeit in small quantities, especially in subsistence farming systems in Asia, Africa and Latin America. Organic materials can improve soil's physical properties such as structure and aggregation and water holding capacity and drainage, biological properties such as increased microbial populations for biological activity and chemical properties such as nutrient holding capacity through increased cation exchange capacity and increased ability to resist changes in soil pH [39]. Improvement in soil physical properties can improve the medium for plant growth especially under well-drained, aerobic condition but less so under submerged paddy field soils, which during land preparation are typically puddled resulting in the breakdown of soil structure [40]. Submergence or flooding tends to buffer pH near neutrality and reduces the decomposition of native soil organic matter (SOM) or mineralization of soil organic nitrogen (SON) as compared to aerobic soils. In addition, puddling of rice soils reduces downward movement of water thereby reducing the need for greater nutrient-holding capacity of soil to reduce loss of nutrients by leaching [40]. Organic materials can also stimulate the activity of aerobic bacteria found in well-drained soils and, to some extent that of anaerobic bacteria found in submerged soils [36].

There are several myths about organic materials/fertilizers and inorganic fertilizers [36,41]. Many of the myths, however, seem to be mostly based on guesses, perceptions, or prejudices, or for political motives, without enough scientific evidences. Some of such myths and associated facts are discussed below:

4.1. Chemical Fertilizers Deteriorate Soil Physical Properties and Degrade Lands

A common myth among the advocates of organic farming is that the chemical fertilizers destroy the soil physical and chemical properties while organic materials or organic fertilizers improve the soil structure and water holding capacity of all soils [36]. Chemical fertilizers are also blamed for soil deterioration through alteration of soil physical properties and making soils acidic [37]. Some policy makers, politicians and even researchers and extension workers perceive that inorganic fertilizers, whether applied in small or large quantities, can degrade soils (structural change and acidification, etc.) and decline soil or crop productivity. There are, however, no scientific evidences demonstrating that the chemical fertilizers, when applied in optimum rates for high yield, destroy soil structure or reduce soil water holding capacity. The reality is that chemical fertilizers per se do not deteriorate soils by changing soil texture or making them acidic. Until and unless fertilizer N acidifies the soil to $pH < 5$, the application of N fertilizers at optimal rate generally has a positive effect on soil biota. It is only when they are applied in excessive amounts they may change soil texture, acidify soil and reduce microbial communities. In most subsistence farming systems practiced in Asia, Africa and Latin America, the use of chemical fertilizers is too low and thus above issues might not be quite important.

Although organic materials, when used as soil cover or mulch, can improve the soil physical properties, such benefits are mostly limited to aerobic soils through improved water retention, reduced soil crusting, increased soil porosity and reduced erosion. In contrast, flooded rice fields are puddled during land preparation destroying the soil structure and hence improvements of soil physical properties are of little significance to such fields. Improvements in soil physical properties however may be of importance for direct-seeded rice established without puddling, or for non-puddled transplanted rice which are now being promoted through conservation agriculture (CA) in South Asia [42,43]. In CA, soil is tilled to a minimum extent and crop residues are retained in the soil to help build up of SOM [43,44].

4.2. Organic Materials Are Available in Adequate Amounts and Have High Nutrient Contents

Advocates of organic fertilizers generally claim that there is enormous amount of organic materials (manures, crop residues, green manures, bio fertilizers, etc.) which contain high amount of essential nutrients to supply the amounts as per the crop demand for high yield. The reality, however, is that organic materials are not universally available in large quantities and contain very minimal macro- and micro-nutrients compared to inorganic fertilizers (Tables 1 and 2).

Table 1. Nutrient contents (%) of some commonly used organic materials in South and SE Asia (adapted from BARC, 2012 and Timsina, 2018, with permission from BARC, Bangladesh, 2012 and Agriculture & Forestry University, Nepal, 2018).

Organic Materials	Nutrient Content (%)			
	N	P_2O_5	K_2O	S
Cow dung (Fresh 60% MC *)	0.50	0.34	0.6	-
Cow dung (Decomposed 30% MC)	2.06	2.29	1.92	0.13
Farm yard manure (70% MC)	1.00	1.90	2.04	0.56
Poultry manure (55% MC)	2.50	1.28	0.9	1.10
Duck manure	2.15	2.59	1.38	-
Goat manure	2.00	3.41	2.94	-
Swine manure	2.76	6.05	1.764	-
Compost (rural 40% MC)	0.75	1.37	1.2	-
Compost (urban 40% MC)	1.50	1.37	1.8	-
Mustard oilcake (15% MC)	5.00	4.12	1.44	-
Linseed oilcake (15% MC)	5.50	3.21	1.44	-
Sesame oilcake (15% MC)	6.20	4.58	1.44	-
Groundnut oilcake	7.00	3.44	1.56	-
Bone meal (raw, 8% MC)	3.50	20.61	-	-
Bone meal (steamed, 7% MC)	1.50	22.90	-	-
Dried blood (10% MC)	11.00	1.10	0.70	-
Fishmeal (10% MC)	7.00	3.50	1.00	-

Source: [41,45]; * MC = Moisture content; - indicates data not available.

Table 2. Nutrient contents (%) of some commonly used green manure crops and crop residues in South and SE Asia (adapted from BARC, 2012 and Timsina, 2018, with permission from BARC, Bangladesh, 2012 and Agriculture & Forestry University, Nepal, 2018).

Green Manure Crops/ Crop Residues	Scientific Name	Moisture (%)	Nutrient Content (%)			
			N	P_2O_5	K_2O	S
Dhaincha	*Sesbania* sp.	80	2.51	0.92	0.92	0.20
Mung bean	*Vigna radiata*	70	0.80	0.46	1.15	0.30
Black gram	*Vigna mungo*	70	0.80	0.46	1.15	0.30
Cowpea	*Vigna unguiculata*	70	0.70	0.34	1.15	-
Pea	*Pisum sativum*	-	1.97	-	-	-
Sun hemp	*Crotolaria juncea*	70	0.70	0.27	1.15	-
Rice straw	*Oryza sativa*	30	0.58	0.23	3.16	-
Wheat straw	*Triticum aestivum*	20	0.50	0.69	2.06	-
Maize stover	*Zea mays*	15.5	0.59	0.71	3.00	-
Sugarcane leaves	*Saccharum officinarum*	20	1.00	1.15	3.21	-
Rice hull	*Oryza sativa*	15	0.31	0.16	0.85	-
Coconut husk	*Cocos nucefera*	-	1.75	0.27	2.06	-
Banana stem	*Musa* sp.	-	1.00	1.05	19.42	-
Leucaena	*Leucaena leucocephala*	-	4.29	0.44	3.14	-
Azolla	*Azolla* sp.	-	3.68	0.46	0.34	-
Acacia	*Acacia Arabica* (leaves)	-	2.61	0.39	2.75	-

Source: [41,45]. – indicates data not available.

Further, nutrient value of organic materials, particularly that of FYM and composts, is highly variable and often more variable, than that of crop by-products such as residues (rice straw or maize stover/hulls/husks, etc.). The animal's diet, the use and type of bedding material, manure age and how it was stored are factors that affect nutrient value of manures. These factors can vary seasonally on and among farms and regionally or on a larger geographic scale. Thus, if different nutrients required for high yields are to be supplied solely through the organic sources, excessively large amounts and volumes of organic materials would be required (Table 3). The exception is that organic materials, especially crop residues (e.g., rice residues), can supply (recycle) considerable potassium (K), sometime even more than crop needs [46,47].

There is poor synchronicity between crop demand and N release from organic manures as N from organic sources could be released during periods without a crop and thus such N could be exposed to leaching when precipitation occurs. Bergstrom and Kirchmann [48,49] demonstrated through two lysimeter studies that leaching of N through NH_4NO_3 was lower compared with animal manures or green manures. Likewise, Aronsson et al. [50] and Torstensson et al. [51] also demonstrated from long-term field studies in Sweden that N losses through leaching were higher in organic than in conventional systems. These results demonstrate that organic N sources are more vulnerable to leaching than inorganic fertilizers because N from organic sources maybe released during periods when there is no crop uptake of N.

Table 3. Quantities of chemical fertilizers, farmyard manure (FYM) and crop residues required (kg ha^{-1}) to attain yield targets of rice, wheat and maize (5, 5 and 10 t ha^{-1}, respectively) for various scenarios of nutrient application (adapted from Timsina, 2018, Agriculture & Forestry University, Nepal, 2018).

Source	Rice	Wheat	Maize
Scenario 1: 100% through chemical fertilizers (kg ha^{-1})			
Urea	159	196	485
TSP	68	64	200
MoP	159	174	400
Scenario 2: 50% through chemical fertilizers; 25% each from FYM and crop residues (kg ha^{-1})			
Urea	79	98	242
TSP	34	32	100
MoP	80	87	200
FYM	1821	2250	5575
Crop residues	1310	1263	3948
Scenario 3: 75% through chemical fertilizers; 12.5% each from FYM and crop residues (kg ha^{-1})			
Urea	119	147	364
TSP	51	48	150
MoP	119	131	300
FYM	913	1125	2788
Crop residues	1547	1940	4806
Scenario 4: 50% each from FYM and crop residues (kg ha^{-1})			
FYM	3650	4500	11150
Crop residues	6186	7759	19224

[1] Author's calculations; Source: [41].

4.3. Organic Fertilizers Undoubtedly Can Produce Quality Products

Promoters of organic farming commonly claim that organic farming or organic fertilizers produce better quality products compared to conventional farming or chemical fertilizers [36]. In fact, a review of multiple studies shows that organic varieties do provide significantly greater levels of vitamin C, iron, magnesium and phosphorus than non-organic varieties of the same foods [52]. Crinnion [53] also reported that organic varieties, while being higher in all these nutrients, are also significantly lower in nitrates and pesticide residues. Meta-analyses based on 343 peer-reviewed publications also indicated that the concentrations of a range of antioxidants were substantially higher in organic crops/crop-based foods than non-organic ones, with higher percentage of phenolic acids, flavanones, stilbenes, flavones, flavanols and anthocyanins [54]. There is also consistent evidence that, in general, organic plant-based foods contain a higher amount of beneficial, health-promoting secondary plant compounds than non-organic plant-based foods. For example, tomatoes grown on fields that have been organically managed for 10 years exhibited respectively 79 and 97% higher quercetin and kaempferol aglycones (i.e., the flavonoid concentrations) than their conventional counterparts [55]. Likewise, a long-term biannual rotation with cauliflower coupled with legume cover crop in an organic system optimized the nutrient fluxes of globe artichoke, suggesting as the most promising approach to foster long-term sustainability for the Mediterranean climate [56]. In a follow-up study in the same environment, polyphenol and Fe and K contents and dihydroxycinnamic and dicaffeoylquinic acids of globe artichoke were higher in organic system than in conventional system [57]. Willer et al. [23] also reported that organically processed products do not contain hydrogenated fats and other additives whose negative health impacts are widely acknowledged. Organic foods are more potent suppressors of the mutagenic action of toxic compounds and inhibit the proliferation of certain cancer cell lines. Clear health benefits from consuming organic dairy products have also been demonstrated regarding allergic dermatitis [53]. Finally, Parrott and Marsden [58] reported an improvement in taste and nutritional content of products by the farmers converted into organic system. Due to high quality of organic products, farmers practicing organic farming can receive higher economic returns due to higher premiums of the products.

While many studies such as above show increase in anti-oxidants and polyphenolics in organically-grown crops or foods, there are also evidences that it is not the application of organic farming alone that results in increase of anti-oxidants, it is when sustainable use of chemical fertilizers but without the use of chemical pesticides can also result in high anti-oxidants.

In fact, some studies have shown that the polyphenol content could be even higher in plants applied with inorganic fertilizers for as long as no pesticides are applied [37]. In the most extreme case, Miller [18] argued that organic foods are less healthy because of the presence of fungi, bacteria and animal manure and provided several examples of organic foods that had dangerous amounts of these substances on them. Thus, it seems unclear from these studies regarding the superiority of organic products over the non-organic ones. More research would be required comparing the performance of organic versus conventional farming or organic versus inorganic fertilizers as the benefits of organic farming/nutrients in terms of product quality or presence of antioxidants is not yet universally accepted. Further, research has shown two important concerns in using organic materials or organic fertilizers. One is that raw organic materials may contain pathogens especially when these are from manures, including human faeces. Another is the level of heavy metals especially when the raw materials are industrial or urban wastes and even household wastes [36]. Hence, bags containing organic materials or organic fertilizers should be properly labelled providing guarantee that these are free of pathogens and that the contents of the heavy metals are within the acceptable levels.

4.4. Organic Fertilizers Are Cheaper and Affordable

One of the widely spread misconceptions by the advocates of organic fertilizers is that organic sources of nutrients are cheaper than the inorganic fertilizers. Research has however shown that, on per unit of nutrient content basis, inorganic fertilizers are cheaper than the organic fertilizers [36]. Inorganic fertilizers contain substantially higher amounts of nutrients, especially macro-nutrients than organic manures. Nutrients from chemical fertilizers are also readily available to plants than that from organic sources. Thus, compared to chemical fertilizers, it can be cost ineffective to purchase, transport and apply organic materials such as FYM and composts with high-moisture and low-nutrient contents.

4.5. Legumes Can Use All N_2 Fixed from Atmosphere

Leguminous plants can fix atmospheric N_2 in the root nodules with the help of aerobic and anaerobic N_2-fixing organisms (Table 4). One of the common misconceptions about green manures, leguminous crops and cover crops and residues and so forth is that all their N content is fixed from the atmosphere and all N is utilized easily by the crops [36]. The reality, however is that the N in green manures and leguminous crops is not necessarily fixed from the atmosphere as a good portion is absorbed from the soil. Also, when green manures or legume residues are incorporated into the soil, not all their N contents are used by the crops as some N is lost during decomposition or mineralization. However, there are exceptions when crops grown in rotation with crops capture nutrient unavailable to crops and recycle the otherwise lost nutrients back to crops. One such case is when crops, weeds, or green manures (grown in rotation with lowland rice) can assimilate nitrate and then recycle the N back to future rice crops through retained biomass. Another case is deep rooting shrubs (such as in agroforestry systems) grown on deep soils, which can capture nutrient from below the rooting depth of crops and recycle them back to future crops (see details about agroforestry systems in a later section below).

Even though legumes or cover crops can fix N_2 from atmosphere they use lands for them to grow at the cost of cropping of main staple crops. In developed countries where mostly monoculture is practiced, inclusion of legumes as a second crop may not be a great issue but in developing countries with small holder farming systems, double or multiple cropping with 200–300% annual cropping intensity is a common phenomenon. For example, rice-wheat or rice-maize systems are

Table 4. Amount of N_2 fixed (kg ha^{-1}) by some common aerobic and anaerobic N_2-fixing organisms and tree legumes (adapted from Akinnifesi et al., 2010, Canadian Center of Science and Education, 2010).

Group	N_2-Fixing Organisms/Legumes	Amount of N_2 Fixed (kg ha^{-1})
Aerobic	*Azospirillium* sp.	20–40 season^{-1}
	Klebsiella	32 year^{-1}
	Anabaena (Cyanobacter/Blue green algae)	15–45 crop^{-1}
	Nostoc (Cyanobacter/Blue green algae)	15–45 crop^{-1}
	Enterobacter	32 year^{-1}
	Achromobacter	32 year^{-1}
	Klebsiella	32 year^{-1}
	Cyanobacteria/Blue green algae	15–45 crop^{-1}
Tree and perennial legumes	*Gliricidia sepium*	212 year^{-1}
	Acacia anguistissima	122 year^{-1}
	Leucaena collinsi	300 year^{-1}
	Cajanus cajan	34–85 crop^{-1}
	Sesbania sesban	84 season^{-1}

Source: [59].

practiced often as double cropping and on many occasions by including a third crop in large areas of South and SE Asia [35,60]. Meeting food security of their people through staple crops (rice, wheat, maize, etc.) is high priority of the governments. Thus, they cannot sacrifice their lands to grow non-staple crops such as legumes instead of staple ones unless replacement of the latter by the former is economically viable without much reduction in total system productivity. Even for the developed countries, there are not enough N_2-fixing cover or legume crops that could fertilize all their crops. Many studies have overestimated the contribution of biological nitrogen fixation (BNF) by legumes. One of such studies is that of Badgley et al. [9] who grossly overestimated the global N supply through BNF, which was immediately disputed by Connor [16] and Dobermann [34].

4.6. Chemical Fertilizers Cannot Supply Micro-Nutrients

One popular misconception about chemical fertilizers is that they provide only a few macro-nutrients and not micro-nutrients. The reality is that while most organic fertilizers contain some micro-nutrients by nature, there are now several commercially-available inorganic fertilizers containing micro-nutrients [36]. Thus, soils deficient in micro-nutrients can now be supplied with smaller amount of inorganic fertilizers containing micro-nutrients rather than large amount of organic materials to supply the same quantity of micro-nutrients required by plants.

4.7. Organic Materials Can Build-Up Large Amount of SOM

Organic crop production has been proposed as a strategy for soil organic carbon (SOC) sequestration. Thus, advocates of organic fertilizers believe that organic materials build up SOM irrespective of the amounts they are applied to the soil. Organic materials no doubt supply nutrients and energy for soil micro-organisms that help in accumulating SOM in soils, their contribution to SOM build-up within a short period of time (e.g., one or two years) is widely misperceived or over-exaggerated [36], as large quantities of organic materials as well as a long time would be required to build up SOM. Moreover, the amount of SOM formed with addition of organic materials depends on the carbon nitrogen ratio (C:N ratio) of the original materials and conditions during decomposition.

Annual carbon input into the soil is the most important factor responsible to build and sequester SOC and crop production practices that result in higher biomass and yields can add more carbon to soil through above- and below-ground crop residues [61]. Since the above-mentioned evidences indicate that crop yields are generally lower in organic farming, it can be hypothesized that the carbon input through crop residues to soil would also be lower resulting in lower SOC sequestration and consequently lower SOM build-up in organic farming. This hypothesis has been proved to be true in many cases. For example, Lutzow and Ottow [62] and Petersen et al. [63] reported lower SOC in organically- than conventionally-managed farms while Burkitt et al. [64] and Leifeld and Fuhrer [65] demonstrated no difference in SOC between organically- and conventionally-managed farms. Thus, it seems clear from these studies that the magnitudes of increases in SOM due to addition of organic materials or organic fertilizers would be far less than what many advocates of organic fertilizers claim.

4.8. Chemical Fertilizers Cannot Build Up SOM

The critics of chemical fertilizers believe that such fertilizers cannot build up organic matter in soil. Some evidences however indicate that inorganic fertilizers, when applied at rates at which maximum yields are achieved, can also result in SOM build-up and microbial biomass by promoting plant growth and increasing the amount of litter and root biomass added to soil. Bijay-Singh [66] reported that only when fertilizer N is applied at rates more than the optimum, it can increase the residual inorganic N accelerating the loss of SOM through mineralization. Fertilizer N application can affect SOM in two ways: (i) it may increase SOM by promoting plant growth and increasing the amount of litter and root biomass added to soil compared with the soil not receiving fertilizer N; and (ii) it may accelerate SOM loss through decay or microbial transformation of litter (leaves, straw, manures) and indigenous forms of organic C already present in the soil [67]. High fertilizer rates however can adversely affect soil microbial biomass (see later).

4.9. Organic Materials Can Be Universally Applied

Advocates of organic fertilizers claim that it is always safe to apply organic materials on every soil, irrespective of amounts and SOM status, including the anaerobic flooded soils. The reality is that excess organic matter could cause zinc and sulphur deficiency especially when the field is continuously flooded [37,40]. In addition, toxicity from products of anaerobic decomposition (such as organic acids and hydrogen sulphide) could also be a concern. Hence, when the SOM in soils is relatively high (>4.0%), organic materials should preferably be applied in dry season or aerobic conditions [36,39].

4.10. Bio-Fertilizers Can Increase Nutrient Content of Soil

Soil organisms (bacteria, fungi, algae, *actinomycetes*, earthworms, etc.) are essential components of the soil, contributing to soil productivity. There are aerobic and anaerobic N_2-fixing bacteria (e.g., *Rhizobia*) and some other bacteria and fungi (e.g., *Trichoderma*), which are effective in decomposing or mineralizing SOM. These microorganisms can be used to dispose farm wastes and to improve soil productivity. Bio fertilizers, which are applied to seeds, soils in seedbed, or to composting materials can increase the number of microorganisms and accelerate certain microbial processes such as atmospheric N_2 fixation, phosphate solubilisation, or cellulose degradation [37]. Advocates of organic fertilizers claim that microbial fertilizers or bio fertilizers, containing organisms such as bacteria, fungi, algae, *actinomycetes* and so forth, contribute significant amount of nutrients to the crop and can be used to any crop or any type of ecosystems [36]. The fact is that bio fertilizers do not directly contribute nutrients but merely make nutrients available from other sources like atmospheric N_2 or SOM [37].

While the role of the bio fertilizers has been recognised, there are evidences regarding their inconsistent effects on crop growth or yield, or not as dramatic as claimed by the advocates of organic fertilizers. Moreover, since most of the microorganisms in bio fertilizers work under aerobic conditions, they may not be effective under anaerobic conditions. Conditions where bio fertilizers are effective are not defined properly to guide extension workers and farmers. Hence, it is important that the bio fertilizer developers indicate the species or strains of organisms present (whether aerobic or anaerobic) and the conditions where the product is effective.

5. Nutrient Supply from Inorganic and Organic Sources

It is a widely recognised fact that small and poor farmers in almost all countries of the world lack resources to purchase high-yielding inputs such as chemical fertilizers and hence rely on the organic inputs in whatever quantities already available in their farm. Organic nutrients are available in varying amounts (from low to high) in soil (i.e., indigenous nutrients) and/or through external sources (i.e., either inorganic or organic). In most cases (except some lowland rice fields), organic nutrients must be supplemented with inorganic fertilizers. Small farmers practising subsistence farming system and with limited income to purchase fertilizers can rely on organic inputs such as FYM, composts, or crop wastes and residues that are available in their farm [68]. However, such inputs contain very low amounts of nutrients which can only support very low-yielding crops. For transitioning from subsistence to commercial agriculture and to achieve high yields and high income, application of inorganic fertilizers is unavoidable.

Erisman et al. [69] reported that over 48% of more than 7 billion people are living today because of increased crop production made possible by applying fertilizer N. Hence, if sufficient amounts of nutrients, especially N, are not applied to plants, high yields will not be possible and transitioning to commercialization of agriculture will be a dream only. However, fertilizers being chemicals can potentially disturb the natural functioning of the soil and may also affect the output of other ecosystem services. The challenge ahead is to manage fertilizers (inorganic and organic) and soil in such a way that not only food demands are continuously met but soil also remains healthy to support adequate food production with minimal environmental impact. As stated earlier, while inorganic fertilizers are crucial to increase crop yields, they ae generally not affordable by small-scale subsistence farmers of developing world. On the other hand, the soil-derived as well as the externally-supplied organic sources of nutrients will not be sufficient to achieve high yield. Hence, depending on their relative availability, nutrients need to be supplied in an integrated manner and in balanced proportions through both inorganic and organic sources.

For illustration purpose, nutrient supply through chemical fertilizers and most common organic sources (FYM and crop residues) for various scenarios involving various combinations of inorganic fertilizers and organic materials to achieve target yields of rice, wheat and maize (5, 5 and 10 t ha^{-1}, respectively) is shown in Table 3. Rice, wheat and maize are chosen because these are the crops grown predominantly in all regions of the world and are globally important especially for achieving the food security of the growing population of the developing countries [32,35,60]. Their sustainable production is necessary in all countries where these are the principal crops. Four scenarios are considered: Scenario 1 is when all nutrients are supplied through 100% chemical fertilizers and with no organic sources; Scenario 2 is when 50% nutrients are applied through chemical fertilizers and 25% each from FYM and crop residues; Scenario 3 is when 75% nutrients are applied through chemical fertilizers and 12.5% each from FYM and crop residues; and finally Scenario 4 is when all nutrients are applied through organic sources (50% each from FYM and crop residues) and with no application of chemical fertilizers. FYM and crop residues are chosen because these are the main sources of organic nutrients in the smallholder crop-livestock or crop-tree-livestock farming systems in tropics and subtropics and contribute to nutrient cycling [68,70–72]. In the example, rice residues are applied to wheat and maize crops and maize residues are applied to the rice crop. The concentrations of N, P_2O_5 and K_2O in urea, TSP and MoP are 46.0%, 46% and 60%, respectively. As stated earlier, the nutrient contents in organic manures are variable and hence the mean values for FYM and crop residues, as shown in Tables 1 and 2, were used for the calculations. Nutrient requirements, predicted by the QUEFTS (Quantitative Evaluation of the Fertility of Tropical Soils) model shows that for rice, wheat and maize, 14.6, 18.0 and 22.3 kg N, respectively, would be required to obtain 1 t of grain yield.

The respective values are 6.2, 5.9 and 9.2 kg P_2O_5 and 19.1, 20.9 and 24.0 K_2O per t grain yield of above crops [67,73–76].

Data in Table 3 reveal that when only chemical fertilizers are used to meet the requirements of high-yielding crops (Scenario 1), only small volumes of chemical fertilizers would be required and hence the handling, storing, transporting and applying the fertilizers in the fields would not be a big issue. This is in contrast to Scenario 4, where very large volumes or amounts of FYM and crop residues would be required to meet crop nutrient requirements and hence all the above issues would be significantly greater. In Scenario 2 and 3, where some fractions of the nutrients are used through organic sources, these issues would still be there but to the much lesser extent than in Scenario 4. The extremely large amounts of organic materials (FYM or other sources of animal manures or crop residues) as required for Scenario 4 and to the lesser extent for Scenario 2, simply would not be available in sufficient amount for organic farming in any country, also due to their multiple uses [68]. Miller [18] also reported that there is simply not enough cow manure in any country to fertilize the organic crops for high yield. Avery [13] also stressed that sewage sources of N would be only about 2% of the synthetic N used to fertilize the crops. The above calculations and the literature review clearly suggest that most countries would need extra lands and water to grow and produce organic materials and feed animals to produce enough quantities of plant biomass and manures to fertilise the soils for achieving high yields if the nutrients were to be supplied through organic sources only.

Our conclusions also agree with many previous workers [2,13,16,19,20,77,78], who all reported that huge amount of extra land would be needed if such sources of N were to be promoted. Except for some countries in Africa, extra lands would not be available in any other countries due to the ever-increasing population and need for housing, industry and other infrastructures. Even if lands would be made available to produce organic inputs, using only organic sources will be highly laborious, costly and impractical unless some novel or innovative practices are developed and used to build on-farm soil fertility and in-situ nutrient application. Further, as mentioned above, nutrient contents in organic materials are highly variable and release nutrients at variable rates. Hence, any assumptions on nutrient contents and release patterns lead to uncertainties in calculations of nutrient recommendations from organic sources. In most developing countries, information on period of nutrient release and on the rates by which nutrients are mineralized are not provided to farmers, further leading to uncertainties in calculations of nutrients supplied through such sources.

6. Need for Site-Specific Nutrient Management

Existing nutrient management recommendations for most crops in most developing countries often consist of one predetermined rate of nutrients for vast areas of production. Such recommendations assume that the need of a crop for nutrients is constant over time and space. However, the nutrient needs for supplemental nutrients for any crop can vary greatly among fields, seasons and years, because of differences in crop-growing conditions, water, nutrient and soil management and climate, resulting in large spatial and temporal variability in soil N supply. Hence, the nutrient management for crops aimed at high yields requires an approach that enables adjustments in nutrient application to accommodate the site- or soil-specific needs of the crop for supplemental nutrients. Site-specific nutrient management (SSNM), a plant-based approach and a form of precision nutrient management, is used to address nutrient differences which exist within and between fields by adjusting the nutrient application through chemical fertilizers or organic sources to match the site, soil, or season differences. SSNM approach for irrigated rice systems for South and SE Asia was developed by International Rice Research Institute (IRRI) in collaboration with National Agricultural Research and Extension Systems (NARES) partners in 1990s [79–81] to address serious limitations arising from blanket fertilizer recommendation for large areas. The approach focused on managing field-specific spatial variation in indigenous N, P and K supply and considering nutrient losses from soil, recovery efficiency of a given fertilizer and nutrient uptake and use efficiencies, temporal variability in plant N status occurring within a growing season and medium-term changes in soil P and K supply resulting from actual nutrient balance.

SSNM or precision nutrient management strategies, based on principles of synchronization of crop demand of nutrients with supply from all sources, including soil, fertilizer and organics, hold great potential for ensuring high yields of crops along with maintenance or improvement in soil health [39,60]. SSNM approaches have the potential to optimize nutrient management for cropping systems as farmers replace crops in their crop rotations. Based on the scientific principles, SSNM recommends nutrients for optimally supplying to crops as and when needed for specific field/soil and cropping season. Scaling-up of nutrient management technologies can be faster if simple computer-based decision support system (DSS) tools can be developed for use by farmers and extension workers from governmental and non-governmental organizations and from the private sector. One of such tools is Crop Manager for rice, maize and wheat developed by International Rice Research Institute (IRRI) (http://cropmanager.irri.org/) and similar other tool is Nutrient Expert for rice, maize, wheat soybean developed by International Plant Nutrition Institute (IPNI) (http://software.ipni.net/article/nutrient-expert). Both tools have been widely evaluated and promoted by IRRI and IPNI in partnership and collaboration with International Centre for Maize and Wheat (CIMMYT) and NARES of several countries in South and SE Asia and Sub-Saharan Africa [82–87]. These tools are available both on-line and off-line in mobile phones and laptops and are interactive and easy-to-use that can rapidly provide nutrient recommendations for an individual farmer's field in the presence or absence of soil testing data. Future approach should give priority for further refinements of simple DSS tools for integration and widespread delivery of improved and integrated nutrient management strategies for diverse agro-ecosystems of Asia, Africa and Latin America.

7. New and Alternative Approaches to Increase On-Farm Soil Fertility and Nutrient Supply

An important question in soil fertility management globally and especially in South and SE Asia and Sub-Saharan Africa, where very low amounts of fertilizers are used, is how crop biomass production can be increased. This is important to enhance surface cover and generate greater quantities of organic nutrients to complement or supplement whatever amounts of inorganic fertilizers a smallholder farmer can afford to apply. The calculations presented in previous section reveal that organic materials (or organic fertilizers) obtained from traditional crop or crop-livestock systems are not enough to improve soil fertility and meet nutrient demand of high-yielding crops. Alternative techniques (or some radical approaches) would be required if the aim was to supply larger proportion of nutrients from organic sources to restore and maintain on-farm soil fertility, obtain high yields and achieve food security for the ever-increasing global population. One of such approaches could be agroforestry system, which is defined as the integration of trees into annual food crop systems, using both perennial and annual species (trees, food and vegetable crops, etc.). In this system, farmers can grow crops and trees in right proportions so that crop residues and tree leaves can provide enough nutrients to build and maintain soil fertility, supply nutrients to plants and can provide green fodder to livestock [88,89]. Sanchez [89] called agroforestry system as "second soil fertility paradigm" which mainly focuses on improved fallow as well as biomass transfer technologies using trees and shrub legumes capable of fixing N_2 through their roots and from the biomass from their leaves and build and maintain soil fertility.

In recent years, more attention has been given to agroforestry system as a possible and sustainable solution to maintain soil fertility. Thus, to promote agroforestry system, a global alliance called Evergreen Agriculture Partnership, has been formed (http://evergreenagriculture.net/). Evergreen Agriculture, an advanced form of agroforestry system, is an approach for maintenance of a green cover on the land throughout the year in the tropical and sub-tropical climate. Such an approach of producing enormous amounts of biomass on-farm does not require extra lands for growing trees as they can be planted in same land together with crops. Depending upon which woody species are used and how they are managed, their cultivation in crop fields can bolster nutrient supply through

N_2 fixation and nutrient cycling, can build-up on-farm soil fertility and provide nutrients to plants as per their demand, enhance suppression of insect pests and weeds, improve soil structure and water infiltration, produce greater amount of food, fodder, fuelwood and fibre and obtain higher income directly from products produced by the intercropped trees and crops [90,91]. Authors suggest that such an intercropped system can enhance carbon storage both above- and below-ground, produce greater quantities of organic matter in soil surface residues, result in more effective conservation of above-and below-ground biodiversity, sequester carbon in trees and soil and thus can mitigate CO_2 emissions and tackle climate change [90,91]. Evergreen Agriculture thus has potential to contribute to integrated soil fertility management and the knowledge to adapt these to local conditions that maximize use efficiencies of chemical fertilizers and organic resources and increase crop productivity. In this respect, the authors [90,91] suggest that the types of intercropped trees can include species whose primary purpose is to provide products or benefits other than soil fertility replenishment alone, such as fodder, fruits, timber and fuel wood. In such cases, the trees are expected to provide an overall value greater than that of the annual crops within the area that they occupy per unit area in the field.

The principles of Evergreen Agriculture have now been widely applied in sub-Saharan Africa where they have been adapted to a diversity of situations, often building successfully on proven indigenous farming technologies and where diversity and polyculture are a common feature of the agricultural systems [91]. For example, in several countries in sub-Saharan Africa, Evergreen Agriculture is practised with conservation farming with *Acacia albida* (or *Faidherbia albida* (Delile) A.Chev.), an indigenous N_2-fixing tree species. *Faidherbia* remains dormant and sheds its foliage during the early rainy season at the time when field crops are being established and re-growing at the end of the wet season, thus exhibiting minimal competition while enhancing yields and soil health. This unique growth habit, known as 'reverse leaf phenology' makes it highly compatible with food crops, since it does not compete with them significantly for light, nutrients or water during the growing season. In contrast, annual crops near *Faidherbia* trees tend to exhibit improved performance and yield [59,92]. Other potential options for sub-Saharan Africa include intercropping maize with *Gliricidia sepium* (Jacq.) Kunth ex Walp., *Tephrosia candida* (Roxb.) DC., *Cajanas cajan* (L.) Millsp., or *Sesbania sesban* (L.) Merr. [59] but can also be recommended for South and SE Asia. For example, research in Africa has revealed that several tons of additional biomass ha^{-1} can be generated annually to accelerate soil fertility replenishment and provide additional livestock fodder and that such systems can result in dramatic increases in maize yield when grown in association with *Faidherbia* of varying age and density, agronomic practices and the weather conditions [92,93]. Akinifessi et al. [59] concluded that fertilizer trees such as *Faidherbia, Gliricidia* and *Leucaena* sp. can add 34–300 kg N ha^{-1} $year^{-1}$ through BNF and that, depending on crops, nutrient contributions from fertilizer tree biomass can reduce the mineral N requirement by up to 75%. This broadens the concept of crop rotations to incorporate the role of fertilizer/fodder trees to more effectively enhance soil fertility and provide needed organic materials to increase crop yield, increase income and achieve food security.

Evergreen Agriculture could also be compatible with crop-tree-livestock integration which is practiced for decades by smallholder farmers in South and SE Asia (for e.g., for example, see Timsina et al., 1991 [68] for a description of crop-livestock and crop-tree-livestock integration for Nepal). This could also be compatible with the three principles of conservation agriculture (CA) (i.e., reduced or no tillage, residue retention on the soil surface and profitable and sustainable rotations) in situations where these are feasible and appropriate [42]. Although some implementation-related issues of CA remain to be addressed, it has now been adapted to many crops and areas in countries of South Asia [42–44] and Africa [94]. Research in Africa has also demonstrated that Evergreen Agriculture by integrating fertilizer trees and shrubs into CA can dramatically enhance both fodder production and soil fertility [91].

8. Conclusions and Research and Policy Implications

A brief review of organic and conventional farming and sources of nutrients in this paper demonstrates that yields of organic agriculture are much less than conventional agriculture and that the current organic sources of nutrients are not enough to increase crop yields required to feed global population. The review also identifies the fact that unless novel and innovative approaches are developed and promoted to build on-farm soil fertility, organic nutrients in the current state of global agriculture are not enough to provide same amount of food that can be produced from conventional agriculture. Integrated and/or site-specific precision nutrient management of inorganic and organic sources is crucial for sustainable soil fertility management and to achieve food security. The application of nutrients in a balanced proportion through organic and inorganic sources and based on SSNM principles can lead to further improvements in soil health and soil fertility and productivity. Based on the available scientific evidences and considering the non-availability of organic materials in sufficient amount in most countries, nutrients from inorganic and organic sources should preferably be applied at 75:25 ratio but the new and alternative concept of Evergreen Agriculture, as discussed in this review, has potential to supply inorganic and organic sources of nutrients at 50:50 or 25:75 ratio.

It is recommended that appropriately-designed field experiments in any country must be conducted to determine the soils and environmental conditions where the organic fertilizers can be effective to better guide and benefit farmers before promoting or spreading the use of organic materials or organic fertilizers. There is also a need to document the long-term fate of organic materials in different cropping systems. Finally, the review strongly recommends that a well-designed agroforestry system for sustainable intensification would be the most effective strategy for integrated soil fertility management and the Evergreen Agriculture, which has been adopted in many countries of Africa, could be introduced and promoted in countries of Asia and Latin America too. Evergreen Agriculture seems to be a sustainable strategy to improve on-farm soil fertility, increase crop yields, provide fodder to livestock and fuel wood to smallholding farmers residing in countries with tropical and sub-tropical climate and finally meet global food demand. In areas where trees are sparse, government policies should aim to increase tree plantation and promote agroforestry and Evergreen Agriculture in those countries. This will encourage farmers to plant trees and will also promote the use of organic materials for sustainable soil fertility management, increase crop yields and feed the ever-increasing global population.

Author Contributions: The author solely conceptualized and wrote the paper.

Acknowledgments: The author is grateful to Saiful Islam, CIMMYT, Bangladesh, for assisting him in the preparation of tables and in reviewing the paper prior to submission.

References

1. Cassman, K.G.; Döbermann, A.D.; Walters, D.T.; Yang, H. Meeting cereal demand while protecting natural resources and improving environmental quality. *Ann. Rev. Environ. Resour.* **2003**, *28*, 315–358. [CrossRef]
2. Kirchmann, H.; Katterer, T.; Bergstrom, L.; Borjesson, G.; Bolinder, M.A. Flaws and criteria for design and evaluation of comparative organic and conventional cropping systems. *Field Crop. Res.* **2016**, *186*, 99–106. [CrossRef]
3. Bruinsma, J. *The Resource Outlook to 2050. By How Much Do Land, Water and Crop Yields Need to Increase by 2050? FAO Expert Meeting on How to Feed the World in 2050*; FAO: Rome, Italy, 2009; p. 33.
4. Alexandratos, N.; Bruinsma, J. *World Agriculture towards 2030/2050: The 2012 Revision; ESA Working Paper No. 12-03*; Food and Agriculture Organization of the United Nations: Rome, Italy, 2012.
5. Searchinger, T.; Hanson, C.; Ranganathan, J.; Lipinski, B.; Waite, R.; Winterbottom, R.; Dinshaw, A.; Heimlich, R. *Creating a Sustainable Food Future. A Menu of Solutions to Sustainably Feed More Than 9 Billion People by 2050. World Resources Report 2013–14: Interim Findings*; World Resources Institute: Washington, DC, USA, 2014.

6. Smil, V. *Feeding the World—A Challenge for the 21st Century*; MIT Press: Cambridge, MA, USA, 2000.
7. Pimentel, D. Green revolution agriculture and chemical hazards. *Sci. Total. Environ.* **1996**, *188*, S86–S98. [CrossRef]
8. Tilman, D.; Cassman, K.G.; Matson, P.A.; Naylor, R.; Polasky, S. Agricultural sustainability and intensive production practices. *Nature* **2002**, *418*, 671–677. [CrossRef] [PubMed]
9. Badgley, C.; Moghtader, J.; Quintero, E.; Zakern, E.; Chappell, J.; Avilés-Vázquez, K.; Samulon, A.; Perfecto, I. Organic agriculture and the global food supply. *Renew. Agric. Food Syst.* **2007**, *22*, 86–108. [CrossRef]
10. Hamer, E.; Anslow, M. 10 reasons why organic farming can feed the world. *The Ecologist*. 1 March 2008. Available online: http:///www.theecologist.org/trial_investigations/268287/10_reasons_why_organic_can_feed_the_world.html (accessed on 25 August 2018).
11. Woese, K.; Lange, D.; Boess, C.; Bogl, K.W. A comparison of organically and conventionally grown foods: Results of a review of the relevant literature. *J. Sci. Food Agric.* **1997**, *74*, 281–293. [CrossRef]
12. Seufert, V.; Ramankutty, N.; Foley, A.E. Comparing the yields of organic and conventional agriculture. *Nature* **2012**, *485*, 229–232. [CrossRef] [PubMed]
13. Avery, D. From saving the planet with pesticides. In *The True State of the Planet*; Baley, R., Ed.; The Free Press: New York, NY, USA, 1995.
14. Borlaug, N.E. Feeding a world of 10 billion people: The miracle ahead. *In Vitro Cell. Dev. Biol. Plant* **2002**, *38*, 221–228. [CrossRef]
15. Goulding, K.W.T.; Trewavas, A.J.; Giller, K.E. *Can Organic Farming Feed the World? A Contribution to the Debate on the Ability or Organic Farming Systems to Provide Sustainable Supplies of Food*; International Fertiliser Society: York, UK, 2009; p. 633.
16. Connor, D.J. Organic agriculture cannot feed the world. *Field Crop. Res.* **2008**, *106*, 187–190. [CrossRef]
17. Connor, D.J. Organically grown crops do not a cropping system make and nor can organic agriculture nearly feed the world. *Field Crop Res.* **2013**, *144*, 145–147. [CrossRef]
18. Miller, J.J. The organic myth. *Natl. Rev.* **2006**, *56*, 35–37.
19. Kirchmann, H.; Kätterer, T.; Bergström, L. Nutrient supply in organic agriculture—Plant availability, sources and recycling. In *Organic Crop Production—Ambitions and Limitations*; Kirchmann, H., Bergström, L., Eds.; Springer: Doordrecht, The Netherlands, 2008.
20. Kirchmann, H.; Bergström, L.; Kätterer, T.; Andrén, O.; Andersson, R. Can organic crop production feed the world? In *Organic Crop Production—Ambitions and Limitations*; Kirchmann, H., Bergström, L., Eds.; Springer: Doordrecht, The Netherlands, 2008; pp. 39–74.
21. Ammann, K. Why farming with high tech methods should integrate elements of organic agriculture. *New Biotech.* **2009**, *25*, 378–388. [CrossRef] [PubMed]
22. Ronald, P.C.; Adamchak, R.W. *Tomorrow's Table: Organic Farming, Genetics and the Future of Food*; Oxford University Press: New York, NY, USA, 2008.
23. Willer, H.; Yussefi-Menzler, M.; Sorensen, N. *The World of Organic Agriculture—Statistics and Emerging Trends 2008*; IFOAM: Bonn, Germany; FiBL: Frick, Switzerland, 2008.
24. NRC. *Toward Sustainable Agricultural Systems in the 21st Century*; The National Academies Press: Washington, DC, USA, 2010; p. 4.
25. Willer, H.; Lernoud, J. (Eds.) *The World of Organic Agriculture: Statistics and Emerging Trends 2017. Research Institute of Organic Agriculture (FIBL)*; Frick; IFOAM-Organic International: Bonn, Germany, 2017.
26. Stanhill, G. The comparative productivity of organic agriculture. *Agric. Ecosys. Environ.* **1990**, *30*, 1–26. [CrossRef]
27. Penning de Vries, F.W.T.; Rabbinge, R.; de Groot, J.J.R. Potential and attainable food production and food security in different regions. *Philos. Trans. R. Soc. Lond B. Biol. Sci.* **1997**, *352*, 917–928. [CrossRef]
28. Lotter, D.W. Organic agriculture. *J. Sustain. Agric.* **2003**, *21*, 59–128. [CrossRef]
29. Ponti, T.D.; Rijk, B.; Ittersum, M.V. The crop yield gap between organic and conventional agriculture. *Agric. Syst.* **2012**, *108*, 1–9. [CrossRef]
30. De Vries, P.; Jansen, D.M.; ten Berge, H.F.M.; Bakema, A. *Simulation of Eco-physiological Processes of Growth in Several Annual Crops*; Simulation monographs 29; Pudoc: Wageningen, The Netherlands, 1989; p. 271, ISSN 0924-8439.
31. Van Ittersum, M.K.; Cassman, K.G.; Grassini, P.; Wolf, J.; Tittonell, P.; Hochman, Z. Yield gap analysis with local to global relevance—A review. *Field Crop. Res.* **2013**, *143*, 4–17. [CrossRef]

32. Timsina, J.; Wolf, J.; Guilpart, N.; van Bussel, L.; Grassini, P.; van Wart, J.; Hossain, A.; Rashid, H.; Islam, S.; van Ittersum, M. Can Bangladesh produce enough cereals to meet future demand? *Agric. Syst.* **2018**, *163*, 36–44. [CrossRef] [PubMed]
33. Cassman, K.G. Editorial response by Kenneth Cassman: Can organic agriculture feed the world—Science to the rescue? *Renew. Agric. Food Syst.* **2007**, *22*, 83–84.
34. Dobermann, A. Getting back to the Field. *Nature* **2012**, *485*, 176–177.
35. Timsina, J.; Connor, D.J. The productivity and management of rice-wheat cropping systems: Issues and challenges. *Field Crop. Res.* **2001**, *69*, 93–132. [CrossRef]
36. Mamaril, C.P.; Castillo, M.B.; Sebastian, L.S. *Facts and Myths about Organic Fertilizers*; Philippine Rice Research Institute (PhilRice): Muñoz, Nueva Ecija, Philippines, 2009.
37. Mamaril, C.P. Organic Fertilizers in Rice: Myths and Facts. FAO: Rome, Italy; Los Banos, Laguna, Philippines, 2004.
38. Dahal, K.R.; Sharma, K.P.; Bhandari, D.R.; Regmi, B.D.; Nandwani, D. Organic Agriculture: A Viable Option for Food Security and Livelihood Sustainability in Nepal. In *Organic Farming for Sustainable Agriculture. Sustainable Development and Biodiversity*; Nandwani, D., Ed.; Springer: Cham, Switzerland, 2016; p. 9.
39. Buresh, R.J.; Dobermann, A. Organic materials and rice. In *Annual Rice Forum 2009: Revisiting the Organic Fertilizer Issue in Rice*; Asia Rice Foundation: College, Laguna, Phiilippines, 2010; pp. 17–33.
40. Ponnamperuma, F.N. The chemistry of submerged soils. *Adv. Agron.* **1972**, *24*, 29–96.
41. Timsina, J. Can organic materials supply enough nutrients to achieve food security? *J. Agric. Forest. Univ.* **2018**, *2*, 9–21.
42. Hobbs, P.; Gupta, R.; Jat, R.K.; Malik, R.K. Conservation agriculture in the indo-gangetic plains of India: Past, present and future. *Exp. Agric.* **2017**, 1–19. [CrossRef]
43. Gathala, M.K.; Timsina, J.; Islam, S.; Rahman, M.; Hossain, I.; Harun-Ar-Rashid; Ghosh, A.K.; Krupnik, T.J.; Tiwari, T.P.; McDonald, A. Conservation agriculture-based tillage and crop establishment options can maintain farmers' yields and increase profits in South Asia's rice-maize systems: Evidence from Bangladesh. *Field Crop. Res.* **2015**, *172*, 85–98. [CrossRef]
44. Gathala, M.K.; Timsina, J.; Islam, S.; Krupnik, T.J.; Bose, T.R.; Islam, N.; Rahman, M.M.; Hossain, M.I.; Harun-Ar-Rashid; Ghosh, A.K.; et al. Productivity, profitability and energetics: A multi-criteria and multi-location assessment of farmers' tillage and crop establishment options in intensively cultivated environments of South Asia. *Field Crop Res.* **2016**, *186*, 32–46. [CrossRef]
45. BARC. *Fertilizer Recommendation Guide-2012. Farmgate, Dhaka-1215*; BARC: Dhaka, Bangladesh, 2012.
46. Timsina, J.; Singh, V.K.; Majumdar, K. Potassium management in Rice-maize systems in South Asia. *J. Soil Sci. Plant Nutr.* **2013**, *176*, 317–330. [CrossRef]
47. Singh, V.K.; Dwivedi, B.S.; Yadvinder-Singh; Singh, S.K.; Mishra, R.P.; Shukla, A.K.; Rathore, S.S.; Shekhawat, K.; Majumdar, K.; Jat, M.L. Effect of Tillage and Crop Establishment, Residue Management and K Fertilization on Yield, K Use Efficiency and Apparent K Balance under Rice-Maize System in North-Western India. *Field Crop Res.* **2018**, *224*, 1–12. [CrossRef]
48. Bergström, L.F.; Kirchmann, H. Leaching of total nitrogen fromnitrogen-15-labeled poultry manure and inorganic nitrogen fertilizer. *J. Environ. Qual.* **1999**, *28*, 1283–1290. [CrossRef]
49. Bergström, L.F.; Kirchmann, H. Leaching of total nitrogen from 15-N-labeledgreen manures and15NH415NO3. *J. Environ. Qual.* **2004**, *33*, 1786–1792. [CrossRef] [PubMed]
50. Aronsson, H.; Torstensson, G.; Bergström, L. Leaching and crop uptake of N, P and K in a clay soil with organic and conventional cropping systems on a clay soil. *Soil Use Manag.* **2007**, *23*, 71–81. [CrossRef]
51. Torstensson, G.; Aronsson, H.; Bergström, L. Nutrient use efficiencies andleaching of organic and conventional cropping systems in Sweden. *Agron. J.* **2006**, *98*, 603–615. [CrossRef]
52. Worthington, V. Nutritional quality of organic versus conventional fruits, vegetables and grains. *J. Altern. Complement. Med.* **2001**, *7*, 161–73. [CrossRef] [PubMed]
53. Crinnion, W.J. Organic foods contain higher levels of certain nutrients, lower levels of pesticides and may provide health benefits for the consumer. *Altern. Med. Rev.* **2010**, *15*, 4–12. [PubMed]
54. Barański, M.; Srednicka-Tober, D.; Volakakis, N.; Seal, C.; Sanderson, R.; Stewart, G.B.; Benbrook, C.; Biavati, B.; Markellou, E.; Giotis, C.; et al. Higher antioxidant and lower cadmium concentrations and lower incidence of pesticide residues in organically grown crops: A systematic literature review and meta-analyses. *Br. J. Nutr.* **2014**. [CrossRef] [PubMed]

55. Mitchell, A.E.; Hong, Y.J.; Koh, E.; Barret, D.M.; Bryant, D.E. Ten-year comparison of the influence of organic and conventional crop management practices on the content of flavonoids in tomatoes. *J. Agric. Food Chem.* **2007**, *55*, 6154–6159. [CrossRef] [PubMed]
56. Deligios, P.A.; Tiloca, M.T.; Sulas, L.; Buffa, M.; Caraffini, S.; Doro, L.; Sana, G.; Spanu, E.; Spissu, E.; Urraci, G.R.; et al. Stable nutrient flows in sustainable and alternative cropping systems of globe artichoke. *Agron. Sustain. Dev.* **2017**, *37*, 54. [CrossRef]
57. Spanu, E.; Deligios, P.A.; Azara, E.; Delogu, G.; Ledda, L. Effects of alternative cropping systems on globe artichoke qualitative traits. *J. Sci. Food Agric.* **2018**, *98*, 1079–1087. [CrossRef] [PubMed]
58. Parrott, N.; Marsden, T. *The Real Green Revolution: Organic and Agro-Ecological Farming in the South*; Greenpeace Environmental Trust: London, UK, 2002; p. 153.
59. Akinnifesi, F.K.; Ajayi, O.C.; Sileshi, G.; Chirwa, P.W.; Chianu, J. Fertilizer tree systems for sustainable food security in the maize-based production systems of East and Southern Africa Region: A review. *J. Sustain. Dev.* **2010**, *30*, 615–629.
60. Timsina, J.; Jat, M.L.; Majumdar, K. Rice-maize systems of South Asia: Current status, future prospects and research priorities for nutrient management. *Pant Soil* **2010**, *335*, 65–82. [CrossRef]
61. Bolinder, M.A.; Janzen, H.H.; Gregorich, E.G.; Angers, D.A.; Vanden Bygaart, A.J. An approach for estimating net primary productivity and annual carbon inputs to soil for common agricultural crops in Canada. *Agric. Ecosyst. Environ.* **2007**, *118*, 29–42. [CrossRef]
62. Lützow, M.; Ottow, J.C.G. Effect of conventional and biological farming on microbial biomass and its nitrogen turnover in agriculturally used Luvisols of the Friedberg plains. *Z. Pflanzenernähr. Bodenk* **1994**, *157*, 359–367.
63. Petersen, C.; Drinkwater, L.; Wagoner, P. *The Rodale Institute Farming 750 Systems Trial: The First 15 Years*; Rodale Institute: Kutztown, PA, USA, 1999.
64. Burkitt, L.L.; Small, D.R.; McDonald, J.W.; Wales, W.J.; Jenkin, M.L. Comparing irrigated biodynamic and conventionally managed dairy farms. 1. Soil and pasture properties. *Aust. J. Exp. Agric.* **2007**, *47*, 479–488. [CrossRef]
65. Leifeld, J.; Fuhrer, J. Organic farming and soil carbon sequestration: What do we really know about the benefits? *Ambio* **2010**, *39*, 585–599. [CrossRef] [PubMed]
66. Bijay-Singh. Are nitrogenous fertilizers deleterious to soil health? *Agronomy* **2018**, *8*, 48. [CrossRef]
67. Recous, S.; Robin, D.; Darwis, D.; Mary, B. Soil inorganic nitrogen availability: Effect on maize residue decomposition. *Soil Biol. Biochem.* **1995**, *27*, 1529–1538. [CrossRef]
68. Timsina, J.; Singh, S.B.; Timsina, D. Integration of crop, animal and tree in rice-based farming systems of hills and Terai of Nepal: Some successful cases. In *Proceeding of Crop-Livestock Integration Workshop, Asian Rice Farming Systems Network*; IRRI: Los Banos, Philippines, 1991.
69. Erisman, J.W.; Sutton, M.A.; Galloway, J.N.; Klimont, Z.; Winiwarter, W. How a century of ammonia synthesis changed the world? *Nat. Geosci.* **2008**, *1*, 636–639. [CrossRef]
70. Bijay-Singh; Shan, Y.H.; Johnson-Beebout, S.E.; Yadvinder-Singh; Buresh, R.J. Crop residue management for lowland rice-based cropping systems in Asia. *Adv. Agron.* **2008**, *98*, 117–199.
71. Thuy, N.H.; Shan, Y.; Bijay, S.; Wang, K.; Cai, Z.; Yadvinder, S.; Buresh, R.J. Nitrogen supply in rice-based cropping systems as affected by crop residue management. *Soil Sci. Soc. Am. J.* **2008**, *72*, 514–523. [CrossRef]
72. Yadvinder-Singh; Bijay-Singh; Timsina, J. Crop residue management for nutrient cycling and improving soil productivity in rice-based cropping systems in the tropics. *Adv. Agron.* **2005**, *85*, 269–407.
73. Chuan, L.; He, P.; Jin, J.; Li, S.; Grant, C.; Xu, X.; Qui, S.; Zhao, S.; Zhou, W. Estimating nutrient requirements for wheat in China. *Field Crop. Res.* **2013**, *146*, 96–104. [CrossRef]
74. Janssen, B.H.; Guiking, F.C.T.; Van der Eijk, D.; Smaling, E.M.A.; Wolf, J.; Van Reuier, H.A. System for quantitative evaluation of the fertility of tropical soils (QUEFTS). *Geoderma* **1990**, *46*, 299–318. [CrossRef]
75. Jiang, W.T.; Liu, X.H.; Qi, W.; Xu, X.N.; Zhu, Y.C. Using QUEFTS model for estimating nutrient requirements of maize in the Northeast China. *Plant Soil Environ.* **2017**, *63*. [CrossRef]
76. Setiyono, T.D.; Walters, D.T.; Cassman, K.G.; Witt, C.; Dobermann, A. Estimating maize nutrient uptake requirements. *Field Crop. Res.* **2010**, *118*, 158–168. [CrossRef]
77. Halberg, N.; Kristensen, I.S. Expected crop yield loss when converting to organic dairy farming in Denmark. *Biol. Agric. Hort.* **1997**, *14*, 25–41. [CrossRef]

78. Quinones, N.A.; Borlaug, N.E.; Dowswell, C.R. A fertilizer-based green revolution for Africa. In *Replenishing Soil Fertility in Africa*; Buresh, R.J., Sanchez, P.A., Calhourn, F., Eds.; SSSA Special Publications No. 51: Madison, WI, USA, 1997; pp. 81–95.
79. Fairhurst, T.H.; Witt, C.; Buresh, R.J.; Dobermann, A. (Eds.) *Rice: A Practical Guide to Nutrient Management*, 2nd ed.; International Rice Research Institute (IRRI): Laguna, Philippines; International Plant Nutrition Institute (IPNI); International Potash Institute (IPI): Singapore, 2007.
80. Witt, C.; Dobermann, A.; Abdulrachman, S.; Gines, H.C.; Wang, G.; Nagarajan, R.; Satawatananont, S.; Son, T.T.; Tan, P.S.; Tiem, L.V.; et al. Internal nutrient efficiencies of irrigated lowland rice and in tropical and sub-tropical Asia. *Field Crop. Res.* **1999**, *63*, 113–138. [CrossRef]
81. Witt, C.; Buresh, R.J.; Peng, S.; Balasubramanian, V.; Dobermann, A. Nutrient management. In *Rice: A Practical Guide to Nutrient Management*; Fairhurst, T.H., Witt, C., Buresh, R., Dobermann, A., Eds.; International Rice Research Institute (IRRI): Laguna, Philippines; International Plant Nutrition Institute (IPNI); International Potash Institute (IPI): Singapore, 2007; pp. 1–45.
82. Banayo, N.P.M.C.; Haefele, S.M.; Desamero, N.V.; Kato, Y. On-farm assessment of site-specific nutrient management for rainfed lowland rice in the Philippines. *Field Crop. Res.* **2018**, *220*, 88–96. [CrossRef]
83. Gupta, S.K.; Ghosh, M.; Kohli, A.; Sharma, S.; Singh, Y.K.; Kumar, S.; Kumar, U.; Lakshman, K. Site-Specific Nutrient Management with Rice-Wheat Crop Manager in South Bihar Alluvial Plain Zone of India. *Int. J. Pure App. Biosci.* **2017**, *5*, 1070–1074. [CrossRef]
84. Islam, S.; Timsina, J.; Salim, M.; Majumdar, K.; Gathala, M.K. Potassium Supplying Capacity of Diverse Soils and K-Use Efficiency of Maize in South Asia. *Agronomy* **2018**, *8*, 121. [CrossRef]
85. Pampolino, M.F.; Witt, C.; Pasuquin, J.M.; Johnston, A.; Fisher, M.J. Development approach and evaluation of the Nutrient Expert software for cereal crops. *Comput. Electron. Agric.* **2012**, *88*, 103–110. [CrossRef]
86. Saito, K.; Sharma, S. *Improving Smallholder Rice Farmers' Yields and Income in Asia and Sub-Saharan Africa. Framework of the 2017 e-Agriculture Call for Good and Promising Practices on the use of ICTs for Agriculture and Rural Development*; FAO: Rome, Italy, 2018.
87. Xu, X.; He, P.; Qiu, S.; Pampolino, M.F.; Zhao, S.; Johnston, A.M.; Zhou, W. Estimating a new approach of fertilizer recommendation across small-holder farms in China. *Field Crop. Res.* **2014**, *163*, 10–17. [CrossRef]
88. Mango, N.; Hebinck, P. Agroforestry: A second soil fertility paradigm? A case of soil fertility management in Western Kenya. *Cogent Soc. Sci.* **2016**, *2*, 121577. [CrossRef]
89. Sanchez, P. Improved fallows come of age in the tropics. *Agrofor. Syst.* **1999**, *47*, 3–12. [CrossRef]
90. Garrity, D.P. Agroforestry and the achievement of the millennium development goals. *Agrofor. Syst.* **2004**, *61*, 5–17. [CrossRef]
91. Garrity, D.; Akinnifesi, F.; Ajayi, O.; Sileshi, G.W.; Mowo, J.G.; Kalinganire, A.; Larwanou, M.; Bayala, J. Evergreen Agriculture: A robust approach to sustainable food security in Africa. *Food Secur.* **2010**, *2*, 197–214. [CrossRef]
92. Barnes, R.D.; Fagg, C.W. *Faidherbia Albida: Monograph and Annotated Bibliography*; Tropical Forestry Papers No 41; Oxford Forestry Institute: Oxford, UK, 2003; p. 281.
93. Kang, B.T.; Akinifessi, F.K. Agroforestry as an alternative land-use production system for the tropics. *Nat. Res.* **2000**, *24*, 137–151. [CrossRef]
94. Pittelkow, C.M.; Liang, X.; Linquist, B.A.; van Groenigen, K.J.; Lee, J.; Lundy, M.E.; van Gestel, N.; Six, J.; Venterea, R.T.; van Kessel, C. Productivity limits and potentials of the principles of conservation agriculture. *Nature* **2014**. [CrossRef] [PubMed]

14

Synergistic Effect of Sulfur and Nitrogen in the Organic and Mineral Fertilization of Durum Wheat: Grain Yield and Quality Traits in the Mediterranean Environment

Francesco Rossini, Maria Elena Provenzano, Francesco Sestili and Roberto Ruggeri *

Department of Agriculture and Forestry Science (DAFNE), University of Tuscia, via San Camillo de Lellis, 01100 Viterbo, Italy; rossini@unitus.it (F.R.); provenzano.mariaelena@gmail.com (M.E.P.); francescosestili@unitus.it (F.S.)

* Correspondence: r.ruggeri@unitus.it

Abstract: In recent years, awareness on sustainable land use has increased. Optimizing the practice of nitrogen fertilization has become crucially imperative in cropping management as a result of this current trend. The effort to improve the availability of organic nitrogen has incurred a bottleneck while seeking to achieve a high yield and quality performance for organic winter cereals. Field experiments were conducted, under rainfed Mediterranean conditions, over a period of two subsequent growing seasons. The objective was to investigate the effect of soil and foliar S application on the performance of three durum wheat cultivars fertilized with either organic or inorganic N. The hypothesis to be verified was if different S fertilization strategies could improve grain yield and quality when coupled with mineral or organic N fertilizer. There were three levels of treatment with mineral N fertilizer (120, 160 and 200 kg ha^{-1}), two levels of organic N fertilizer (160 and 200 kg ha^{-1}), two levels of S fertilizer applied to the soil (0 and 70 kg ha^{-1}), and two levels of foliar S application at flag leaf stage (0 and 5 kg ha^{-1}). Cultivars were Dylan, Iride and Saragolla. Analyzed traits were grain yield, yield components and quality features of grain. Overall, at the same N rate, grain yield and quality were markedly higher for mineral than organic N source. Cultivar × Year × N treatment interactions significantly affected grain yield and quality indices. Iride showed a high yield stability throughout the mineral N rates in the most favorable year (2011) and, in the same year, was the top performing cultivar in organic N treatments. Dylan was the top performing cultivar for protein content, while Saragolla for the SDS sedimentation test. Soil S fertilization had no effect on grain quality, whereas it significantly increased grain yield (+ 300 kg ha^{-1}) when coupled with organic rather than a mineral N source. However, foliar S application at flag leaf stage did not affect grain yield, but it significantly enhanced quality indices such as test weight (81 vs. 79.9 kg hL^{-1}), protein content (13.7% vs. 12.9 %) and SDS value (72.5 vs. 70.5 mm). A rate of 160 kg ha^{-1} of N (both mineral and organic) determined the optimal response for both grain yield and quality. Finally, soil and foliar application of S may help to contain the large yield and quality gap that still exists between mineral and organic fertilization of durum wheat.

Keywords: durum wheat; mineral N; organic N; S fertilization; grain quality; grain yield

1. Introduction

Durum wheat (*Triticum turgidum* L. subsp. *durum* (Desf.) Husn.) is an economically important crop cultivated worldwide. Europe-28 is by far the largest world durum wheat producer. In 2017, it was grown on 2.7 million hectares only in the European Union (EU), providing an output of about 9 million tons. The cultivation area of durum wheat in Europe is mostly concentrated in the

Mediterranean region: Italy, Spain and France together account for 80% of total EU production [1]. Italy is the top EU producer country and a traditional durum wheat growing region as it dedicates half of the total EU durum wheat area to this crop, thus accounting for 45% of the entire EU production, with a yield of about 3.2 t ha^{-1}.

Grain quality has become one of the most important goals for the breeders and growers [2,3], because it is essential in obtaining premium prices and meeting markets needs for high-quality end-products of durum wheat such as pasta, couscous and burghul [4].

There is no simple and complete definition for the quality of durum wheat [4,5]. Grain protein content, color and gluten strength are considered the most important features needed for use in pasta and bread production. Grain protein content is known to be influenced by climatic parameters, genetic factors, nitrogen fertilizer rate, time of nitrogen application, residual soil nitrogen and available moisture during grain filling [6–9]. The yellow color is due to the carotenoid pigment content in the whole kernel, and it is commercially identified as the yellow index in semolina. Besides their role as an important aesthetic parameter, the carotenoids have important nutritional and health characteristics [10]. While yellow index was found to be affected by weather conditions, cultivar and N rate and timing [10–12], less is known about the effect of N source and S fertilization on this quality index.

Gluten strength contributes to the ability of dough to rise and maintain its shape as it is baked. Gluten strength is commonly estimated using the sodium dodecyl sulfate (SDS) sedimentation test that, depending on the protein quality, provides a good indicator of pasta cooking quality [13–15]. Ercoli et al. (2011) [15] found that SDS value increases with the increase of inorganic N (from 120 to 180 kg ha^{-1}) and S (from 0 to 60 kg ha^{-1}) rate.

The use of nitrogen is normally considered a key factor in cereal crops and numerous studies on the best N fertilization rates and timing have been conducted. In fact, if on the one hand it has been proved that nitrogen positively affects grain yield and quality, on the other hand N fertilizer management is pivotal to avoiding N losses caused by leaching, runoff, denitrification or volatilization [11,16,17].

After taking all this into account, the use of organic N fertilizers may be a further option, together with other cropping management and practices, to reduce nitrate pollution and improve the environmental sustainability of conventional farming systems [12]. Thus, another feature can be added to the definition of the quality of wheat products [18].

Moreover, fertility management was the identified key factor in limiting both yield and grain protein content in the organic wheat management [12,19,20]. The results on common wheat (*Triticum aestivum* L.) emphasize the importance of a sufficient supply of soils with organic fertilizers as well as the need to improve the availability of organic nitrogen [19,21]. This latter option might be accomplished by trying to regulate the degradation and mineralization of organic matter (OM) in the soils, which is the traditional role assigned to heterotrophic microbes [22]. The number of these microorganisms, and in particular those which oxidize sulfur (S), was found to be: (i) greater in some rizospheres (e.g., canola and wheat) than in bulk soil controls [23]; and (ii) stimulated by S fertilization and soil OM [24]. A recent study conducted in the Canadian prairie showed that common wheat biomass production in organic systems was positively related (among other factors) to the plant tissue S concentration [21]. Thus, S application to the soil might have a synergistic effect with organic N fertilization of durum wheat, determining higher yields and better quality. This hypothesis could be particularly verified in those agroecosystems that extend along the Mediterranean coast, in which soil temperature and water availability during the winter season do not drastically reduce mineralization capacity by the soil biota.

Sulfur is an essential element for all organisms since it is present in many molecules (amino acids, oligopeptides, vitamins and many secondary metabolites) and it is involved in several biochemical processes. Plants absorb S as sulfate ion (SO_4^{2-}) from soil solution and use it in key steps of their metabolism [25]. Furthermore, findings provide evidence for the uptake and metabolization of elemental S also at the leaf level [26,27]. However, the fact that symptoms of deficiency appear earlier in young leaves than mature ones suggests that S is relatively immobile in mature leaves and that the re-distribution from vegetative tissues to wheat kernels is noticeably less than that of N and P[28]. These are significant findings when considering foliar S application from flag leaf emergence to anthesis, aiming to reach greater efficiency in the S fertilization [29,30].

The importance of S in plant nutrition is highlighted by the fact that a limited availability of this element causes both direct (biomass reduction) and indirect production loss [28,31–33]. The indirect effects on plants productivity are attributable to the role that S plays in the synthesis of several metabolites such as Sulfur-containing Defense Compounds (SDC) involved in the physiological response to biotic and abiotic stresses [34–36]. Such considerations have led to studying grain yield and quality responses to S fertilizer and thus developing improved N and S fertilization strategies [11,16,37,38].

Although many studies have been conducted on the influence of N and S fertilization on common wheat characteristics such as growth, yield, quality and technological properties [28–30,38–40], still very little is known about the effect of S and N nutrition on grain yield and quality of durum wheat.

Studies conducted in the Mediterranean basin show different results leading to different conclusions. Garrido-Lestache et al. (2005) [11] in a three-year field experiment in southern Spain found that soil or leaf application of S had no effect on quality indices, with the exception of ash content. Conversely, Lerner et al. (2006) [41] in Argentina described a positive effect of sulfur fertilization on wheat quality traits. A similar result was observed in Southern Italy [42] and by Ercoli et al. (2011) [15] in Central Italy, even if they did not find a significant effect of S fertilization on grain protein content. Moreover, to the best of our knowledge, no studies have been conducted on durum wheat aiming to test simultaneously the interaction of S fertilization type with both mineral and organic N source.

Thus, the aims of the present study were: (1) to evaluate the effect of different N-S fertilization rates and types on grain yield and quality of three durum wheat cultivars representative of the Mediterranean region; and (2) to verify the hypothesis that soil S fertilization has a synergetic effect with organic N fertilization, on improving grain yield and quality of durum wheat.

2. Materials and Methods

2.1. Site and Experimental Design

A field experiment was set up in Tarquinia, Central Italy (42°12′ N, 11°45′ E; altitude: 22 m a.s.l.), during the 2010–2011 and 2011–2012 growing seasons. The area has a Mediterranean climate, with a mean air temperature of 15.5 °C and a mean annual precipitation of 658 mm. The weather data were retrieved from a meteorological station located in Tarquinia, at a short distance from the site. Meteorological data were characterized by a consistent difference between the growing seasons, particularly in terms of precipitation, so that in 2010–2011 rainfall was 38% higher compared to the 20-year average rainfall, while in 2011–2012 it was lower by 58%. Mean monthly temperature and total rainfall during 2010–2011 and 2011–2012 growing seasons are shown in Figure 1.

Temperatures were similar in the 2010–2011 and 2011–2012 growing seasons and, compared to the 20-year averages, were higher by 2 °C over the period from sowing to tillering stage and lower by 5.5 °C on average from February until grain ripening. Rainfall were under the average for more than five months in 2012.

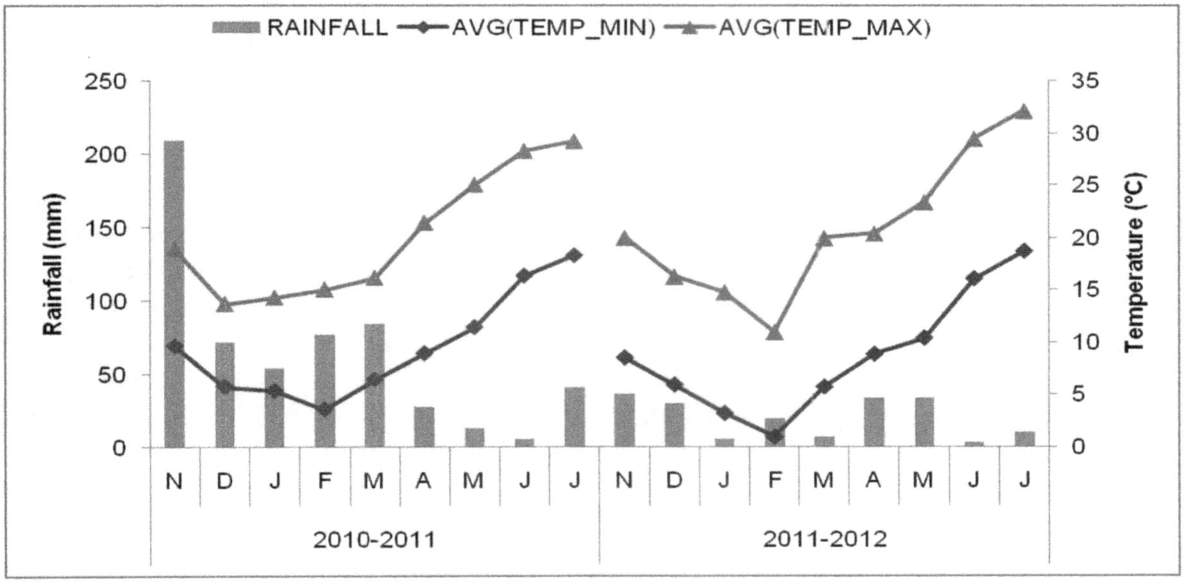

Figure 1. Minimum and maximum temperatures and rainfall recorded during the growing season (November–July) in 2010–2011 and 2011–2012.

Soil samples were collected from fields in both years before sowing. Samples, taken at 0–40 cm depth, were oven dried, grounded and then analyzed for texture and chemical analysis: pH, OM, sulfur content, total N, and total carbonate content. The soil was classified as clay according to the International Soil Science Society (ISSS) classification. The relevant soil characteristics are presented in Table 1.

Table 1. Soil characteristics of the experimental sites.

Parameter	Unit	2010–2011	2011–2012
Clay (Ø < 2 μm)	%	45.2	55.9
Silt (2.0 < Ø < 20 μm)	%	20.5	19.5
Sand (2.0 > Ø > 0.02 mm)	%	34.3	24.6
pH		7.4	7.2
Organic matter	%	1.8	1.6
Total $CaCO_3$	%	3.8	0.4
Total N	%	0.1	0.1
Available P	mg kg^{-1}	11.4	10.1
Exchangeable K	mg kg^{-1}	488.0	452.0
Available S	mg kg^{-1}	8.2	4.16

A split-split plot design with three replications was used: nitrogen fertilization levels was the main treatment, sulfur soil and foliar fertilization were the sub-treatments and varieties were the sub-sub-treatment. N-fertilization was arranged in five main plots while soil sulfur rates in two subplots in each main plot, as well as foliar sulfur rates and the three varieties were arranged in three sub-subplots in each subplot.

At the end of summer, the experimental field was ploughed at 30 cm depth and then divided into plots and subplots with three replicates for a total of 180 sub-subplots. The area of each sub-subplot was 180 m^2. Plots were sown on 20 January 2010 and 23 December 2011 at a seeding rate of 350 viable seeds m^{-2}. Three durum wheat varieties (Dylan, Iride and Saragolla) were chosen as representative of the cultivation area. They are widely adapted to different Mediterranean environments and characterized by high and constant productivity and good grain quality. Iride and Saragolla are early maturing and medium size varieties, while Dylan is medium-late maturing having a medium-taller size. All of them are relatively new varieties, released and registered in the Italian register of varieties since 1996 (Iride), 2004 (Saragolla) and 2002 (Dylan). The preceding crop was tomato (*Solanum lycopersicum* L.) in the first season and melon (*Cucumis melo* L.) in the second season.

Different nitrogen fertilizers (organic and mineral) and rates were applied: for organic fertilization 160 and 200 kg ha^{-1} of N (hereafter referred to as NO160 and NO200, respectively) and for the mineral fertilization 120, 160 and 200 kg ha^{-1} of N (hereafter referred to as NM120, NM160 and NM200, respectively). All NM plots received 92 kg ha^{-1} of P_2O_5 before sowing as diammonium phosphate. Nitrogen doses were determined by considering the minimum crop requirement of 3 kg of N per 100 kg of grain produced [43] and the more common yields recorded for the above mentioned cultivars in that environment (4–6 t ha^{-1}). In addition, subplots were treated with four combinations of sulfur fertilization: nil (hereafter referred to as SS0 or FS0), granular soil-sulfur fertilization (70 kg ha^{-1} of elemental S, hereafter referred to as SS70), foliar fertilization (5 kg ha^{-1} of S, hereafter referred to as FS5) and soil and foliar fertilization. Foliar S fertilization was applied at flag leaf emergence stage and soil S fertilization before sowing, as well as organic nitrogen fertilizer. This latter was a pelletized organic NP fertilizer (6% N; 3% P_2O_5; 30% C) derived from the fermentation of organic materials such as feather meal, bone meal, manure, etc. Nitrogen mineral fertilization was split as follows: (i) 36 kg ha^{-1} as diammonium phosphate for all rates at sowing; and (ii) 42–42, 62–62 and 82–82 kg ha^{-1} in the form of ammonium nitrate at early tillering stage and flag leaf emergence for NM120, NM160 and NM200, respectively. Weeds and diseases were chemically controlled.

2.2. Sampling and Measurements

Grain yield was determined at 13% moisture content, harvesting 15 m^2 sampling areas. At the same time, one square meter of plants was cut and then processed to obtain the following grain yield components: number of spikes, number of kernels per spike, and mean kernel weight. From each main sample, a sub-sample of grains was taken for the following measurements: test weight, vitreousness, thousand kernels weight, protein content, sodium dodecyl sulfate (SDS) sedimentation test and yellow index. Grain test weight was measured by the Schopper chondrometer. To determine protein content and yellow index and perform the SDS sedimentation test, samples were ground and analyzed using a Foss NIR System 6500 monochromator (Foss NIR Systems Inc., Silver Spring, Laurel, MD, USA), equipped with a sample transport module and a small ring cup. Prior to taking the measurements, the instrument was validated according to the diagnostic procedure of Win ISI II (InfraSoft International, LLC., Port Mathilda, PA, USA) software. The calibration equation developed at the Società Produttori Sementi Spa (Bologna, Italy) in accordance with the NIR guidelines for prediction model development [44] was used in this study. The range of wavelength used for analyses was set from 400 to 2500 nm and recorded at 2-nm intervals as log (1/R), where R represents decimal fraction transmittance [45]. Each sample was analyzed twice.

2.3. Data Analysis

All data were processed using R statistic software (R Development Core Team, 2006). Statistically significant differences among means were detected by the least significant difference test [46] after analysis of variance (ANOVA). The main effect of Year, N rate, S application and variety and their interactions were tested. Means were separated at 95% probability level. Each set of data was checked for normality and appropriate transformations were used, when necessary, prior to the ANOVA to improve normality [47].

3. Results

3.1. Yield and Yield Components

The analysis of variance for yield, protein content, test weight, SDS test, yellow index, vitrousness, and yield components are presented in Table 2. Treatments differently affected the measured traits.

Fourth and fifth order interaction was never significant. Third order interactions were significant only for V × N × Y.

For the same nitrogen treatment, grain yield was significantly lower in 2012 as compared to 2011 in all tested cultivars (Table 3). This may be largely attributed to the dramatic reduction of rainfall amount in 2012 which had a negative effect on the number of kernels per spike and mean kernel weight. A decrease in the number of kernels per spike were detected for Dylan in NO160 (−37%), Iride in both NO160 and NO200 (−36% and −40%, respectively) and Saragolla in NO200 treatment (−47%). On the contrary, the mean kernel weight showed a lower variation in the NO treatments (5–10%) than in the NM ones (18–20% in NM200) for all cultivars. Considering these results and the number of spikes per unit area, more yield decreases were registered for Iride in NM120 and NO200 plots (−52.6% and−52.5%, respectively) and for Saragolla in NM160 plots (−53.4%). In contrast, the lower differences in grain yield between 2011 and 2012 were found for Dylan in both NM160 and NO160 plots (−29.8%). Iride showed a high yield stability throughout the NM rates in the most favorable year (2011) and, in the same year, was the top performing cultivar in NO treatments. However, when limiting weather conditions occurred (2012), Dylan had a grain yield significantly higher than that of Iride and Saragolla, both in NO160 (+21% and +19%, respectively) and NO200 (+23% and +19%, respectively). In general, at the same N rate, yield responses were dramatically higher for mineral than organic N because of a significantly higher number of both spikes per unit area and kernels per spike. Cultivars differently responded to the increase of N rate, both for mineral and organic form. Particularly, 2011 grain yield significantly increased in Dylan by 11% from NM160 to NM200, while it did not change significantly between NO rates. By contrast, grain yield did not significantly vary in Iride through NM rates but it increased significantly from NO160 to NO200 (+11%). Finally, Saragolla showed the best yield performance at 160 kg ha^{-1} of N, both mineral and organic form, since no significant increases were detected at the higher rate. NM160 was also the best solution in 2012 with the exception of Saragolla which yielded significantly higher with NM200 (+25%).

Regarding the effect of S fertilization, SS × N interaction significantly affected grain yield, highlighting a positive effect of S soil application in both organic nitrogen fertilization rates, while no significant effect was detected for mineral nitrogen (Table 4). This was due to a significant increase in both the number of spike per unit area and the number of kernels per spike (this latter significant only for NO160).

S foliar treatment had no significant effect on yield (Table 5), as it increased the number of kernels per spike (+3.4%) but decreased the mean kernel weight (−2.4%).

3.2. *Quality Traits*

As expected from such an assorted collection of cultivars and fertilization treatments, quality characteristics of grain varied considerably.

As for yield, V × N × Y interaction was significant for all quality traits.

For the same nitrogen treatment, test weight significantly decreased in 2012 in all tested cultivars (Table 6). This was due to kernel shriveling caused by the severe drought and heat stress which occurred in the second year during grain filling. NM200 treatment resulted in the highest test weight decrease between the two years for all cultivars (from 14.3% of Dylan to 18.2% of Saragolla). On the contrary, organic fertilization showed the lowest differences in the test weight values between the two years (from 4.3% of Dylan in NO200 to 8.7% of Iride in NO160). Moreover, Dylan was the only cultivar which accomplished market request for test weight in 2012, overcoming 80 kg hL^{-1} in NO treatments. Soil sulfur fertilization had no significant effect on the test weight (data not shown), whereas foliar S application slightly increased this trait from 79.9 to 81 kg hL^{-1} (Table 5).

Table 2. Results of the ANOVA on different grain yield and quality traits.

Main Effect	Significance								
	Grain Yield	Grain Protein	Test Weight	SDS Test	Yellow Index	Vitreous Kernels §	Spikes m^{-2}	Kernels $Spike^{-1}$	Kernel Weight
Year (Y)	***	**	**	***	***	***	***	***	***
Nitrogen (N)	***	***	***	***	***	***	***	***	***
Variety (V)	***	***	***	***	***	***	***	***	***
Soil sulfur (SS)	n.s.	n.s.	n.s.	n.s.	n.s.	***	n.s.	*	***
Foliar sulfur (FS)	n.s.	***	***	***	n.s.	***	n.s.	***	***
Two-way interactions									
N × Y	***	***	***	***	n.s.	***	***	***	***
V × Y	***	***	***	***	***	***	**	***	**
V × N	n.s.	***	*	***	n.s.	**	***	***	*
SS × N	***	n.s.	*	n.s.	n.s.	*	***	***	***
FS × SS	*	n.s.	n.s.	*	n.s.	n.s.	n.s.	n.s.	***
Three-way interactions									
V × N × Y	***	***	**	***	*	**	***	***	**

ANOVA signif. codes: '***' < 0.001; '**' < 0.01; '*' < 0.05; '.' < 0.1; n.s.: not significant. Other interactions are not reported since they are not significant. § transformed data were used to perform the ANOVA.

Table 3. Yield related traits. Variety × Nitrogen × Year interaction.

Yield Related Traits		NM120			NM160			NM200			NO160			NO200		
		Dylan	Iride	Saragolla	Dylan	Iride	Saragolla	Dylan	Iride	Saragolla	Dylan	Iride	Saragolla	Dylan	Iride	Saragolla
Grain yield (t ha^{-1})	2011	5.95	6.46	5.81	6.07	6.47	6.83	6.75	6.79	6.72	5.26	5.54	5.19	5.53	6.06	5.32
	2012	3.55	3.06	3.38	4.26	4.43	3.18	4.13	3.91	3.96	3.69	3.06	3.11	3.54	2.88	2.98
	LSD	0.36 ***														
Kernels spike^{-1} (n)	2011	39.22	41.05	38.10	41.95	43.16	42.92	41.95	41.97	39.20	35.72	40.13	36.49	38.72	39.18	41.20
	2012	28.02	26.34	26.01	36.30	35.37	30.36	35.88	30.93	32.62	22.51	25.55	26.47	27.88	23.35	21.71
	LSD	1.97 ***														
Spikes m^{-2} (n)	2011	405.2	464.1	441.9	437.4	477.2	497.8	484.1	543.6	520.3	403.9	401.5	415.8	398.8	455.5	378.8
	2012	458.8	468.4	484.9	473.8	495.6	510.3	492.6	544.4	509.2	462.7	438.6	409.4	405.9	442.2	472.6
	LSD	13.44 ***														
Mean kernel weight (mg)	2011	47.65	42.32	43.88	46.35	40.56	43.99	44.61	41.28	43.76	51.18	46.14	47.18	51.00	46.35	48.35
	2012	39.37	37.75	40.06	39.10	36.40	34.47	35.62	33.68	35.32	48.19	42.50	44.82	46.81	43.78	43.68
	LSD	1.83 ***														

ANOVA signif. codes: '***' < 0.001; LSD: least significant difference ($p < 0.05$).

Table 4. Yield related traits. Soil Sulfur × Nitrogen interaction.

Trait		NM120	NM160	NM200	NO160	NO200
Grain yield (t ha^{-1})	SS0	4.63 cd	5.28 ab	5.34 a	4.17 e	4.23 e
	SS70	4.77 c	5.13 b	5.41 a	4.45 d	4.54 d
Kernels spike^{-1} (n)	SS0	32.13 d	39.46 a	36.60 b	30.22 e	31.72 d
	SS70	34.12 c	37.23 b	37.58 b	32.07 d	32.30 d
Spikes m^{-2} (n)	SS0	455.4 c	477.2 b	516.5 a	411.6 e	417.2 e
	SS70	452.3 c	486.8 b	514.9 a	432.4 d	434.0 d
Mean kernel weight (mg)	SS0	42.25 d	40.24 e	39.38 ef	48.33 a	46.35 b
	SS70	41.43 d	40.05 e	38.71 f	45.01 c	46.97 b

For each trait, numbers followed by the same letter are not significantly different at $p < 0.05$.

Table 5. Yield and quality related traits. Foliar sulfur fertilization mean values.

Treatment	Grain Yield (t ha^{-1})	Kernels Spike^{-1} (*n*)	Spikes m^{-2} (*n*)	Mean Kernel Weight (mg)	Test Weight (kg hL^{-1})	Protein Content (%)	SDS Test (mm)	Yellow Index
FS0	4.76	33.77	457.7	43.39	79.88	12.86	70.53	24.81
FS5	4.83	34.91	462.0	42.35	80.97	13.69	72.53	24.72
ANOVA signif.	n.s.	***	n.s.	***	***	***	***	n.s.

ANOVA signif. codes: '***' < 0.001; n.s.: not significant.

Table 6. Quality related traits. Variety × Nitrogen × Year interaction.

Quality Related Traits		NM120			NM160			NM200			NO160			NO200		
		Dylan	Iride	Saragolla	Dylan	Iride	Saragolla	Dylan	Iride	Saragolla	Dylan	Iride	Saragolla	Dylan	Iride	Saragolla
Test weight (kg hL^{-1})	2011	85.35	85.05	84.43	84.69	85.47	84.86	84.60	86.13	84.95	85.43	84.98	83.79	85.15	85.33	84.68
	2012	75.09	74.82	73.74	78.11	73.71	71.50	72.53	71.94	69.46	80.94	77.61	78.83	81.49	79.70	78.39
	LSD	1.61 ***														
Protein content (%)	2011	14.01	12.00	11.59	14.52	12.72	13.30	15.01	12.49	13.49	11.28	10.23	10.44	11.49	10.54	11.55
	2012	14.80	14.67	15.22	16.29	16.58	16.86	16.45	15.58	16.35	12.69	12.29	11.85	11.19	11.23	11.44
	LSD	0.72 ***														
SDS (mm)	2011	71.28	67.88	70.78	69.55	68.07	74.60	74.76	66.64	75.47	58.59	59.65	66.79	62.19	62.22	71.45
	2012	76.91	73.36	82.19	73.04	79.36	85.50	81.27	83.15	90.75	64.58	67.85	72.68	60.42	64.91	70.00
	LSD	3.69 ***														
Yellow index	2011	23.52	21.66	23.78	23.59	21.49	23.87	23.78	20.98	23.92	22.84	21.10	23.77	22.85	21.10	23.46
	2012	27.84	25.44	27.06	27.43	25.79	27.83	28.02	26.17	27.09	27.13	25.40	26.79	27.23	24.96	27.05
	LSD	0.57 ***														

ANOVA signif. codes: '***' < 0.001; LSD: least significant difference ($p < 0.05$).

Grain protein content was significantly higher in 2012 than 2011 for all cultivars and N treatments with the exception of NO200, in which it remained substantially unchanged (Table 6). This increase between the two years was due to the fact that in 2012 grains were shriveled, with low starch accumulation, resulting in higher protein concentration [4]. Dylan showed a protein content significantly higher than Iride and Saragolla for each N fertilization treatment in 2011 with the exception of NO200. Even with lower rates, mineral nitrogen fertilization always showed a significantly higher protein content with respect to NO treatments, with the exception of NM120, where the protein content was similar to that of NO200 for Saragolla in 2011.

Even though grain protein content was not affected when sulfur was applied to the soil (data not shown), foliar S fertilization had a positive effect, increasing this trait significantly from 12.9% to 13.7% (Table 5).

As for grain protein content, the sedimentation values (SDS test) increased markedly in 2012 over 2011, with the exception of NO200 treatments (Table 6). Generally, mineral fertilization showed higher values than NO and the best performing cultivar for this trait was Saragolla in both years.

Similar to grain protein concentration, S fertilization applied to the soil had no effect on the sedimentation values (data not shown), while foliar application significantly increased this trait, especially when coupled with soil application (Figure 2).

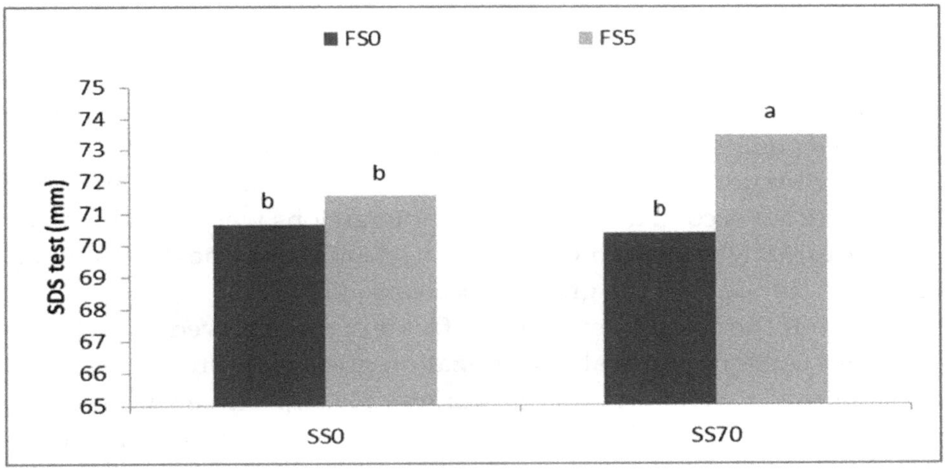

Figure 2. SDS sedimentation test. Foliar and soil sulfur interaction. Bars sharing the same letter are not significantly different ($p < 0.05$).

Values for the yellow index from 2012 were significantly higher than those from 2011 for all cultivars and N treatments (Table 6). In general, Dylan and Saragolla showed similar results for each N treatment while Iride always highlighted a significantly lower yellow index. Moreover, the increase of N rate had a significant effect on this trait just in two circumstances (from NM120 to NM200 for Iride and from NM120 to NM160 for Saragolla, both in 2012).

Sulfur fertilization had no effect on this trait.

Similar to the other quality traits, vitrousness was also affected by N × V × Y interaction, showing significantly higher values in 2012 than 2011 for all cultivars and N treatments (Figure 3). Compared to the other cultivars, Dylan showed a significantly higher percentage of vitrous kernels in 2011 NM treatments, while it performed similar to Iride and Saragolla in 2011 NO200 fertilization and in all 2012 N treatments. Mineral nitrogen fertilization always caused a significantly higher percentage of vitrous kernels than NO treatments, with the exception of NM120 compared with NO 200 for Saragolla in 2011.

Consistent with the result on grain protein concentration, vitrousness of kernels increased with foliar sulfur application from 80.8% to 83.5% (data not shown).

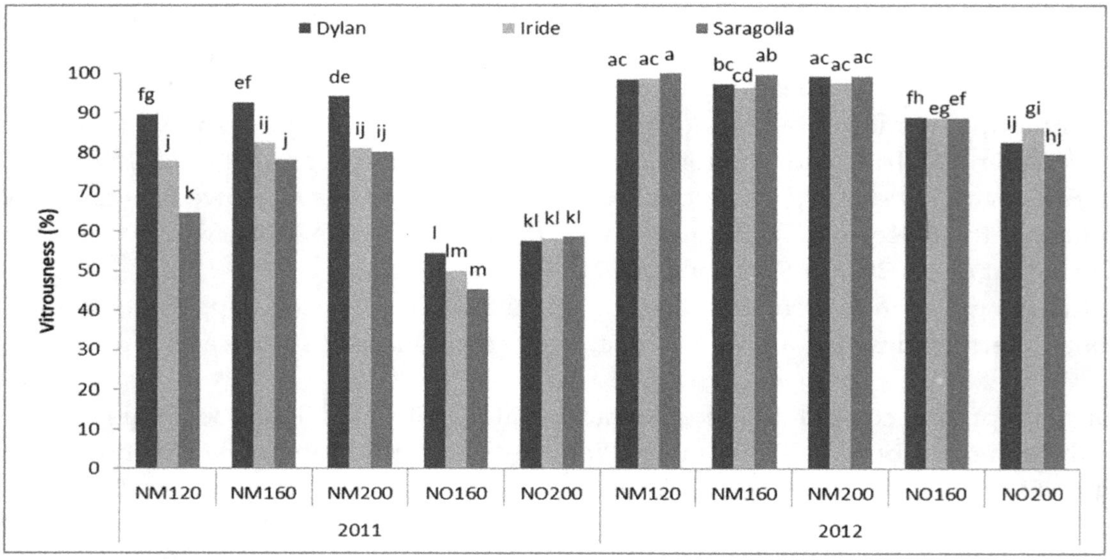

Figure 3. Grain vitrousness. Nitrogen × variety × year interaction. Bars sharing the same letter are not significantly different ($p < 0.05$).

4. Discussion

Results of this study demonstrate that fertilization type and rate have a strong influence on durum wheat yield, yield components and quality characteristics of grain. However, as also found by other authors [6,11,17], crop growth, yield and quality traits are mainly a function of environmental conditions. In fact, significant second and third order interactions were found which confirm the year on year variations for durum wheat production. This variability may be due to changes in the rainfall amount and distribution throughout the growing seasons [48].

Overall, organic fertilization was a determining factor in the observed reduction of grain yield and quality. As other studies confirm, mineral N fertilization gives better results as compared to organic fertilization, either in terms of yield and protein content [49,50]. These studies reported that winter wheat receiving organic fertilization had yields up to 19% lower than that fertilized with mineral N, on average. In our study, considering only the N effects and the same N rate, mineral nitrogen fertilization gained significantly higher grain yield than NO, ranging from +21% to +23%.

Similarly, protein content and SDS test values were higher for mineral fertilization as compared to the organic one, and they increased with the increase of nitrogen rate. These results are consistent with those by other authors and corroborate the issue of N availability in the case of organic fertilization, which is limited during the crop reproductive phases and always lower when compared to mineral nitrogen [12,50]. Systems that are based on organic fertilization usually have very different seasonal N cycle and availability than those that use mineral fertilizers. Thus, reliance on organic N sources requires an understanding of organic N mineralization–immobilization and turnover patterns in relation to crop N demands and N loss pathways. Besides the dependence of mineralization on pedoclimatic conditions [51], it usually takes many years to mineralize past organic fertilizer and support crops with appropriate N availability [52]. Moreover, even if a balance is reached, winter crops generally suffer significant yield reduction due to slow mineralization during their growth cycle. Even though commercial organic fertilizer, such as that used in this study, contains this reduction due to its low C/N ratio and higher nutrient availability, the yield gap was found to be significant even after six years [53]. However, a long-term study would better clarify if a multi-year application of an organic fertilizer may reduce this yield gap, thanks to the positive effects that organic matter has on physical, chemical and biological properties of the soil.

In this study, yield gain for NM fertilization treatments was at 14.5% between 120 and 200 kg ha^{-1} while it was just at 3% from 160 to 200 kg ha^{-1}. On the contrary, no significant increase was detected with the increase of NO rate (from 160 to 200 kg ha^{-1}). Ercoli et al. (2011) [15] found an increase of 20% (from 120 to 180 kg ha^{-1} of NM) in similar climatic conditions. Other authors in Spain reported no grain yield response to NM rates of up to 100 kg ha^{-1} [11,54,55]. Such a large difference is probably because wheat yield is influenced by N rate only when the amount of rainfall exceeds 450 mm during the growing season, as reported by Lopez-Bellido et al. (1996) [48] in a long-term experiment. In fact, in the wetter year (2011), we obtained significantly different yields at each N rates (both NM and NO), whereas in the drier year (2012), NM160 and NM200 yields were similar. In 2012, only the NM120 yield was significantly lower, supporting the finding that the crop fertilized with 120 kg ha^{-1} of N, often has sub-optimal yield performance [15].

Concerning varieties, all the tested genotypes responded similarly to the year-on-year variations of climatic conditions, but were differently sensitive to N fertilization in each year. Particularly, all the three cultivars were markedly sensitive to water shortage which especially reduced the number of kernels per spike and the mean kernel weight. This behavior was also verified by Ercoli et al. (2011) [15] in medium and late-maturing varieties (Claudio and Creso, respectively), while in early or medium-early varieties they did not find any yield difference between wet and dry season. However, other authors also found short-cycle cultivars decreasing grain yield with the decrease of the rainfall amount in the Mediterranean environment [11,56]. The different behavior of cultivars in response to climatic fluctuations is of crucial importance for Mediterranean environments, because of the high year-to-year variability in rainfall and temperature pattern existing in the climate.

In our study, the yield performance of the cultivar Dylan in the drier year was unexpected. This medium-late variety was expected to yield poorly in the most limiting environmental conditions of the second year, while it performed similarly to early-maturing cultivars with NM treatments and even better with NO fertilization. This finding is consistent with the results from the Italian durum wheat network, which show that, in the last seven years (from 2011 to 2017) and in an environment comparable to that of this study, Dylan yielded similarly to Iride and/or Saragolla in five different seasons (2012, 2013, 2014, 2015, and 2017). Although Iride had the same yielding performance of Dylan, it did not have the same quality of grains, showing a significantly lower protein content, yellow index, test weight and grain vitrousness. Saragolla was the lowest yielding variety but that with the highest SDS value.

Concerning the effect of sulfur, it is known that elemental sulfur (ES) has to be oxidized to SO_4^{2-} before it is available for plants and that the response to sulfur fertilization can be very variable in wheat. Sulfur uptake and metabolization depends on soil N and S balance, water supply, timing and rates of N and S application [28,57,58]. With regard to grain yield, we found that, for a given N rate, the S application to the soil had a synergistic effect with organic rather than mineral N fertilization. Yield gains obtained from S fertilization ranged between 280 and 310 kg ha^{-1} (+7%) within a same NO rate and were caused by a higher number of both spikes per unit area and kernels per spike. Consistently with our findings, several studies reported a similar grain yield increase and suggested that S deficiency leads to a reduction of the number of spikelets or to an increase of floret mortality [28,39,59,60]. The synergistic effect of S with NO fertilization may be attributed to the higher rates of OM degradation achieved by the improved activity of the heterotrophic S-oxidizing microorganisms and the resulting release of other nutrients [61,62]. In fact, there is evidence which shows that organic amendments to the soil promote ES oxidation rates and that some specific rhizosphere (e.g., wheat and canola) may stimulate the proliferation of heterotrophic ES oxidizing microorganisms and arbuscular mycorrhyzal fungi [21,23,24,63–66]. Moreover, in a recent study, regression analysis showed that initial soil pH was the most important factor affecting ES oxidation, followed by OM content [67]. Specifically, in soils with pH above 6.65 and higher S and OM content, the ES oxidation rate was found to be significantly higher than that of the other soils. In our study, pH after fertilization treatments may have played an important role in the ES oxidation dynamic, considering that ammonium nitrate (used to fertilize the NM plots) has soil acidifying potential while organic fertilization often resulted in an increased soil pH [68].

Even though some authors found that S deficiency may have a significant effect on the synthesis and accumulation of proteins [29,69], generally, the S nutrition of wheat has a marked effect on the composition of the seed storage protein, rather than the concentration of total proteins in seeds [28,70]. Although we did not find any effect of S application to the soil on grain protein concentration and SDS test [15,71,72], it must be observed that foliar S application significantly increased both these traits. This effect could be due to a better assimilation of N and S, as previously reported by Tea et al. (2007) [30]. In that same study, the authors demonstrated that S applied by foliar spray was mainly assimilated in the grain and here it may favor N accumulation. Furthermore, many studies, as reviewed by Zhao et al. (1999) [28], demonstrated that the higher S accumulation in grain determines an increase in disulfide groups (polymeric glutenins), which are related to a higher gluten strength that means a higher SDS sedimentation value. Our results are consistent with those by Ercoli et al. (2011) [15], who showed that SDS sedimentation values and alveograph W were the quality indices highlighting the highest correlation with S concentration. The same authors argued that a high S concentration in grain is a key factor in obtaining a high quality pasta.

5. Conclusions

Our results evidenced the clear advantage of the mineral nitrogen fertilization when compared with the organic one in a relatively short time frame (two-year study). Nevertheless, for those cropping systems that base soil fertility on organic sources, soil S fertilization may be a winning strategy to improve grain yield in the Mediterranean environment. Even though future long-term studies should be conducted under strict organic conditions, also involving organic weed and disease control, results from our study may help to contain the large yield gap that farmers usually experience during the transition from a conventional system to an organic one or when they aim to improve the environmental sustainability of conventional farming systems. In general, we did not find any effect of soil S fertilization on the quality indices of durum wheat, whereas foliar S application proved to be a key factor in determining higher protein content and SDS value. Further studies are needed to verify if this finding results in improved rheological characteristics of grain. Concerning varieties, Iride and Dylan yielded similarly under both mineral and organic fertilization but Dylan had higher quality indices. However, all tested genotypes responded similarly to the year-on-year climatic fluctuations, which remain the most limiting factors for yield and quality performance of durum wheat in the Mediterranean region. Finally, from our results, a rate of 160 kg ha^{-1} of N (both mineral and organic) determined the optimal response in terms of both grain yield and quality.

Author Contributions: Conceptualization and supervision, F.R.; Data Curation, M.E.P., R.R. and F.S.; Writing—Original Draft Preparation and Review & Editing, R.R.

Acknowledgments: The authors gratefully acknowledge PANTANO cooperative (Tarquinia, Italy) for the collaboration during the project and Società Produttori Sementi Spa (Bologna, Italy) for NIR analysis.

References

1. Europoean Commission: Agriculture and Rural Development EU Crops Market Observatory—Cereals. Available online: https://ec.europa.eu/agriculture/market-observatory/crops/cereals/statistics_en (accessed on 20 Jul 2018).
2. Ceoloni, C.; Kuzmanović, L.; Ruggeri, R.; Rossini, F.; Forte, P.; Cuccurullo, A.; Bitti, A. Harnessing genetic diversity of wild gene pools to enhance wheat crop production and sustainability: Challenges and opportunities. *Diversity* **2017**, *9*, 55. [CrossRef]

3. Camerlengo, F.; Sestili, F.; Silvestri, M.; Colaprico, G.; Margiotta, B.; Ruggeri, R.; Lupi, R.; Masci, S.; Lafiandra, D. Production and molecular characterization of bread wheat lines with reduced amount of α-type gliadins. *BMC Plant Biol.* **2017**, *17*, 1–11. [CrossRef] [PubMed]
4. Troccoli, A.; Borrelli, G.M.; De Vita, P.; Fares, C.; Di Fonzo, N. Durum wheat quality: A multidisciplinary concept. *J. Cereal Sci.* **2000**, *32*, 99–113. [CrossRef]
5. Dexter, J.E.; Marchylo, B.A. Recent Trends in Durum Wheat Milling and Pasta Processing: Impact on Durum Wheat Quality Requirements. In *Proceedings International Workshop on Durum Wheat, Semolina and Pasta Quality: Recent Achievements and New Trends*; Abecassis, J., Autran, J., Feillet, P., Eds.; INRA: Montpellier, Franch, 2000; pp. 139–164.
6. Campbell, C.A.; Davidson, H.R.; Winkleman, G.E. Effect of Nitrogen, Temperature, Growth Stage and Duration of Moisture Stress on Yield Components and Protein Content of Manitou Spring Wheat. *Can. J. Plant Sci.* **1981**, *61*, 549–563. [CrossRef]
7. Rao, A.C.S.; Smith, J.L.; Jandhyala, V.K.; Papendick, R.I.; Parr, J.F. Cultivar and Climatic Effects on the Protein Content of Soft White Winter Wheat. *Agron. J.* **1993**, *85*, 1023–1028. [CrossRef]
8. Uhlen, A.K.; Hafskjold, R.; Kalhovd, A.-H.; Sahlström, S.; Longva, Å.; Magnus, E.M. Effects of Cultivar and Temperature During Grain Filling on Wheat Protein Content, Composition, and Dough Mixing Properties. *Cereal Chem. J.* **1998**, *75*, 460–465. [CrossRef]
9. Rharrabti, Y.; Villegas, D.; Del Moral, L.F.G.; Aparicio, N.; Elhani, S.; Royo, C. Environmental and genetic determination of protein content and grain yield in durum wheat under Mediterranean conditions. *Plant Breed.* **2001**, *120*, 381–388. [CrossRef]
10. Ficco, D.B.M.; Mastrangelo, A.M.; Trono, D.; Borrelli, G.M.; De Vita, P.; Fares, C.; Beleggia, R.; Platani, C.; Papa, R. The colours of durum wheat: A review. *Crop Pasture Sci.* **2014**, *65*, 1–15. [CrossRef]
11. Garrido-Lestache, E.; López-Bellido, R.J.; López-Bellido, L. Durum wheat quality under Mediterranean conditions as affected by N rate, timing and splitting, N form and S fertilization. *Eur. J. Agron.* **2005**, *23*, 265–278. [CrossRef]
12. Fagnano, M.; Fiorentino, N.; D'Egidio, M.G.; Quaranta, F.; Ritieni, A.; Ferracane, R.; Raimondi, G. Durum Wheat in Conventional and Organic Farming: Yield Amount and Pasta Quality in Southern Italy. *Sci. World J.* **2012**, *2012*, 1–9. [CrossRef] [PubMed]
13. Dexter, J.E.; Matsuo, R.R.; Kosmolak, F.G.; Leisle, D.; Marchylo, B.A. The Suitability of the sds-sedimentation test for assessing gluten strenght in durum wheat. *Can. J. Plant Sci.* **1980**, *60*, 25–29. [CrossRef]
14. Carter, B.P.; Morris, C.F.; Anderson, J.A. Optimizing the SDS sedimentation test for end-use quality selection in a soft white and club wheat breeding program. *Cereal Chem.* **1999**, *76*, 907–911. [CrossRef]
15. Ercoli, L.; Lulli, L.; Arduini, I.; Mariotti, M.; Masoni, A. Durum wheat grain yield and quality as affected by S rate under Mediterranean conditions. *Eur. J. Agron.* **2011**, *35*, 63–70. [CrossRef]
16. Alcoz, M.M.; Hons, F.M.; Haby, V.A. Nitrogen Fertilization Timing Effect on Wheat Production, Nitrogen Uptake Efficiency, and Residual Soil Nitrogen. *Agron. J.* **1993**, *85*, 1198–1203. [CrossRef]
17. Barraclough, P.B.; Howarth, J.R.; Jones, J.; Lopez-Bellido, R.; Parmar, S.; Shepherd, C.E.; Hawkesford, M.J. Nitrogen efficiency of wheat: Genotypic and environmental variation and prospects for improvement. *Eur. J. Agron.* **2010**, *33*, 1–11. [CrossRef]
18. Mäder, P.; Hahn, D.; Dubois, D.; Gunst, L.; Alföldi, T.; Bergmann, H.; Oehme, M.; Amadò, R.; Schneider, H.; Graf, U.; et al. Wheat quality in organic and conventional farming: Results of a 21 year field experiment. *J. Sci. Food Agric.* **2007**, *87*, 1826–1835. [CrossRef]
19. Bilsborrow, P.; Cooper, J.; Tétard-Jones, C.; Średnicka-Tober, D.; Barański, M.; Eyre, M.; Schmidt, C.; Shotton, P.; Volakakis, N.; Cakmak, I.; et al. The effect of organic and conventional management on the yield and quality of wheat grown in a long-term field trial. *Eur. J. Agron.* **2013**, *51*, 71–80. [CrossRef]
20. Mayer, J.; Gunst, L.; Mäder, P.; Samson, M.-F.; Carcea, M.; Narducci, V.; Thomsen, I.K.; Dubois, D. Productivity, quality and sustainability of winter wheat under long-term conventional and organic management in Switzerland. *Eur. J. Agron.* **2015**, *65*, 27–39. [CrossRef]
21. Dai, M.; Hamel, C.; Bainard, L.D.; Arnaud, M.St.; Grant, C.A.; Lupwayi, N.Z.; Malhi, S.S.; Lemke, R. Negative and positive contributions of arbuscular mycorrhizal fungal taxa to wheat production and nutrient uptake efficiency in organic and conventional systems in the Canadian prairie. *Soil Biol. Biochem.* **2014**, *74*, 156–166. [CrossRef]

22. Kirchman, D.L. Degradation of organic material. In *Processes in Microbial Ecology*; Oxford University Press: New York, NY, USA, 2012.
23. Graystone, S.; Germida, J. Influence of crop rhizospheres on populations and activity of heterotrophic sulfur-oxidizing microorganisms. *Soil Biol. Biochem.* **1990**, *22*, 457–463. [CrossRef]
24. Germida, J.J.; Janzen, H.H. Factors affecting the oxidation of elemental sulfur in soils. *Fertil. Res.* **1993**, *35*, 101–114. [CrossRef]
25. Scherer, H.W. Sulphur in crop production—Invited paper. *Eur. J. Agron.* **2001**, *14*, 81–111. [CrossRef]
26. Legris-Delaporte, S.; Ferron, F.; Landry, J.; Costes, C. Metabolization of elemental sulfur in wheat leaves consecutive to its foliar application. *Plant Physiol.* **1987**, *85*, 1026–1030. [CrossRef] [PubMed]
27. Tea, I.; Genter, T.; Naulet, N.; Morvan, E.; Kleiber, D. Isotopic study of post-anthesis foliar incorporation of sulphur and nitrogen in wheat. *Isot. Environ. Health Stud.* **2003**, *39*, 289–300. [CrossRef] [PubMed]
28. Zhao, F.J.; Hawkesford, M.; McGrath, S. Sulphur assimilation and effects on yield and quality of wheat. *J. Cereal Sci.* **1999**, *30*, 1–17. [CrossRef]
29. Tea, I.; Genter, T.; Naulet, N. Effect of foliar sulfur and nitrogen fertilization on wheat storage protein composition and dough mixing properties. *Cereal* **2004**, *81*, 759–766. [CrossRef]
30. Tea, I.; Genter, T.; Naulet, N.; Lummerzheim, M.; Kleiber, D. Interaction between nitrogen and sulfur by foliar application and its effects on flour bread-making quality. *J. Sci. Food Agric.* **2007**, *87*, 2853–2859. [CrossRef]
31. Blake-Kalff, M.M.A.; Hawkesford, M.J.; Zhao, F.J.; McGrath, S.P. Diagnosing sulfur deficiency in field-grown oilseed rape (*Brassica napus* L.) and wheat (*Triticum aestivum* L.). *Plant Soil* **2000**, *225*, 95–107. [CrossRef]
32. Zhao, F.J.; Wood, A.P.; McGrath, S.P. Effects of sulphur nutrition on growth and nitrogen fixation of pea (*Pisum sativum* L.). *Plant Soil* **1999**, *212*, 207–217. [CrossRef]
33. Zhao, F.; Evans, E.J.; Bilsborrow, P.E.; Syers, J.K. Influence of sulphur and nitrogen on seed yield and quality of low glucosinolate oilseed rape (*Brassica napus* L.). *J. Sci. Food Agric.* **1993**, *63*, 29–37. [CrossRef]
34. Chan, K.X.; Wirtz, M.; Phua, S.Y.; Estavillo, G.M.; Pogson, B.J. Balancing metabolites in drought: The sulfur assimilation conundrum. *Trends Plant Sci.* **2013**, *18*, 18–29. [CrossRef] [PubMed]
35. Nazar, R.; Iqbal, N.; Masood, A.; Syeed, S.; Khan, N.A. Understanding the significance of sulfur in improving salinity tolerance in plants. *Environ. Exp. Bot.* **2011**, *70*, 80–87. [CrossRef]
36. Rausch, T.; Wachter, A. Sulfur metabolism: A versatile platform for launching defence operations. *Trends Plant Sci.* **2005**, *10*, 503–509. [CrossRef] [PubMed]
37. Luo, C.; Branlard, G.; Griffin, W.B.; McNeil, D.L. The effect of nitrogen and sulphur fertilisation and their interaction with genotype on wheat glutenins and quality parameters. *J. Cereal Sci.* **2000**, *31*, 185–194. [CrossRef]
38. Moss, H.; Wrigley, C.; MacRichie, R.; Randall, P. Sulfur and nitrogen fertilizer effects on wheat. II. Influence on grain quality. *Aust. J. Agric. Res.* **1981**, *32*, 213. [CrossRef]
39. Ramig, R.; Rasmussen, P.; Allmaras, R.; Smith, C. Nitrogen-Sulfur Relations in Soft White Winter Wheat. I. Yield Response to Fertilizer and Residual Sulfur. *Agron. J.* **1975**, *67*, 219–224. [CrossRef]
40. Salvagiotti, F.; Miralles, D.J. Wheat development as affected by nitrogen and sulfur nutrition. *Aust. J. Agric. Res.* **2007**, *58*, 39–45. [CrossRef]
41. Lerner, S.E.; Seghezzo, M.L.; Molfese, E.R.; Ponzio, N.R.; Cogliatti, M.; Rogers, W.J. N- and S-fertiliser effects on grain composition, industrial quality and end-use in durum wheat. *J. Cereal Sci.* **2006**, *44*, 2–11. [CrossRef]
42. Pompa, M.; Giuliani, M.M.; Giuzio, L.; Gagliardi, A.; di Fonzo, N.; Flagella, Z. Effect of sulphur fertilization on grain quality and protein composition of durum wheat (*Triticum durum* Desf.). *Ital. J. Agron.* **2009**, *4*, 159–170. [CrossRef]
43. Borghi, B. Nitrogen as determinant of wheat growth and yield. In *Wheat: Ecology and Physiology of Yield Determination*; Satorre, E.H., Slafer, G.A., Eds.; Haworth Press Inc.: Binghmanton, NY, USA, 1999; pp. 67–84. ISBN 1-56022-874-1.
44. AACC (American Association of Cereal Chemists) AACC Method 39-00 1999. Available online: http://methods.aaccnet.org/summaries/39-00-01.aspx (accessed on 20 June 2018).
45. Botticella, E.; Sestili, F.; Ferrazzano, G.; Mantovani, P.; Cammerata, A.; D'Egidio, M.G.; Lafiandra, D. The impact of the SSIIa null mutations on grain traits and composition in durum wheat. *Breed. Sci.* **2016**, *66*, 572–579. [CrossRef] [PubMed]
46. Steel, R.; Torrie, J.; Dickey, D. *Principles and Procedures of Statistics: A Biometrical Approach*; McGraw-Hill: New York, NY, USA, 1997.

47. Smith, J.E.K. Data transformations in analysis of variance. *J. Verbal Learn. Verbal Behav.* **1976**, *15*, 339–346. [CrossRef]

48. López-Bellido, L.; Fuentes, M.; Castillo, J.; López-Garrido, F.; Fernández, E. Long-Term Tillage, Crop Rotation, and Nitrogen Fertilizer Effects on Wheat Yield under Rainfed Mediterranean Conditions. *Agron. J.* **1996**, *88*, 783–791. [CrossRef]

49. Černý, J.; Balík, J.; Kulhánek, M.; Časová, K.; Nedvěd, V. Mineral and organic fertilization efficiency in long-term stationary experiments. *Plant Soil Environ.* **2010**, *56*, 28–36. [CrossRef]

50. Tosti, G.; Farneselli, M.; Benincasa, P.; Guiducci, M. Nitrogen fertilization strategies for organic wheat production: Crop yield and nitrate leaching. *Agron. J.* **2016**, *108*, 770–781. [CrossRef]

51. Dawson, J.C.; Huggins, D.R.; Jones, S.S. Characterizing nitrogen use efficiency in natural and agricultural ecosystems to improve the performance of cereal crops in low-input and organic agricultural systems. *Field Crops Res.* **2008**, *107*, 89–101. [CrossRef]

52. Drinkwater, L.E.; Letourneau, D.K.; Workneh, F.; van Bruggen, A.H.C.; Shennan, C. Fundamental Differences Between Conventional and Organic Tomato Agroecosystems in California. *Ecol. Appl.* **1995**, *5*, 1098–1112. [CrossRef]

53. Sacco, D.; Moretti, B.; Monaco, S.; Grignani, C. Six-year transition from conventional to organic farming: Effects on crop production and soil quality. *Eur. J. Agron.* **2015**, *69*, 10–20. [CrossRef]

54. Garrido-Lestache, E.; López-Bellido, R.J.; López-Bellido, L. Effect of N rate, timing and splitting and N type on bread-making quality in hard red spring wheat under rainfed Mediterranean conditions. *Field Crops Res.* **2004**, *85*, 213–236. [CrossRef]

55. López-Bellido, L.; López-Bellido, R.; Castillo, J.; López-Bellido, F. Effects of Tillage, Crop Rotation, and Nitrogen Fertilization on Wheat under Rainfed Mediterranean Conditions. *Agron. J.* **2000**, *92*, 1054–1063. [CrossRef]

56. Flagella, Z.; Giuliani, M.M.; Giuzio, L.; Volpi, C.; Masci, S. Influence of water deficit on durum wheat storage protein composition and technological quality. *Eur. J. Agron.* **2010**, *33*, 197–207. [CrossRef]

57. Ercoli, L.; Arduini, I.; Mariotti, M.; Lulli, L.; Masoni, A. Management of sulphur fertiliser to improve durum wheat production and minimise S leaching. *Eur. J. Agron.* **2012**, *38*, 74–82. [CrossRef]

58. Rasmussen, P.; Ramig, R.; Allmaras, R.; Smith, C. Nitrogen-Sulfur Relations in Soft White Winter Wheat. II. Initial and Residual Effects of Sulfur Application on Nutrient Concentration, Uptake, and N/S Ratio. *Agron. J.* **1975**, *67*, 224–228. [CrossRef]

59. Withers, P.J.A.; Zhao, F.J.; McGrath, S.P.; Evans, E.J.; Sinclair, A.H. Sulphur inputs for optimum yields of cereals. In *Aspects of Applied Biology 50, Optimising Cereal Inputs: Its Scientific Basis*; Gooding, M., Shewry, P., Eds.; The Association of Applied Biologists: Wellesbourne, UK, 1997; pp. 191–198.

60. Archer, M. A sand culture experiment to compare the effects of sulphur on five wheat cultivars (*T. aestivum* L.). *Aust. J. Agric. Res.* **1974**, *25*, 369–380. [CrossRef]

61. Attia, K.; El-Dosuky, M. Effect of elemental sulfur and inoculation with Thiobacillus, organic manure and nitrogen fertilization on wheat. *Assiut. J. Agric. Sci.* **1996**, *27*, 191–206.

62. Badawy, F.; Ahmed, M.; El-Rewainy, H.; Ali, M. Response of wheat grown on sandy calcareous soils to organic manures and sulfur application. *Egypt. J. Agric. Res.* **2011**, *89*, 785–807.

63. Lawrence, J.; Germida, J. Relationship between microbial biomass and elemental sulfur oxidation in agricultural soils. *Soil Sci. Soc. Am. J.* **1988**, *52*, 672–677. [CrossRef]

64. Pepper, I.; Miller, R. Comparison of the oxidation of thiosulfate and elemental sulfur by two heterotrophic bacteria and *Thiobacillus thiooxidans*. *Soil Sci.* **1978**, *126*, 9–14. [CrossRef]

65. Skiba, U.; Wainwright, M. Oxidation of elemental-S in coastal-dune sands and soils. *Plant Soil* **1984**, *77*, 87–95. [CrossRef]

66. Janzen, H.; Bettany, J. Measurement of sulfur oxidation in soils. *Soil Sci.* **1987**, *143*, 444–452. [CrossRef]

67. Zhao, C.; Degryse, F.; Gupta, V.; McLaughlin, M.J. Elemental Sulfur Oxidation in Australian Cropping Soils. *Soil Sci. Soc. Am. J.* **2015**, *79*, 89–96. [CrossRef]

68. Clark, M.S.; Horwarth, W.R.; Shennan, C.; Scow, K.M. Changes in soil chemical properties resulting from organic and low-input farming practices. *Agron. J.* **1998**, *90*, 662–671. [CrossRef]

69. Castle, S.; Randall, P. Effects of Sulfur Deficiency on the Synthesis and Accumulation of Proteins in the Developing Wheat Seed. *Funct. Plant Biol.* **1987**, *14*, 503–516. [CrossRef]

70. Klikocka, H.; Cybulska, M.; Barczak, B.; Narolski, B.; Szostak, B.; Kobiałka, A.; Nowak, A.; Wójcik, E. The effect of sulphur and nitrogen fertilization on grain yield and technological quality of spring wheat. *Plant Soil Environ.* **2016**, *62*, 230–236. [CrossRef]
71. Wieser, H.; Gutser, R.; Von Tucher, S. Influence of sulphur fertilisation on quantities and proportions of gluten protein types in wheat flour. *J. Cereal Sci.* **2004**, *40*, 239–244. [CrossRef]
72. Karamanos, R.; Harapiak, J.; Flore, N. Sulphur application does not improve wheat yield and protein concentration. *Can. J. Soil Sci.* **2013**, *93*, 223–228. [CrossRef]

Permissions

All chapters in this book were first published in MDPI; hereby published with permission under the Creative Commons Attribution License or equivalent. Every chapter published in this book has been scrutinized by our experts. Their significance has been extensively debated. The topics covered herein carry significant findings which will fuel the growth of the discipline. They may even be implemented as practical applications or may be referred to as a beginning point for another development.

The contributors of this book come from diverse backgrounds, making this book a truly international effort. This book will bring forth new frontiers with its revolutionizing research information and detailed analysis of the nascent developments around the world.

We would like to thank all the contributing authors for lending their expertise to make the book truly unique. They have played a crucial role in the development of this book. Without their invaluable contributions this book wouldn't have been possible. They have made vital efforts to compile up to date information on the varied aspects of this subject to make this book a valuable addition to the collection of many professionals and students.

This book was conceptualized with the vision of imparting up-to-date information and advanced data in this field. To ensure the same, a matchless editorial board was set up. Every individual on the board went through rigorous rounds of assessment to prove their worth. After which they invested a large part of their time researching and compiling the most relevant data for our readers.

The editorial board has been involved in producing this book since its inception. They have spent rigorous hours researching and exploring the diverse topics which have resulted in the successful publishing of this book. They have passed on their knowledge of decades through this book. To expedite this challenging task, the publisher supported the team at every step. A small team of assistant editors was also appointed to further simplify the editing procedure and attain best results for the readers.

Apart from the editorial board, the designing team has also invested a significant amount of their time in understanding the subject and creating the most relevant covers. They scrutinized every image to scout for the most suitable representation of the subject and create an appropriate cover for the book.

The publishing team has been an ardent support to the editorial, designing and production team. Their endless efforts to recruit the best for this project, has resulted in the accomplishment of this book. They are a veteran in the field of academics and their pool of knowledge is as vast as their experience in printing. Their expertise and guidance has proved useful at every step. Their uncompromising quality standards have made this book an exceptional effort. Their encouragement from time to time has been an inspiration for everyone.

The publisher and the editorial board hope that this book will prove to be a valuable piece of knowledge for researchers, students, practitioners and scholars across the globe.

List of Contributors

Dorota Wichrowska
Department of Microbiology and Food Technology, University of Science and Technology, Kaliskiego 7 Street, 85-796 Bydgoszcz, Poland

Małgorzata Szczepanek
Department of Agronomy, UTP University of Science and Technology, Kaliskiego 7 Street, 85-796 Bydgoszcz, Poland

Gerhard Moitzi and Helmut Wagentristl
Experimental Farm Groß-Enzersdorf, Department of Crop Sciences, University of Natural Resources and Life Sciences, Vienna (BOKU), Schloßhofer Straße 31, 2301 Groß-Enzersdorf, Austria

Reinhard W. Neugschwandtner and Hans-Peter Kaul
Department of Crop Sciences, Institute of Agronomy, University of Natural Resources and Life Sciences, Vienna (BOKU), Konrad-Lorenz-Straße 24, 3430 Tulln an der Donau, Austria

Saiful Islam and Mahesh K Gathala
International Maize and Wheat Improvement Centre, Bangladesh Office, House-10/B, Road-53, Gulshan-2, Dhaka 1212, Bangladesh

Muhammad Salim
Department of Agronomy, Bangladesh Agricultural University, Mymensingh 2022, Bangladesh

Kaushik Majumdar
International Plant Nutrition Institute, South Asia Office, Palm Drive, B-1602, Golf Course Ext Road, Sector-66, Gurgaon 122001, Haryana, India

Sheetal Sharma
International Rice Research Institute-India Office, 1st Floor, CG Block, NASC Complex, DPS Marg, Pusa, New Delhi 110012, India

Rajeev Padbhushan
International Rice Research Institute-Odisha Office, Plot no. 340/C, School St., Saheed Nagar, Bhubaneswar, Odisha 751007, India

Upendra Kumar
National Rice Research Institute (NRRI), Bidyadharpur, Cuttack, Odisha 753006, India

Timothy J. Boring
Michigan Agri-Business Association, East Lansing, MI 48824, USA

Kurt D. Thelen
Plant, Soil and Microbial Sciences, Michigan State University, East Lansing, MI 48824, USA

James E. Board
Emiritus, Department of Plant, Environmental and Soil Science, Louisiana State University, Baton Rouge, LA 70803, USA

Jason L. De Bruin
DuPont Pioneer, Macomb, IL 61455, USA

Chad D. Lee
Department of Plant and Soil Sciences, University of Kentucky, Lexington, KY 40546, USA

Seth L. Naeve
Department of Agronomy and Plant Genetics, University of Minnesota, St. Paul, MN 55108, USA

William J. Ross
Department of Crop, Soil and Environmental Science, University of Arkansas, Little Rock, AR 72204, USA

Wade A. Kent
Becks Hybrids, Urbandale, IA 55032, USA

Landon L. Ries
DuPont Pioneer, Algona, IA 50511, USA

Khin Thuzar Win and Aung Zaw Oo
Faculty of Agriculture, Tokyo University of Agriculture and Technology, Saiwaicho 3-5-8, Fuchu, Tokyo 183-8509, Japan

Naoko Ohkama-Ohtsu and Tadashi Yokoyama
Institute of Agriculture, Tokyo University of Agriculture and Technology, Saiwaicho 3-5-8, Fuchu, Tokyo 183-8509, Japan

Marta Aranguren, Ana Aizpurua, Ander Castellón and Gerardo Besga
NEIKER-Basque Institute for Agricultural Research and Development, Berreaga, 1, 48160 Derio, Biscay, Spain

Nerea Villar
Department of Plant Biology and Ecology, The University of the Basque Country, Barrio Sarriena s/n, 48940 Leioa, Biscay, Spain

Bijay-Singh
Department of Soil Science, Punjab Agricultural University, Ludhiana 141 004, India

Lu Yang
Graduate School, Chinese Academy of Agricultural Sciences, Beijing 100081, China
Key Laboratory of Plant Nutrition and Fertilizer, Ministry of Agriculture and Rural Affairs/Institute of Agricultural Resources and Regional Planning, Chinese Academy of Agricultural Sciences, Beijing 100081, China

Jinshun Bai, Naohua Zeng and Weidong Cao
Key Laboratory of Plant Nutrition and Fertilizer, Ministry of Agriculture and Rural Affairs/Institute of Agricultural Resources and Regional Planning, Chinese Academy of Agricultural Sciences, Beijing 100081, China

Jia Liu
Institute of Soil & Fertilizer and Resources & Environment, Jiangxi Academy of Agricultural Sciences, Nanchang 330200, China

Vinod K. Singh, Brahma S. Dwivedi, Pravin K. Upadhyay and Kapila Shekhawat
ICAR-Indian Agricultural Research Institute, New Delhi-110 012, India

Rajendra P. Mishra and Azad S. Panwar
ICAR-Indian Institute of Farming Systems Research, Modipuram-250 110, India

Arvind K. Shukla
ICAR-Indian Institute of Soil Science, Bhopal-462 038, India

Dale Loussaert, Josh Clapp, Nick Mongar, Dennis P. O'Neill and Bo Shen
DuPont Pioneer, 7250 NW 62nd Ave, Johnston, IA 50113, USA

Hirak Banerjee
Bidhan Chandra Krishi Viswavidyalaya, Regional Research Station (CSZ), Kakdwip 743347, South 24 Parganas, West Bengal, India

Krishnendu Ray
Ramakrishna Mission Vivekananda Educational and Research Institute, Sasya Shyamala Krishi Vigyan Kendra, Narendrapur, Kolkata 700103, West Bengal, India

Sudarshan Kumar Dutta
International Plant Nutrition Institute, South Asia (East India and Bangladesh) Program, 36 Gorakshabasi Road, Kolkata 700028, West Bengal, India

Talatam Satyanarayana
International Plant Nutrition Institute, South Asia Program, 354, Sector 21, Gurgaon 122016, Haryana, India

Jagadish Timsina
Soils and Environment Research Group, Faculty of Veterinary and Agricultural Sciences, University of Melbourne, Melbourne, VIC 3010, Australia
Agriculture & Forestry University, Rampur, Chitwan 44209, Nepal

Francesco Rossini, Maria Elena Provenzano, Francesco Sestili and Roberto Ruggeri
Department of Agriculture and Forestry Science (DAFNE), University of Tuscia, via San Camillo de Lellis, 01100 Viterbo, Italy

Index

Printed in the USA
CPSIA information can be obtained
at www.ICGtesting.com
LVHW080049211023
761596LV00010BA/1987